T0291674

CAMBRIDGE LIBRARY COLLECTION

Books of enduring scholarly value

Life Sciences

Until the nineteenth century, the various subjects now known as the life sciences were regarded either as arcane studies which had little impact on ordinary daily life, or as a genteel hobby for the leisured classes. The increasing academic rigour and systematisation brought to the study of botany, zoology and other disciplines, and their adoption in university curricula, are reflected in the books reissued in this series.

The Ferns (Filicales)

Frederick Orpen Bower (1855–1948) was a renowned botanist best known for his research on the origins and evolution of ferns. Appointed Regius Professor of Botany at the University of Glasgow in 1885, he became a leading figure in the development of modern botany and the emerging field of paleobotany, devising the interpolation theory of the life cycle in land plants. First published between 1923 and 1928 as part of the Cambridge Botanical Handbook series, *The Ferns* was the first systematic classification of ferns according to anatomical, morphological and developmental features. In this three-volume work Bower analyses the major areas of comparison between different species, describes primitive and fossil ferns and compares these species to present-day fern species, providing a comprehensive description of the order. Volume 2 describes, analyses and classifies primitive and fossil ferns.

Cambridge University Press has long been a pioneer in the reissuing of out-of-print titles from its own backlist, producing digital reprints of books that are still sought after by scholars and students but could not be reprinted economically using traditional technology. The Cambridge Library Collection extends this activity to a wider range of books which are still of importance to researchers and professionals, either for the source material they contain, or as landmarks in the history of their academic discipline.

Drawing from the world-renowned collections in the Cambridge University Library, and guided by the advice of experts in each subject area, Cambridge University Press is using state-of-the-art scanning machines in its own Printing House to capture the content of each book selected for inclusion. The files are processed to give a consistently clear, crisp image, and the books finished to the high quality standard for which the Press is recognised around the world. The latest print-on-demand technology ensures that the books will remain available indefinitely, and that orders for single or multiple copies can quickly be supplied.

The Cambridge Library Collection will bring back to life books of enduring scholarly value (including out-of-copyright works originally issued by other publishers) across a wide range of disciplines in the humanities and social sciences and in science and technology.

The Ferns (Filicales)

Treated Comparatively with a View to their Natural Classification

Volume 2:
The Eusporangiatae and Other Relatively Primitive Ferns

F. O. Bower

CAMBRIDGE UNIVERSITY PRESS

Cambridge, New York, Melbourne, Madrid, Cape Town, Singapore,
São Paolo, Delhi, Dubai, Tokyo, Mexico City

Published in the United States of America by Cambridge University Press, New York

www.cambridge.org
Information on this title: www.cambridge.org/9781108013178

This edition first published 1926
This digitally printed version 2010

ISBN 978-1-108-01317-8 Paperback

Cambridge Botanical Handbooks

Edited by A. C. Seward

THE FERNS

VOLUME II

CAMBRIDGE UNIVERSITY PRESS
LONDON : FETTER LANE, E.C. 4

LONDON: H. K. LEWIS & CO., LTD.,
136, Gower Street, W.C. 1
LONDON: WHELDON & WESLEY, LTD.
2-4, Arthur Street, New Oxford Street, W.C. 2
NEW YORK : THE MACMILLAN CO.
BOMBAY, CALCUTTA, MADRAS } MACMILLAN & CO., LTD.
TORONTO : THE MACMILLAN CO. OF CANADA, LTD.
TOKYO : MARUZEN-KABUSHIKI-KAISHA

Royal Ferns (*Osmunda Regalis* L.) in the Kibble House, Botanic Garden, Glasgow

THE FERNS
(FILICALES)

TREATED COMPARATIVELY WITH A VIEW TO THEIR NATURAL CLASSIFICATION

VOLUME II

THE EUSPORANGIATAE AND OTHER RELATIVELY PRIMITIVE FERNS

BY

F. O. BOWER, Sc.D., LL.D., F.R.S.

EMERITUS PROFESSOR OF BOTANY
IN THE UNIVERSITY OF GLASGOW

CAMBRIDGE
AT THE UNIVERSITY PRESS
1926

PRINTED IN GREAT BRITAIN

PROLOGUE TO VOLUME II

ἔστην θεατής, πύργον εὐαγῆ λαβών.
ὁρῶ δὲ φῦλα τρία τριῶν στρατευμάτων.

EURIPIDES, *Suppliants* 653-4.

"On a far-looking tower I stood to watch,
And three tribes I beheld, of war-bands three."

Way's translation, 1912.

IN offering this Second Volume on the Ferns, the author may fitly quote these words of the messenger in the Greek Play. By establishment of the twelve criteria of comparison detailed in the first Volume we have taken our place upon a tower of vision. Thence we may now witness the phyletic advance. As the armies in the play were seen to be formed in three distinct columns, each moving independently, so also the three main phyla of Ferns, which our comparative study will disclose, may be held to have progressed independently in their evolutionary march, their separate movements being discernible by the observer from his point of vantage. It is immaterial that on both sides of our comparison the number is three. Later writers on Ferns may recognise some different number. The point is that each phylum takes its own course: in fact the evolutionary movements are polyphyletic.

An impressive feature that will emerge from further phases of this study of Ferns is that the lines of development, previously distinct, converged in character as their evolution progressed. Their constituent genera and species thus assumed features so similar that it may often baffle the student to trace their phyletic origin. As in the battle, graphically described by the messenger in the play, the several columns finally merged in an inextricable mêlée, so in the later phases of the evolution of Ferns it becomes difficult or even impossible to segregate completely the several phyla of descent according to their detailed features. But this problem of convergent evolution will be reserved for later treatment: the present Volume deals with the evolutionary progressions of earlier geological time.

The author desires to acknowledge with gratitude the continued assistance given by the Carnegie Trustees, in the form of a grant towards the cost of illustration of this Volume. By such means they promote the advancement of science in a most practical way.

CONTENTS

INTRODUCTION TO VOLUME II

As some years have necessarily elapsed since the publication of the First Volume of this Work, and since the volumes will be on sale separately, it seems desirable to offer a brief summary of the contents of the First Volume which shall serve as an introduction to the Second, and so to link the whole Work together into a coherent whole. It has been shown in Chapter III of Vol. I that those arrangements of the constituent Families of the Filicales which have their place in current Botanical Literature appear to be marked by chance rather than by considered method, and are often wanting in suggestion of phyletic, that is evolutionary, sequence. This is believed to have been a natural consequence of the insufficiency of the foundations upon which such systems have hitherto been based. Accordingly, a definite attempt has been made to extend those foundations by using wider comparisons, and by seeking new criteria for that purpose. The result has been that twelve leading characteristics of Ferns have been selected to serve as a broad basis for their comparison with a view to the phyletic seriation of the Class—this being the end aimed at in any Natural Classification. The number of such criteria used by different authors may vary, and will probably be added to by later writers. Those to be used here are these:

(1) The external morphology of the shoot.
(2) The initial constitution of the plant-body as indicated by segmentation.
(3) The architecture and venation of the leaf.
(4) The vascular system of the shoot.
(5) The dermal appendages.
(6) The position and structure of the sorus.
(7) The indusial protections.
(8) The characters of the sporangium, and of the spores.
(9) The spore-output.
(10) The morphology of the prothallus.
(11) The position and structure of the sexual organs.
(12) The embryology of the sporophyte.

Each of these criteria has been examined critically in Ferns at large, and the scope of variation noted in respect of each. The extreme variants have been checked according to the fossil record, so that it should become possible with some degree of confidence to assert for each criterion which of its variants are to be held as relatively primitive, and which as relatively derivative. The consequences of this procedure in respect of the several criteria have been found to work out as follows:

(1) The simple shoot, composed of an axis and an acropetal succession of leaves, is the unit of construction of the sporophyte. In primitive types it is commonly unbranched, upright, and radial. The prone position with dorsiventral symmetry is probably derivative. The branching of the shoot gives distinctive features. All the distal branchings (exclusive of adventitious buds) may be referred to dichotomy. Shoots which dichotomise equally may be held as primitive in that feature, and any departure from equality, so as to produce some form of dichopodium, may be held as derivative (Chapter IV, Vol. I).

(2) Comparison as regards cellular constitution of the apices of stem, leaf, or root, or of the wings of the leaf, or of the sporangia, shows variation from types having regular segmentation of a single initial cell, as in Leptosporangiate Ferns, to those where the segmentation is not referable to a single initial, but to several, as in the Eusporangiate Ferns. Comparison with related fossils shows clearly that the latter type is relatively primitive. Hence the provisional hypothesis is entertained that the more robust Eusporangiate Ferns appeared first in descent, and that the less robust Leptosporangiate Ferns are derivative and specialised forms, and appeared later in descent. In fact that there has been a progression from a more robust and less exact to a less robust and more exact type of organisation in Ferns (Chapter VI).

(3) The leaf is traceable in origin to an elongated rachis with a dichotomous distal region; and in primitive forms it may have basal stipular growths. In various ways sympodial development of the distal dichotomy may give dichopodia, the pinnae being essentially branches of the distal region. But in advanced types where a phyllopodium is strongly established the lower pinnae may arise monopodially, the later only arising by dichotomous branching. In primitive types the segments may be all separate, each with a single vein. All webbed expansions are held to be derivative. The primitive venation is always open with free endings: looping of the veins in various ways to form a reticulum is held to be derivative (Chapter V).

(4) The primitive vascular structure of the axis was the protostele. Those Ferns which in the adult state are structurally nearest to being protostelic are held to be relatively primitive in respect of that feature: those which have departed from it, showing solenostely, dictyostely, polycycly and perforation, are held to be relatively advanced. Similarly the primitive leaf-trace is an integral strand: the undivided horse-shoe curve of the Osmundaceous type which follows is probably the prototype of later Fern-petioles, while in yet more advanced types the meristele breaks up into parts still disposed in the original curve. The marginal pinna-trace is relatively primitive, and the extra-marginal derivative (Chapters VII to X).

(5) The dermal appendages may take the form of simple hairs or scales.

As a result of simple induction from the observed facts it may be stated generally that simple hairs are a primitive feature, and that branched hairs and in particular flattened scales are an indication of advance from that simple state. Evidence of this advance parallel with other indications of advance in other characters may be found in various phyletic sequences, and this fully substantiates the general correctness of the conclusion thus stated (Chapter XI).

(6) Comparison shows that the sorus is not a constant entity for Ferns at large. It is liable to vary in position, individuality, and constitution, and the differences afford material for phyletic comparison. The marginal position, more frequent in early than in late Fern-types, is shown to have passed in several distinct phyletic series into the superficial as the area of the leaf-blade increased. This makes it seem probable that the latter is a derivative state wherever it occurs. The individuality of the sorus is lost in many sequences, whether by fissions or by fusions: or the sori may be obliterated by spreading the sporangia generally over the leaf-surface. These states appear to be all derivative. The most important variations are, however, those of constitution of the sorus. Three types have been distinguished, viz. the Simple, Gradate, and Mixed types. The first is characteristic of Palaeozoic Ferns, though it survives to the present day. The last is the type which prevails at the present day but does not appear in Palaeozoic Ferns. The Gradate is an intermediate type in many but not in all phyletic lines. Thus the soral characters provide ample material for phyletic comparison (Chapter XII).

(7) Palaeozoic Ferns have no indusial protection, nor have those characteristic of modern conditions which are included in the old comprehensive genera *Polypodium* and *Acrostichum*. But between such extremes protection by various means has been achieved, notably by different types of indusium. An indusiate sorus is then a later and derivative type: on the other hand there is evidence that the modern non-indusiate state has often resulted from the abortion of an indusium previously present. Such facts afford ample material useful for phyletic seriation (Chapter XII).

(8) The characters of the sporangium are more important than those of the sorus. The sporangium shows consistent reduction in size, with increasing specialisation of the mechanism of dispersal, as we progress from the Palaeozoic to modern types. Moreover the form, length of stalk, and the position of the annulus and stomium vary in close relation to the constitution of the sorus. The ends of the series, viz. the Eusporangiate and the advanced Leptosporangiate types, are strongly dissimilar, but they grade one into the other by most gentle steps, which give plentiful phyletic material (Chapter XIII).

(9) At the back of this gently graded series are the facts of spore-output from each sporangium. The gradual diminution of the spore-number from

many thousands to a single one gives a numerical index of the progressive simplification that cannot fail to be impressive (Chapter XIII).

(10) The vegetative features of the prothallus are chiefly negative, and for phyletic purposes stand far behind those of the sporophyte. The structure is uniformly parenchymatous, and the form is plastic under varied conditions of lighting and moisture. The persistence of the cordate type is noteworthy, but it may under certain conditions take a filamentous form, while the latter is characteristic in certain genera. The massive mycorhizic type may very well be a special modification following on fungal nutrition. The Eusporangiate types have relatively massive prothalli as a rule, but it is difficult to found any consecutive phyletic argument on the vegetative characters of the prothallus (Chapter XIV).

(11) The sexual organs of Eusporangiate Ferns are apt to be more deeply sunk than those of Leptosporangiate Ferns, which habitually project from the prothallus. In this they compare with the sporangia. The archegonia appear highly standardised for the Class, presenting little of comparative value in Ferns. But it is not so with their antheridia. These are not only deeply sunk in the Eusporangiate Ferns, but are also massive: while in Leptosporangiate Ferns they project and are relatively delicate. The sperm-numbers in each antheridium run roughly parallel with the spore-numbers in the corresponding sporangia, the primitive Ferns having relatively large numbers in each of these distinct organs. This parallel cannot be pursued into strict numerical detail, but along the lines thus indicated the sexual organs provide useful confirmatory material (Chapter XIV).

(12) The embryology of Ferns took a new value for comparison when a suspensor was discovered in certain Ophioglossaceae and Marattiaceae. It seems probable that this is an archaic feature stamping a primitive character of the plants in which it occurs. In all others it is absent, and they may be held as derivative in this feature. Its existence in these and other plants suggests that the embryo was originally a spindle-like structure, one pole of which is the suspensor, the other the apex of the shoot: while the root is an accessory organ of lateral origin. From a phyletic point of view the presence or absence of a suspensor appears to be the most important comparative feature relating to the embryo (Chapters XV, XVII).

The strength of the field of comparison thus widened by new criteria, and supplemented by those already in use, does not lie simply in the number of the lines upon which the comparisons are based, but upon the degree in which those lines run parallel. In so far as they do so they mutually support one another. This may even be held as a test of the validity of the criteria themselves. As the application of the method proceeds in the present volume it will appear that the parallelism is impressively uniform. But it cannot be assumed that progressions in respect of all the criteria will necessarily march together. Not uncommonly a Fern may be advanced in respect of certain features, and still retain one or more primitive characters: or Ferns essentially primitive may show some single feature of advance. An example of the latter is seen in *Christensenia*, in which the anatomical and soral features are clearly Marattiaceous but the venation is reticulate. The like appears in *Anemia* §*Anemidictyon*, and in *Lygodium* §*Hydroglossum*: here, while the primitive characters of the Schizaeaceae are retained, the venation is reticulate, though in the rest of those genera it is open. It may be presumed that in such examples there has been an advance in the type of venation while other features retained their primitive state. Many species of *Eu-Gleichenia* bear dermal scales, though the rhizome is protostelic, and the sorus primitive in its structure and in its sporangia. A very notable example of discrepancy is that of *Cheiropleuria*, in which a protostelic axis, a singularly primitive leaf-trace, and unbranched dermal hairs are associated with a reticulate venation and a mixed and Acrostichoid sorus. The fact that such examples are exceptional makes them stand out prominently from the rest as witnesses to the fact that the progressions in respect of the several characters march as a rule parallel.

The criteria thus recognised and evaluated were at the close of Volume I abstracted, and the most primitive aspects of them combined into a verbal specification of an ideal plant which should embody them all. The archetype which was thus presented before the mind came out as a plant not unlike the type of vegetation characteristic of the Devonian Rhynie Flora. This need not in itself be held as evidence of actual ancestry. But at least the fact that plants, comparable to this archetype extracted by analysis of Ferns at large, did exist at a very early period, is in itself evidence that the method pursued in Volume I has led to conclusions not in any way improbable.

But a much more severe test will be the application of this method of analysis to the Ferns at large with a view to their seriation in time. If it appears that the phyletic grouping thus arrived at for modern Ferns harmonises with the appearance of the various families in successive geological ages, then it may be confidently felt that we possess a morphological weapon that has real value. This is the test that will be made in the succeeding Volumes. The order of presentment will be roughly in accordance with probable evolutionary sequence. But it is of course impossible in such arrangements to do more than avoid the more gross misrepresentations: for actual affinities are from the very first doubtful, and probably complex. As we progress to the more modern types the sequences of probable affinity become clearer, and the arrangement of the material will be such as to lead readily to their recognition. But the most modern present a fresh difficulty, for owing to parallel development in them the latest types become so standardised that it becomes difficult to assign them severally to their phyletic source.

The positive facts of Palaeontology relating especially to the Ferns have been summarised critically by Seward (*Fossil Plants*, ii). They indicate the following sequence of appearance for those Classes of Ferns which a general comparison has indicated as being relatively primitive. The first Ferns of which any record exists were the **Coenopteridaceae**[1], including the Botryopterideae, and Zygopterideae, with which the Anachoropterideae may also be associated. They are known as fossils only, and are represented from the Culm and Lower Carboniferous, extending onwards to the Permian (Seward, p. 432). They appear then to have died out, and to have left no direct representatives. Perhaps we are right in looking upon the Ophioglossaceae as the most nearly related of living Filicales to them (Seward, p. 427).

The **Marattiaceae**, as represented by the Psaronieae, have been recognised from the Upper Carboniferous and Lower Permian, while *Ptychocarpus unitus* with its Marattiaceous sorus (Fig. 407, p. 115) occurs in the Middle Coal Measures of France (Seward, p. 412). Some other Carboniferous fossils may well have been really of Marattiaceous affinity, but this family appears to have attained its maximum development in Mesozoic times. Its modern representatives are the living Marattiaceae.

Next in the palaeontological succession come the **Osmundaceae**, the reliable record of which opens with the Upper Permian of Russia. The type has persisted through the Mesozoic Age, and it is represented by the Osmundaceae of the present day (Seward, p. 325). Numerous sporangia exist in Carboniferous Rocks which in section are almost indistinguishable from sections of the sporangia of living Osmundaceae (Bower, *Ann. of Bot.* Vol. v, 1891, p. 109). It has been pointed out, however, that favourable sections of a sporangium of *Botryopteris* might yield a similar appearance. On the other hand, Kidston has summed up his own wide experience of Fern sporangia, as seen from without, thus: "In the great majority of cases where Carboniferous plants have borne annulate sporangia, these have most frequently been described as formed of a single row of cells: but in some cases at least the annulus is composed of two rows of cells[2]." This is seen in his genus *Boweria* (Fig. 310). The fact seems to be that in Carboniferous times the annulus as an opening mechanism was in its experimental stage: and the Botryopterid or the early Osmundaceous types of sporangium gave a favourable field for experiment.

The Ferns above mentioned were undoubtedly of Palaeozoic origin. In the Mesozoic Period a great evolutionary outburst of Ferns took place. Several of the families which were well represented then have been traced, though with doubtful credentials, to the Palaeozoic Period. This remark

[1] I prefer this title introduced by Seward, as expressing a generalised Filical type, to that of "Primofilices" suggested by Arber, which appears to convey more than is desirable. But I have ventured to change the form from Coenopterideae to the more comprehensive title Coenopteridaceae.

[2] *Fossil Plants of the Carboniferous Rocks*, Vol. ii, Part IV, 1923, p. 278

applies to the **Gleicheniaceae**: but the reference of *Oligocarpia* to this family may be held as doubtful. Seward sums up the case thus (*l.c.* p. 352): "Despite an agreement between *Oligocarpia* and *Gleichenia* as regards the form of the sori and the number of sporangia, it is not certain that the existence of a typical Gleicheniaceous annulus has been proved to occur in any Palaeozoic sporangia." But in Triassic and Jurassic times their presence is proved, and they survive in the modern Gleicheniaceae.

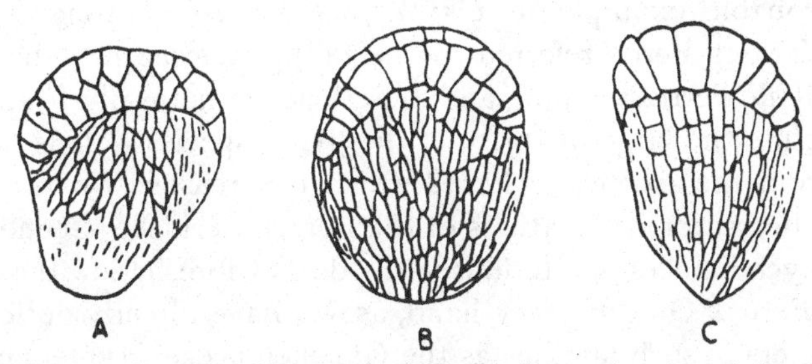

Fig. 310. *Boweria schlatzlarensis* Kidston. Sporangia enlarged about 50 times. *A*, from Clifton, Lancashire; *B*, *C*, from Monckton Main Colliery, near Barnsley. (After Kidston.)

The **Schizaeaceae** are also early Ferns in geological time. But according to Seward (*l.c.* p. 347), the first confident recognition of the Family is in the older Jurassic Rocks. It is true that earlier records have been quoted: in particular *Senftenbergia elegans* described by Corda from the Carboniferous of Bohemia was referred to this Family. But it seems now more probable that this and other early fossils, while bearing a resemblance to Schizaeaceae, are really generalised types foreshadowing lines of evolution rather than already representing the actually differentiated Family.

The **Matonineae** were undoubtedly represented in Mesozoic times by *Laccopteris* and *Matonidium*: and the type survives to the present day in *Matonia*.

The **Dipteridinae** also appear represented by numerous fossil types in the Mesozoic Age, while the family still exists in the living genus *Dipteris*, and it is probably represented also in a large number of those Polypodiaceae which may be looked upon as Dipterid-derivatives.

With regard to the **Cyatheaceae** (incl. **Dicksonieae**) Seward remarks that "we have as yet no satisfactory evidence of the existence of the Cyatheaceae in Palaeozoic Floras. It is not until we reach the Jurassic Period that trustworthy data are obtained" (*l.c.* p. 366). But then they appear to have been fairly well represented, and the type has survived in a large number of living species.

Certain early fossils have been referred to the **Hymenophyllaceae**, in particular *Hymenophyllum weissii* from the Coal Measures of Saarbrücken, and *Hymenophyllites quadridactylites* Gutbier from the French Coal Measures.

But after considering these carefully Seward concludes (*l.c.* p. 365) that "there is no evidence which can be adduced in favour of regarding the Hymenophyllaceae as Ferns of great antiquity which played a prominent part in the Floras of the past."

With regard to the **Polypodiaceae** Seward remarks (*l.c.* p. 375) that "we have as yet no satisfactory evidence of the existence of true Polypodiaceae in the Palaeozoic Era." As regards their Mesozoic existence, however, he quotes a probable example in *Onychiopsis mantelli* (Brongn.) from the Wealden. Various Ferns referred to the Polypodiaceae have been derived from the British Tertiary strata, and these lead on naturally to those Ferns which constitute the large majority of the living species. "It is noteworthy that apart from the absence of Ferns which can reasonably be included in this family, the anatomical features of the Botryopterideae (Coenopterideae) and of the Cycadofilices or Pteridosperms do not foreshadow those of Polypodiaceous Ferns. On the other hand, as we have already noticed, anatomical characters of such families as the Gleicheniaceae, Hymenophyllaceae, and Schizaeaceae are met with in certain generalised Palaeozoic types. These facts are perhaps of some importance as supplying collateral evidence in favour of the relatively more recent origin of the dominant family of ferns in modern floras" (Seward, *l.c.* p. 376).

From this brief summary, drawn from a reliable source, it will be gathered that the general trend of the results of Palaeontological enquiry runs parallel with that of Comparative Morphology. Together the two methods justify the recognition of certain Families as the most ancient. They are the **Coenopteridaceae** and **Osmundaceae**, which have had undoubted Palaeozoic existence, and with them may perhaps be associated the **Marattiaceae.** With the first of these the **Ophioglossaceae** may be ranked on comparative grounds, though they have not yet been definitely recognised as early fossils. These four Families include all the Eusporangiate Ferns now living.

The remainder of the Ferns, though prefigured in some measure by fossils of the Palaeozoic Period, first began to assume their ordinal characters in the Mesozoic Period, or later. They will be found to show more or less clear relationship to two relatively primitive Families, viz. the **Schizaeaceae** and the **Gleicheniaceae**, both of which became firmly established in early Jurassic times: one, the Schizaeaceae, is characterised by bearing its sori on the margins of the leaves; the other, the Gleicheniaceae, bears them superficially. The two series thus consistently defined have been designated respectively the **Marginales**, and the **Superficiales.** Both are regarded as having not improbably originated from some common earlier source. The order of treatment of the several Families will follow the lines thus indicated, the most ancient fossil types being taken first. But since the Class of the Filicales shows many indications of polyphylesis, the descriptions cannot possibly follow naturally in any simple linear sequence. The succession in which they are taken up must therefore be somewhat artificial. Short series may be suggested, and built up. But the actual evolutionary relation of those series to one another will still remain highly problematical. The time has not yet come for any confident reconstruction of an "evolutionary tree," which shall represent the phylesis of the Filicales as a series of connected branches springing from a common trunk, even if there ever was one. Nevertheless the relative position of those branches is a legitimate subject for discussion, and definite views may even be entertained regarding it.

In any Family of related organisms, excepting the very simplest, difficulties must be expected in grouping them phyletically. The more complex the organisms are, and the more numerous the criteria of their comparison, the greater those difficulties may be expected to be: for there is no reason to anticipate that evolution will have expressed itself equally in respect of all the features. And so it comes about that it appears to be sometimes impossible to sum up the results of comparison in respect of all the features into a conclusion, that genus or species (A) is more primitive than genus or species (B). Such difficulties naturally present themselves when any attempt is made to seriate any group of organisms phyletically: and any one who fully grasps the problem will place his own value upon the phyletic seriation of any group submitted for his judgment. Some may say, "If that is true, why attempt phyletic seriation at all?" The proper reply to that question is, "Why study Evolution at all?" Phyletic seriation is merely a form of recording tentative evolutionary conclusions: it has, however, the advantage of raising questions which otherwise might very likely remain dormant.

CHAPTER XVIII

COENOPTERIDACEAE

THE plants included under this heading are all fossils, and include the Botryopterideae, the Zygopterideae, and the Anachoropterideae: but it is possible that others may be added as knowledge of the fossils increases. They are all Palaeozoic types, and are distinct from any living Ferns. Still there is no reason to doubt their Fern-nature, though they appear as generalised rather than specialised types. Their recognition as Ferns is based upon the external characters of the shoot with pronounced megaphyllous character, while the usual circinate vernation of the large alternate leaves has been seen in some of them. But more particularly it rests upon the anatomical details of axis and leaf, and upon the fact that in some of them sporangia containing numerous homosporous spores have been found, borne upon the distal region of their relatively large branched sporophylls. Finally, Scott has shown in *Stauropteris oldhamia* that the spores possessed the capacity for germination within the sporangium, a feature seen in some modern Ferns (*Todea*) (*New Phyt.* v, 1906). As the three orders are somewhat divergent in their characters it will be well to describe them separately.

A. BOTRYOPTERIDEAE

To this family the genera *Grammatopteris*, *Tubicaulis*, and *Botryopteris* are referred. They were mostly plants of relatively small size with stems probably upright or straggling, which in some cases were short, bearing crowded leaves (*Grammatopteris*, *Tubicaulis*, *Botryopteris ramosa* and *hirsuta*, Fig. 311): in others the stem was slender with dichotomous branching, and it appears to have borne leaves at distant intervals (*Botryopteris cylindrica*, Fig. 312). The leaves are imperfectly known: they have never been seen to bear a well-developed lamina. They were repeatedly branched, with narrow fleshy pinnules upon which the sporangia were inserted (*B. forensis*). In the young state the circinate vernation has been seen. Stiff conical hairs are found upon the leaves having peculiar structure (*B. forensis*, Vol. I, Fig. 187). In some instances the plants bore numerous adventitious roots with diarch structure (*Tubicaulis*, *B. ramosa*, Fig. 311). Though the habit of these plants is still imperfectly known, their Fern-nature seems clearly established, while their anatomy, which is better understood, confirms that conclusion.

The vascular system of the stem is protostelic, giving off monodesmic leaf-traces to the several leaves. The roots which are often numerous also

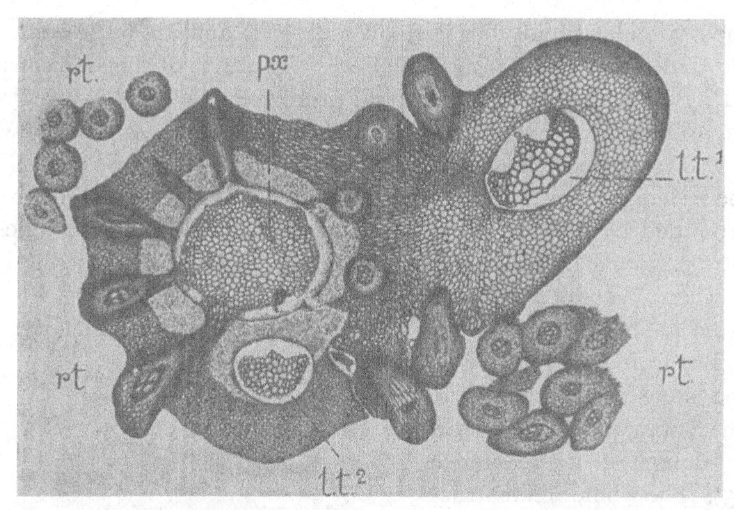

Fig. 311. *Botryopteris ramosa*. Transverse section of stem and leaf-bases. *px*, protoxylem-group in stele; *l.t.*[1], leaf-trace in attached petiole; *l.t.*[2], leaf-trace in cortex; *rt*, roots, some in connection with stem, and others free. (× about 8.) S. Coll. 2314 (G.T.G.). (After Scott.)

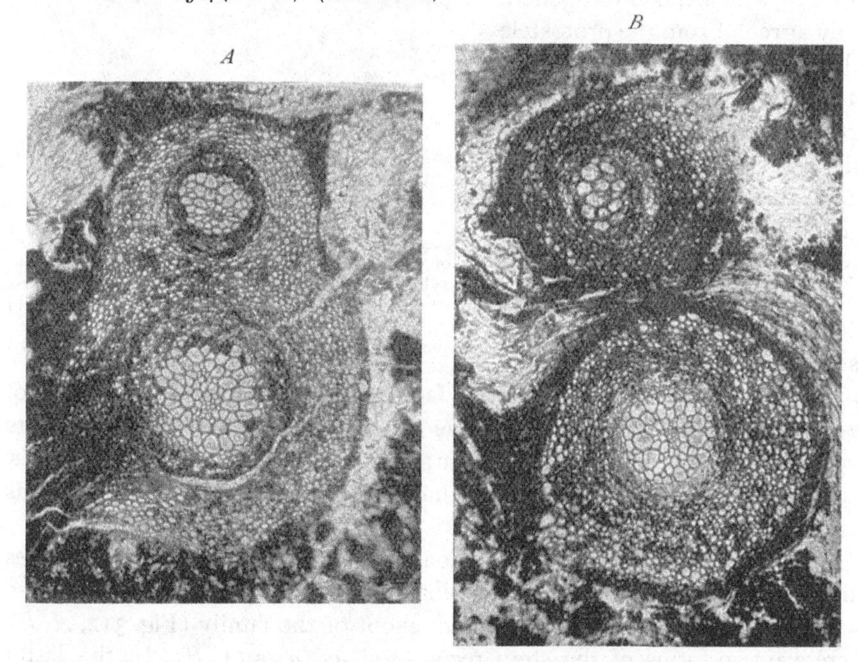

Fig. 312. *Botryopteris cylindrica*, α form. *A*, Transverse section, showing unequal dichotomy. In the smaller branch the double protoxylem is still very excentric: in the larger branch there are three protoxylems. A root is being given off from this branch. *B*, Transverse section of stem and petiole. In the stem there are either two or three internal protoxylems. In the petiole there are two, almost confluent, on the lower (originally adaxial) side of the bundle. In both specimens the delicate outer cortex is almost lost. (× about 24.) S. Coll. 1906. (From photographs by Mr W. Tams. After Scott.)

derive their vascular supply from it, and their stele appears to be essentially similar to that of the roots of modern Ferns. The stele of the axis consisted of a solid xylem-core with endarch spiral protoxylem in one or more groups (Fig. 312). The wider tracheides of the metaxylem are mostly pitted with a transition to scalariform structure. There is no pith in any of these Ferns, nor is the xylem differentiated into external and internal zones. The xylem was surrounded by a broad band of phloem in which the sieve-tubes are sometimes particularly well preserved (Fig. 312). In many cases it is not clear whether an endodermis was present: nor is this surprising if it had been of that ill-specialised type seen in *Helminthostachys* (see Vol. I, Fig. 173). But Dr Bancroft has recognised it in the stem of *B. cylindrica*, and in the young petiole there can be little doubt of its existence (Fig. 313). The whole stelar structure suggests a primitive state. The character of the stele is the feature in which the three genera chiefly agree. From the protostele of the stem of *Botryopteris hirsuta* and *ramosa* the monodesmic leaf-traces passed off with a phyllotaxis of two-fifths. The form of the trace was approximately oval at its departure, but it soon took a very characteristic form. As it leaves the stele the single protoxylem points inwards; as the trace passes into the petiole three prominent points appear upon the adaxial face of it, and to the apex of each of these a protoxylem-group passes. The degree of projection of these points varies in the species, while their orientation relative to the axis also varies, doubtless in accordance with the habit, as is seen in modern Ferns (Fig. 311, *l.t.*).

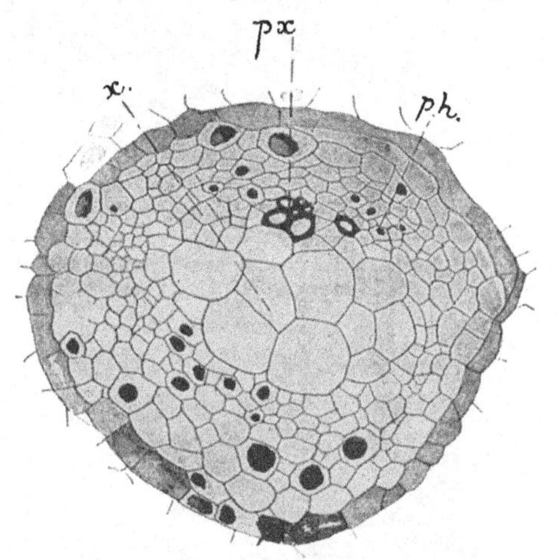

Fig. 313. *Botryopteris hirsuta.* Transverse section of vascular bundle of young petiole, showing the xylem in course of differentiation. *px*, lignified protoxylem; *x*, thin-walled xylem not yet lignified; *ph*, phloem-zone. The dark layer may be the endodermis. (× 150.) S. Coll. 564 (R.S.). (After Scott.)

In *B. cylindrica* from the Lower Coal Measures, with its isolated leaves and long internodes, both stem and leaf have been carefully examined, so that it is now probably the best known shoot of the family (Fig. 312, *A*, *B*). There are two forms of the stem recognised as (*a*) and (*β*). In the latter the stele is simpler and the cortex more bulky with superficial hairs, giving the impression either of a rhizome or of an aquatic stem. In the former the vascular tissue is better developed, and it gives important features for comparison. In the first place the axis dichotomises, giving rise to branches which may be either equal or unequal (Fig. 312, *A*). Secondly, it gives rise to leaves of which the trace closely resembles that of an unequal dichotomy. At the point of origin the trace is oval in outline, with a single external pole. A little higher the presence of a few adaxial metaxylem elements may indicate a relic of ancestral structure such as is seen in *B. antiqua.* Still higher the single protoxylem may divide into two or into three, as in the tridentate petioles. In *B. antiqua* described by Kidston from the Petticur Beds (*Trans. Roy. Soc. Edin.* Vol. 46, Part II, 1908) the form of the leaf-trace is even more condensed, giving an oval outline with internal protoxylem; but it becomes superficial when the petiole is reached, and sometimes divides into two, though not always; so that some petioles are diarch, others monarch (Fig. 314). Dr Bancroft suggests that we may regard *B. antiqua*, *cylindrica*, *racemosa*, and *hirsuta* as forming a series in which the foliar traces show progression from a simple structure in the oldest species (*B. antiqua*) to a more complex development in the later species. In *B. ramosa* from the Lower Coal Measures the tridentate type is well shown, while in *B. forensis* from the French Permo-Carboniferous the most extreme development of the three xylem-arms is seen. The simplest trace of all, that of *B. antiqua* from the Petticur Beds, links readily in point of structure with what is seen in the simplest Osmundaceous trace.

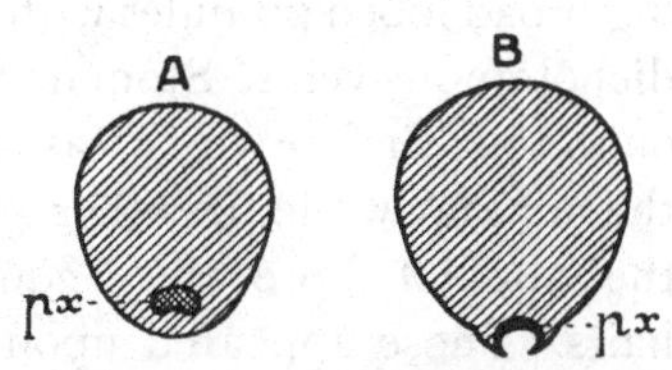

Fig. 314. *Botryopteris antiqua* Kidston. *A*, foliar trace as it leaves the stem, with an internal pole of protoxylem. *B*, a foliar trace in the primary petiole, with the pole cup-shaped at the inner boundary. *px* = protoxylem. (After Dr M. Benson, from Bertrand.)

The peculiar "equisetoid" hairs have already been described in Vol. I, p. 199, Fig. 187. It will be noted that notwithstanding their complex appearance they are essentially simple conical hairs transversely septate. Such hairs are seen in *B. cylindrica* with simple transverse septa, and those of *B. forensis* may be held as a peculiar variant on that type, which is repeated in the hairs of some modern Ferns, e.g. *Dicksonia squamosa* (see Williams, *Ann. of Bot.* 1925, p. 655).

The genera *Grammatopteris* and *Tubicaulis* are chiefly known by their stems, which bore crowded leaf-bases, and show a more complicated phyllotaxis than *Botryopteris.* As before, the stele is protostelic and the monodesmic leaf-trace has a simple form at first, but widens upwards into a broad ribbon. In *Grammatopteris*, which was described by Renault from the Permo-Carboniferous of Autun, each petiolar bundle appears as a straight tangential band. But in *Tubicaulis*, of which specimens are recorded from the Coal Measures (*T. Sutcliffii* Stopes), from the Permian of Saxony (*T. solenites* Cotta), and from the Permo-Carboniferous of Autun (*T. Bertieri* Bertrand), the petiolar trace was shaped like a horse-shoe, with the concavity outwards, as characteristic of the "Inversicatenales" of P. Bertrand (Fig. 315). Since the leaves and sporangia of these fossils are unknown, their position in the Botryopterideae is an uncertain one.

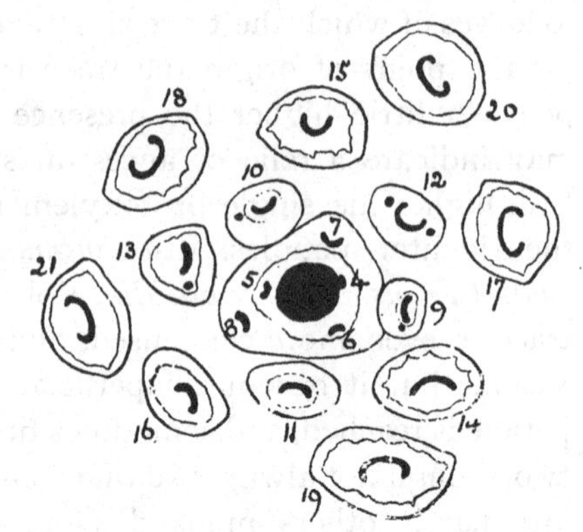

Fig. 315. Transverse section of the shoot of *Tubicaulis solenites* Cotta, after Stenzel, showing stem with protostele, and the last four leaf-traces traversing the cortex. The leaves are numbered in their succession. The drawing is simplified by the omission of roots, etc.

The leaves of *Botryopteris* appear to have been branched, with alternate pinnae. In the French species, *B. forensis*, they ran into fine branches, bearing broad lobed pinnules with prominent dichotomous veins. Stomata were borne on the one surface which was presumably the adaxial, while the other, presumably the abaxial, bore the "equisetiform" hairs. These appeared upon the outer surface of a young leaf still with circinate vernation, thus confirming the presumed orientation. The sporangia have been found attached to branches of the fertile rachis showing the characteristic features of *B. forensis*. In other species similar sporangia are associated with the parts preserved, but not attached to them. They were of small size and pear-shaped, with a wall composed when ripe of a single layer of cells. An annulus ran up one side only of the sporangium, composed of thick-walled cells, forming a broad band. The spores were evidently very numerous (Fig. 316). The sporangia associated with the British *B. hirsuta* and *ramosa* are smaller. It seems probable that many of the sporangia from British coal-balls previously described as being of an Osmundaceous type may really have been sporangia of *Botryopteris*. The comparison of these with the sporangia of the Osmundaceae may still hold good, though not perhaps the definite reference to that family (*Ann. of Bot.* v, 1891, p. 109).

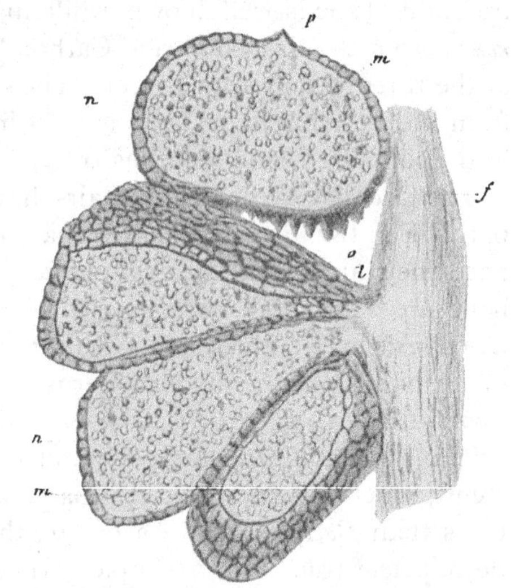

Fig. 316. *Botryopteris forensis*. Group of sporangia, *m*, *m*, inserted on rachis; *l*, pedicel of sporangium; *n*, wall of sporangium; *a*, multiseriate annulus. The uppermost sporangium is in nearly transverse section: *p*, stomium, or place of dehiscence. (× 35.) (From Renault, after Scott.)

A particular interest attaches to the sporangia found by Scott in Petticur material closely associated with a rachis of *Botryopteris antiqua* (Fig. 317)[1]. Not only do they show the sporangia cut transversely with multiseriate annuli, and very numerous spores (probably at least 500—1000 in each), but the sporangia appear to be naturally grouped round a centre, with the thinner region of the wall, where dehiscence would occur, directed outwards. Between the sporangia lies what Scott describes as an "indusium-like" structure. This suggests comparison with those irregular laciniate growths seen at the distal end of the sporangiophores of *Helminthostachys*. Except for the elongation of the stalk of the sporangiophore and the absence of an annulus in *Helminthostachys*, the similarity is very striking. (Compare Bauer and Hooker's *Genera Filicum*, Pl. 47, Fig. *B*, 3–8.)

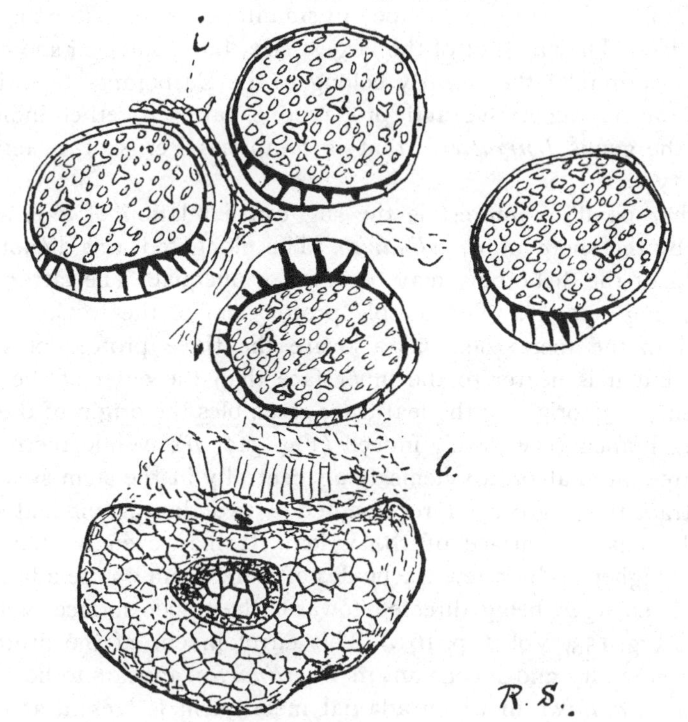

Fig. 317. Group of four annulate sporangia in close association with the rachis of *Botryopteris antiqua*. *i*, indusium-like structure between two sporangia. *l*, represents a foreign body. Drawn by Mrs D. H. Scott, F.L.S. (× about 100.) Scott Collection 2496.

Comparison

The characters thus shown by the Botryopterideae, their megaphylly, the circinate vernation, their anatomy, the vestiture of hairs, the numerous sporangia borne upon the large leaves, their sporangial structure, and homosporous spores, all point to them as undoubted Ferns. On the other hand,

[1] Scott, *Ann. of Bot.* xxiv, 1910, p. 819.

there are many features which indicate for them a relatively primitive position in that Class, such as would fully accord with their early geological history. This is shown by the protostelic structure of the stem, whether short with crowded leaves or elongated with leaves inserted at long intervals. They share also with primitive Ferns the radial construction of the shoot as a whole, and the monodesmic leaf-trace, which shows at the base of the adult leaf a structure such as is seen only in some of the most primitive of living Ferns. The much-branched distal region of the leaf and its narrow segments with dichotomous branching again afford features that are primitive. The hairs though apparently of complex structure are actually of the simple transversely septate type, characteristic of the earlier Ferns. The sporangia of *B. forensis* are grouped in simultaneous sori after the type of the Simplices. They are not of the largest size: but from their spore-output and complex annulus they clearly belong to the Eusporangiate series. All these characters, vegetative and propagative, taken together indicate, at least for the genus *Botryopteris*, that it comprises Ferns of an extremely primitive type.

A further point of interest is the suggestive relation of axis and leaf afforded by the shoot of *B. cylindrica*. The elongated axis dichotomises: the shanks of the dichotomy may be equal or unequal. The latter is illustrated by Fig. 312, *A*. The double protoxylem of the weaker shank is immersed in the metaxylem, as also are the three protoxylems of the stronger; but it is nearer to the inner face than the outer of the slightly oval xylem. The origin of the leaf-trace resembles the origin of the supply to a weak branch very closely indeed (Fig. 312, *B*). While there may be two or three internal protoxylems lying centrally in the stem-stele, in the petiolar trace the protoxylem (represented by two almost confluent strands) lies at the adaxial surface of the xylem. It may even be temporarily enveloped higher up by a few tracheides of metaxylem on the adaxial side, but later it emerges, being directed towards the upper surface of the leaf. (Compare Fig. 153, Vol. I, p. 161.) This envelopment of the protoxylem, which is temporary and inconstant in *B. cylindrica*, appears to be a regular feature of *B. antiqua*, in which adaxial metaxylem is present at the level where the trace separates from the stele of the stem (Fig. 314). Thus in this most ancient species the leaf-trace at its base resembles very closely the vascular supply passing to the weaker shank of a dichotomy of the axis. At the same time its structure as seen in *B. antiqua* resembles that of the extreme base of the leaf in *Thamnopteris*, a primitive member of the Osmundaceous series (Fig. 155, 2, Vol. I, p. 162).

The similarity of anatomical structure thus seen in *B. cylindrica* between the leaf-base and a weak branch points towards the conclusion that the leaf of a Fern may actually be phyletically a shank of a dichotomy, as was originally suggested in 1884 (Bower, *Phil. Trans.* 1884, Part II, p. 565). The suggestion was then based only upon analogy derived from a comparative study of Fern-leaves. The results of that study have been re-stated and amplified in Vol. I (Chapter V, on Leaf-Architecture). The question was further discussed in Vol. I, Chapter XVII. In Chapter IX it has been argued (p. 175) that the most valuable features for comparison are found in the basal region of the leaf, and in the leaf-trace itself. If this principle be put in practice here, the argument in favour of the dichopodial origin of leaf and axis in Ferns acquires added strength. It has further been shown that the embryological facts for Ferns would accord with this view; in particular it has been seen that a dichotomising embryo, such as the hypothesis would demand, actually exists in *Tmesipteris*, with that equal or unequal development of the shanks that the hypothesis presupposes. In view of the cumulative effect of these various facts the hypothesis may now be entertained that the shoot in the genus *Botryopteris*, and perhaps also of Ferns generally, originated as a dichopodial development from a primitively equal dichotomy, such as is actually seen in the adult form of the Psilophytales.

B. ZYGOPTERIDEAE

This family of fossil Ferns includes a number of generic types (about twelve), which are best known by the anatomical features of the leaf and sometimes also of the stem. But in some of them the axis has never been seen (*Stauropteris*). The leaf-form and general habit, as well as the sporangia, are only known in a few of them. Accordingly the tracing of relationships between themselves, and their comparison with other families of Ferns, must be held as provisional. They have been found to include some forms with massive upright stems and crowded leaves (*Asterochlaena, Asteropteris*): in others the axis appears to have been elongated and rhizomatous with isolated leaves (*Diplolabis, Metaclepsydropsis*): others again had a thin slightly elongated stem, and these are believed to have had a straggling or a climbing habit (*Ankyropteris*). In fact in these early Fern-types those differences of habit are already to be found which characterise Ferns of the present day. The plants were attached to the soil by numerous adventitious roots.

A feature which is common to them all, and distinguishes them from the Botryopterideae, is the greater complexity of their vascular system whether of the axis or of the petiole. This goes along with larger dimensions of both, as will be shown later. It may be a question how far the increased complexity is a direct consequence of the greater size: but whatever view may be taken on this point its diagnostic value may be accepted provisionally.

Stratigraphically the Zygopterideae were already represented in the Upper Devonian, as seen in *Asteropteris noveboracensis* from the Portage Beds of the State of New York. They extended through the Carboniferous

Series to the Permian, where the large upright stems of *Asterochlaena laxa* appear in the Lower Permian of Saxony, and two species of *Ankyropteris* have been found in the Permian of Autun and of Bohemia. After this Period they appear to have died out.

VEGETATIVE STRUCTURE

The conformation of the shoot in the Zygopterideae, and its internal structure as well, are probably the best known in *Ankyropteris,* which appears to have consisted of small plants with thin stems, and an upright, perhaps straggling or even climbing habit. The stems branched, sometimes by equal distal dichotomy as in *Ankyropteris corrugata* (Fig. 318), sometimes unequally so that the weaker axis lying between the stronger and the base

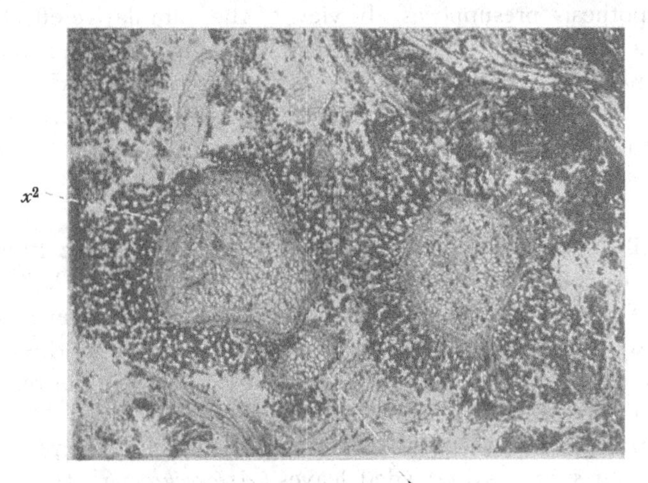

Fig. 318. *Ankyropteris corrugata.* Transverse section of a dichotomous stem, showing the two steles. The stele on the left has a band of secondary wood at x^2; *l.t.*, leaf-trace. (× about 8.) From a photograph by Mr W. Tams. S. Coll. 2715. (After Scott.)

of a subtending leaf appears as an axillary shoot. This is seen in *A. Grayi* (Fig. 319), and it occurs also in other species (Scott, p. 295). In view of the fact that the axillary shoot repeats on a simpler scale the anatomy of the relative main axis, the conclusion may be accepted that this is its real nature. The condition thus seen appears to have resulted from development of a dichotomy derived primarily from equal dichotomy such as is seen in *A. corrugata,* but with one of the shanks arrested. It is to be noted that equal dichotomy is also a character of *Botrychioxylon, Dineuron,* and some others.

The leaves of the Zygopterids were relatively large, and probably upright in their habit. They consisted of a stout rachis traversed by a meristele

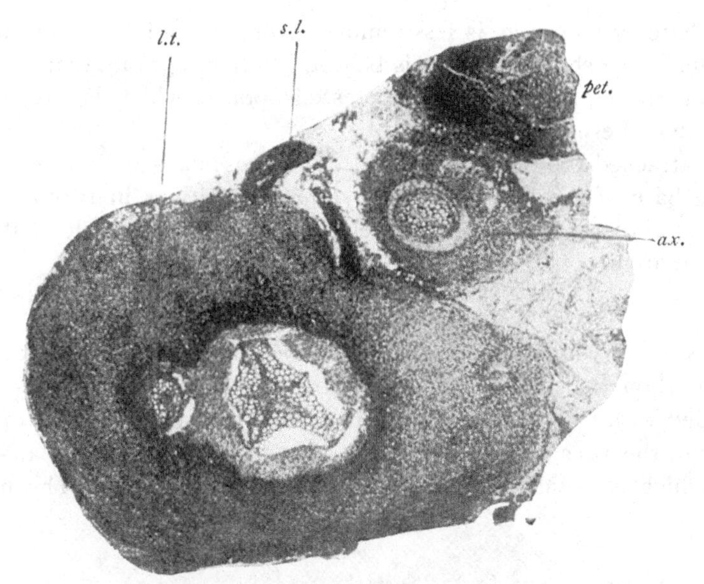

Fig. 319. *Ankyropteris Grayi.* Transverse section of stem, with axillary shoot, and part of petiole. In the middle of the stem the 5-angled stele is seen. *l.t.*, leaf-trace, about to give off the axillary strand; *s.l.*, "scale-leaf"; *ax.*, axillary shoot of next node below; *pet.*, part of petiole of the subtending leaf. The small strand in the cortex to the right belongs to a scale-leaf. (× about 4.) From a photograph by Mr L. A. Boodle. Will. Coll. 1919 A. (After Scott.)

singularly complex in many of its forms, from which pinna-traces were given off to supply the pinnae. These were arranged either in two orthostichies (Clepsydroideae, including *Ankyropteris* (?), *Clepsydropsis*, *Asterochlaena*, and *Asteropteris* (?)): or in four orthostichies (Dineuroideae, including *Dineuron*, *Metaclepsydropsis*, *Diplolabis*, *Botrychioxylon* (?), and *Etapteris* (Fig. 320).

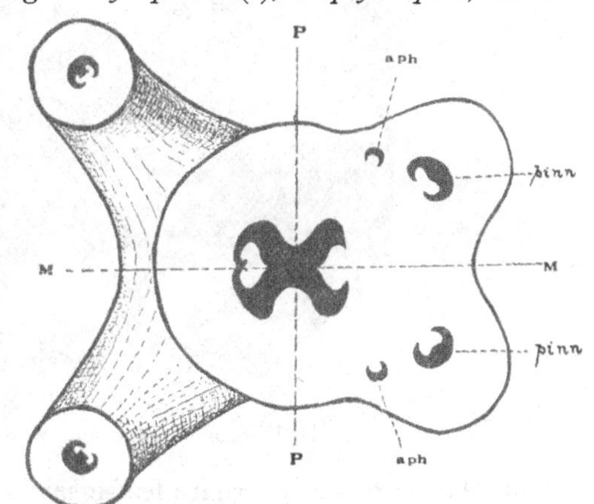

Fig. 320. *Diplolabis Römeri* Solms. Diagrammatic drawing showing the ramification of the primary petiole, as it might be seen in a thick transverse section. *P*, *P* = principal plane of symmetry; *M*, *M* = median plane; *aph.* = aphlebia-trace; *pinn.* = pinna-trace. (From P. Bertrand. After Gordon.)

In either case it appears that the orientation of the pinnae was not in the plane including the rachis, as in modern Ferns. They were expanded in planes transverse to this, giving to the whole upright leaf an almost radial symmetry, and a shoot-like aspect. This is an important point in its bearing on foliar theory: for it is clear that such an orientation, whatever its origin, is less removed from the radial dichotomy from which the whole shoot of Ferns is believed to have sprung, than is the leaf with bilateral symmetry deeply impressed upon it, which is characteristic of all modern Ferns (Vol. I, Chap. V).

Roots attached the plant to the soil. Their vascular supply arises usually from the base of the leaf-trace. They are adventitious in much the same way as in Ferns at large. Their stelar structure is as a rule that of the ordinary diarch roots of Ferns.

THE AXIS AND LEAF-TRACE

The axis as a whole in *Ankyropteris Grayi* may be as much as 22 mm. in diameter, though usually less; that shown in Fig. 319 measures 18 mm. Embedded within a broad cortex is the stele, with an outline as of a five-rayed star, the rays being of unequal length. Its form follows that of its xylem, which is usually the best preserved part (Fig. 321). The five rays

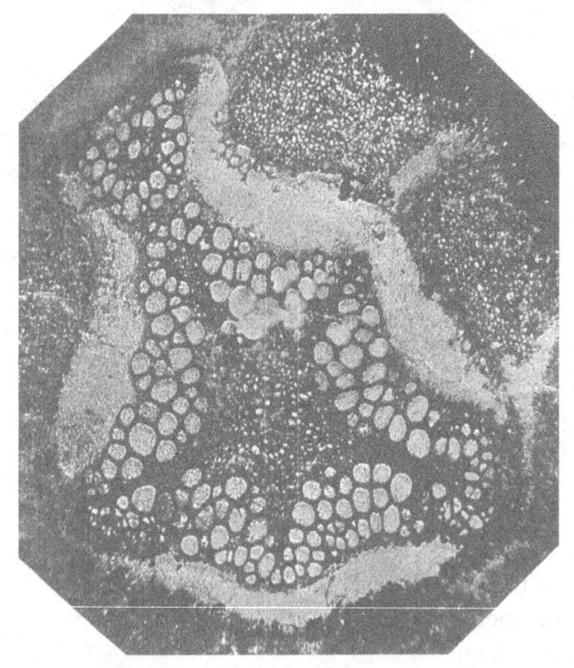

Fig. 321. *Ankyropteris Grayi.* Stele, from a section in Dr Kidston's collection. (× 18.) (After Seward.)

are in relation to the alternate leaf-insertion, which is on a two-fifths divergence. They are unequal, the least prominent having just parted with its leaf-trace, while the most prominent is just about to do so. The leaf-trace is a strand whose xylem takes a crescentic form with the concave side directed outwards. It is nipped off from the margin of the ray, and passes out through the cortex into the petiole. The vascular supply to the axillary bud resembles structurally that of the main stele, but on a smaller scale: it separates from the inner face of the trace on its course through the cortex, in a manner very similar to that seen in the living Hymenophyllaceae (Vol. I, p. 70, Fig. 63). The roots draw their vascular supply directly from the main system of the shoot. Finally small isolated vascular strands may be given off from the leaf-trace either below or above its actual separation from the stele: each enters a small scale-like appendage called an aphlebia, and dividing into a few strands supplies its small expanse. The aphlebiae may be held as peculiar basal pinnae, partly on the ground of their form and position, partly on that of their vascular supply.

The stele of *Ankyropteris* is protostelic, but it differs from the simplest of these by its corrugated form, and by the differentiation of the xylem. Two regions are clearly distinguishable, the peripheral or outer-xylem has relatively large scalariform tracheides which form arcs following the incurved outer surfaces of the stele, but deepening those curves: the inner-xylem is composed of narrower tracheides, mostly scalariform, mixed with parenchyma, and it forms a narrow five-rayed star. From each of its rays a band of small tracheides containing protoxylem-elements extends outwards up the middle of each arm to each several leaf-trace. Phloem surrounded the xylem of the stele, having a single or even a double series of sieve-tubes. There is no pith.

The leaf-trace of *Ankyropteris Grayi* at its departure from the stele spreads out laterally, and consists also of central and peripheral regions: the latter supplies three groups of protoxylem: one group retains a median position, and gives off two lateral groups each enclosed by narrow tracheides with parenchyma, which distending forms a loop at either cusp of the crescentric trace. These enlarge as the trace passes into the petiole, still enclosed by the phloem. Each cusp spreading into adaxial and abaxial curves, the result is the H-shaped meristele so characteristic of the Zygopterideae: indeed this was the leading character of the old genus *Zygopteris* (Fig. 326). The cortex was indurated, and hairs were borne on the outer surface. Those about the base of the petiole were developed as stiff bristles (Kidston Coll. 1985). In *A. corrugata* they were not only transversely septate, but longitudinal walls also appeared in their cells, as is seen in modern Ferns, e.g. *Dipteris*. (See Vol. I, p. 198. Fig, 185, 2, 3.)

The stems of other Zygopterideae show variants on this structure, but still with differentiated xylem. In *Diplolabis Römeri* the stele is cylindrical (diam. 3·5 mm.), and there is no parenchyma in the xylem. In *Metaclepsydropsis duplex* the stele contains an oval mass of xylem composed of a dense band of outer wide tracheides surrounding an inner-xylem of the nature of a "mixed pith" (diam. 2·2 mm.). In this plant the stem was an elongated creeping rhizome (see Vol. I, Fig. 122, p. 128). Many of the Zygopterideae appear to have had a lax form of shoot. Dr M. Benson has observed in *Metaclepsydropsis* how the same creeping rhizome, followed through 22 inches of its course, bore only five leaves within that distance. Its habit must have been very like that of *Pteridium*.

On the other hand, where the axis is massive and upright, with short internodes and crowded leaf-bases, a different structure may be found which is, however, related to that corrugated form seen in *Ankyropteris*. An extreme example of it has been described in detail with beautiful plates by Prof. P. Bertrand in *Asterochlaena laxa* Stenzel, a plant of Permian Age. The upright axis increased conically upwards to 5 cm. in diameter. It was covered by the bases of the closely crowded petioles together with roots.

Fig. 322. *Asterochlaena laxa*. General transverse section, showing the stellate stele, with bilobed or trilobed arms, and the leaf-traces, but only the inner petioles. The Roman numerals indicate the leaves of two successive whorls. *R*=roots; *S*=pinna-traces. (Slightly magnified.) (After Dr P. Bertrand. From Scott.)

The phyllotaxis was on a complicated plan with a tendency towards verticillation (Fig. 322). Centrally in the transverse section is a large stellate stele measuring 5 mm. or more in diameter, with some 7 or 8 diverging rays, each more or less lobed at the periphery. It is from these lobed arms of the star that the leaf-traces are abstricted, each lobe giving rise to two or three traces but occasionally to only one. It thus appears that the number of arms of the star bears no constant relation to the number of the leaf-traces. This condition is comparable with what is seen in *Lycopodium*, where the small leaf-traces do not correspond in number or in arrangement to the rays of the stellate xylem (Jones, *Trans Linn. Soc.* Vol. vii, Part II, 1905).

The details of this remarkable fossil are seen in Fig. 323. The stellate stele consists of xylem which is all primary, showing as in *Ankyropteris* a distinction of external tracheides of wide, and internal tracheides of narrower lumen. Outside this there is phloem with large sieve-tubes, surrounded

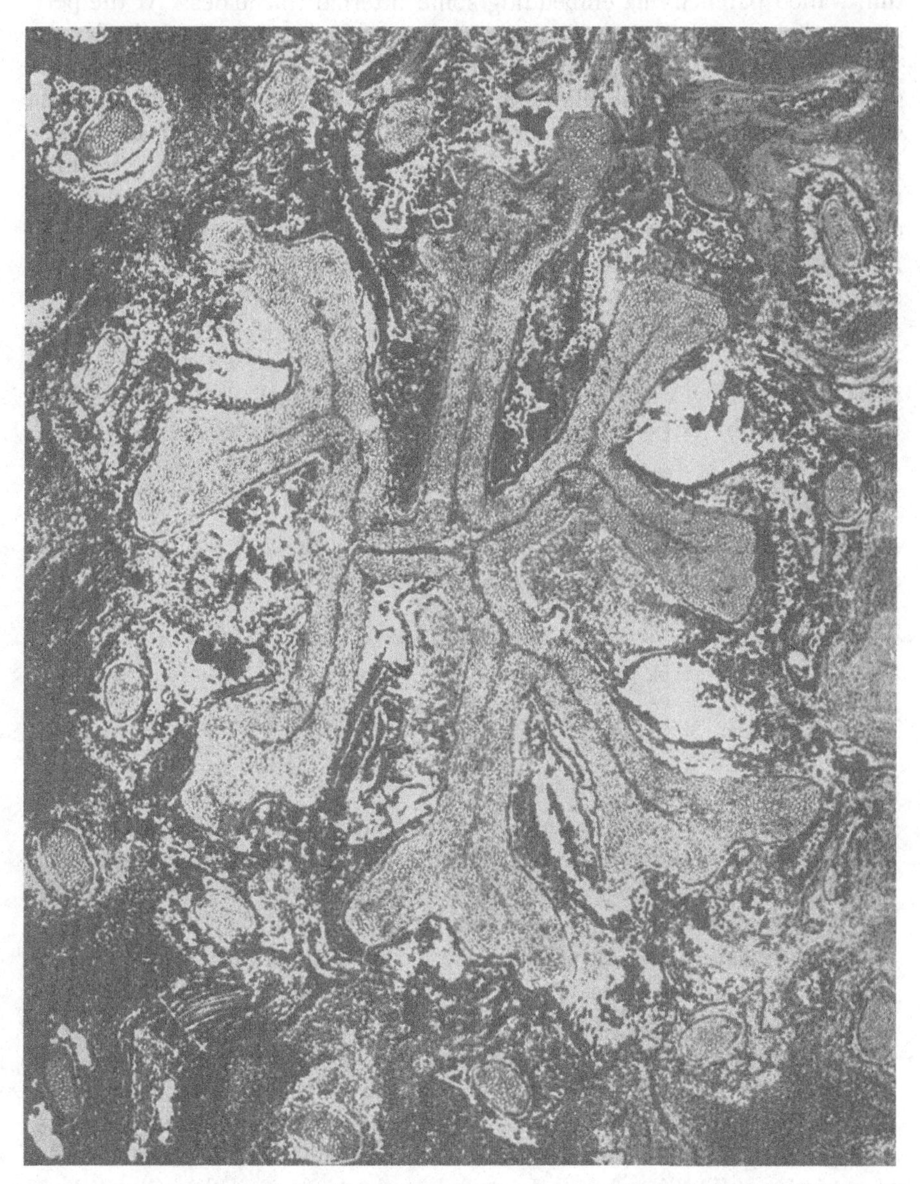

Fig. 323. *Asterochlaena laxa*: specimen showing in transverse section the whole stellate stele, with the neighbouring leaf-traces. Prep. 484 *c*, 2 *a* from the Solms-Laubach collection. The foliar traces are verticillate, with alternating sequence, ten in each whorl. (After P. Bertrand.)

again by three or four layers of cells with more resistant walls, forming a sheath. The internal xylem forms a narrow median band traversing each arm, and consisting of tracheides of small bore, presumably including protoxylem. At the centre of the star these widen into a small mass of thin-walled parenchyma embedding some internal tracheides. At the periphery of the arms a single protoxylem-group passes into each leaf-trace. The trace at its origin is small (1 mm. diam.) in proportion to the whole stele (5·5 mm.), as is also the petiole (6·5 mm.) in proportion to the whole stem (5 cm.). As it passes outwards the trace widens, and the single protoxylem branches into two which pass to the poles of the Clepsydroid trace, so that as it enters the petiole it measures 2 mm. diam. as against the 6·5 mm. diam. of the relatively small petiole. From the margins of the meristele the supply passes out to the two rows of pinnae. Thus it would appear that the type of *Asterochlaena* offers near analogies with that of *Ankyropteris* of which it may be held as a variant. Perhaps the comparison would lie better with *Clepsydropsis australis*, the stem of which has recently been described by Mrs Osborn (*Brit. Ass. Report*, 1915). It appears to have been an upright, fairly large Tree-Fern of the Osmundaceous type.

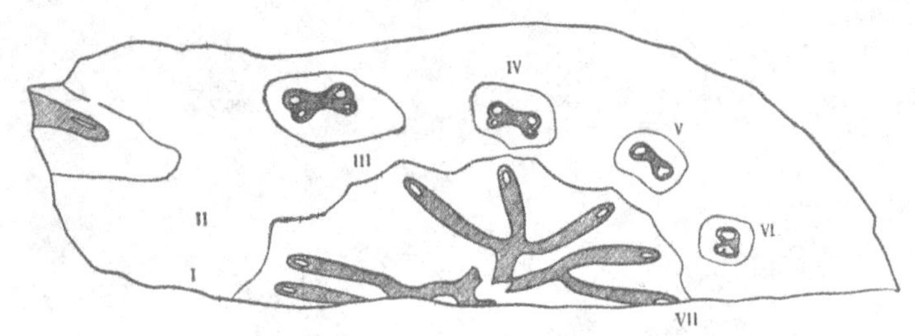

Fig. 324. *Asteropteris noveboracensis*. Incomplete transverse section of stem, showing part of the star-shaped stele and several leaf-traces I–VII. The fracture at the middle of the stele is accidental. (× nearly 3.) (From Dr Bertrand. After Scott.)

If *Asterochlaena* were the only example of a stellate stele in these plants, its stele might be held to be of a late derivative type, since the genus is recorded only from the Permian. But a stellate stele has been described by Dawson from the Portage Beds of New York State of Devonian age, in *Asteropteris noveboracensis*, which is the oldest of the known Zygopterideae. The diameter of the stem is here also considerable (2·5 cm.), and it bears numerous though less crowded leaves. The transverse section shows a xylem-star with about 12 arms, each with a single peripheral loop. Each arm gives off a single trace. The leaves are in verticils which are superposed (Fig. 324). The xylem is simpler than in the previous examples, being homogeneous, without either pith or any distinctive internal system of tracheides. The frond is unknown, but the trace is at first Clepsydroid, passing later into a more complex form.

All the vascular structures hitherto described are primary, but occasionally secondary wood of cambial origin has been seen in *Ankyropteris corrugata* (Scott, *Studies*, i, p. 303, Fig. 136), and indications of secondary growth have also been observed in *Metaclepsydropsis* by Gordon (*l.c.* p. 319). But in these plants cambial activity never took such firm hold as it appears to have done in *Botrychioxylon*, where secondary developments are normally present (Scott, *Trans. Linn. Soc.* vii, 1912, p. 373). The plant showed equal dichotomy of its rhizomatous axis, which bore scattered leaves, forked spines or aphlebiae, and adventitious roots. The stele was cylindrical; it had a central "mixed pith," and protoxylem-groups lay at its outer border. Outside this was a broad zone of radially disposed wood, with cambium at its periphery, and a narrow phloem. The leaf-trace departs in the usual way, and the foliar bundle is that of an ordinary Zygopterid, but at its extreme base the trace shows the unusual feature of secondary structure. Such facts indicate that in the Zygopterideae secondary thickening existed, though it was never turned to full account. Their condition in this respect appears to have been similar to that of the Ophioglossaceae, where cambial activity appears in *Botrychium* so similar as to have suggested to Scott the name of *Botrychioxylon* for this Zygopterid. Though sporadic in each of these families the existence of cambial activity is at least a feature which they hold in common, though it is unusual in Ferns.

The most prominent characteristic of the Zygopterideae is found in the vascular system serving their peculiarly constructed upright leaves. The origin of the leaf-trace from the stele has been described above in several examples. At first it is monarch, but the single protoxylem soon divides in simple cases into two (*Clepsydropsis*), and the two groups then diverge towards the poles of the trace, which has an elliptical outline in transverse section (Fig. 325).

A central type round which the whole series may be grouped has been recognised by Kidston and Gwynne-Vaughan in their "assumed primitive type" (Fig. 325, 6), which was oval in outline, with two protoxylems at the foci of the elliptical solid mass of xylem. It has been seen in *Asterochlaena* that these originated from a single protoxylem by fission. This structure is seen with very slight modification in the "Clepsydroid" trace characteristic of *Clepsydropsis antiqua* from the Lower Carboniferous (Fig. 328), and it appears in *Asteropteris*, the earliest recorded of the family (Fig. 324). Kidston and Gwynne-Vaughan show that from their "assumed primitive type" all the Zygopterid traces may be held to have been derived by modifications in two directions. (1) The two ends of the xylem-strand become more or less extended at right angles to the longer axis of transverse section, taking in simple examples a dumb-bell shape, but in more complex forms growing into long arms, more or less curved; (2) Islands of parenchyma appear in relation to the protoxylems, which also become extended in the same plane as the ends, each protoxylem dividing into two which pass to the anterior and posterior ends of their particular islands.

In applying their theory it must be borne in mind that there are two types of Zygopterid leaves: first those with two rows of pinnae, one on each side of the rachis (Clepsydroideae,

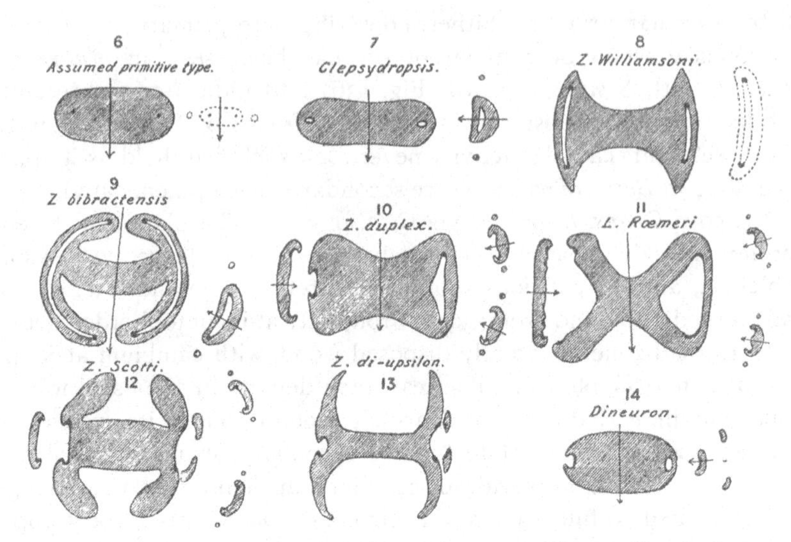

Fig. 325. Diagrammatic comparison of the various types of Zygopteridean leaf-trace. (After Kidston and Gwynne-Vaughan.) (*Trans. Roy. Soc. Edin.* vol. xlvii, p. 472.) For a detailed description see text.

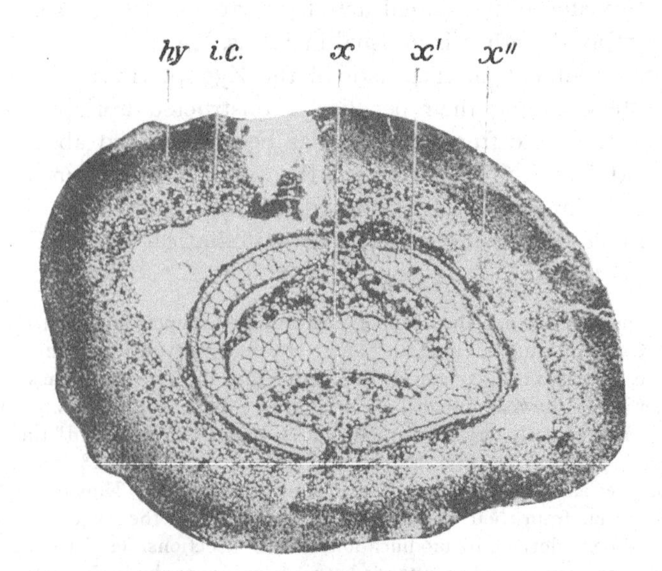

Fig. 326. *Ankyropteris westphalensis*. Transverse section of a petiole, showing the double-anchor form of stele. x, middle band of stele ("a-polar"); x', the main lateral bands ("antennae"); x'', the small-celled external arcs of xylem ("filaments"); the protoxylem lies between x' and x''; *i.c.*, inner cortex; *hy*, sclerenchymatous hypoderma. (× 7.) From a photograph by Mr L. A. Boodle. S. Coll. 914. (After Scott.)

Sahni); secondly, those with four rows, two being on each side of it (Dineuroideae, Sahni). Also a further distinction exists between those in which at the departure of a pinna-trace the xylem-strand of the pinna forms a closed ring (Fig. 325, 7–9), and those in which it has the form of a slightly curved band, which divides into halves (Fig. 325, 10, 11); or the division may actually precede the separation of the trace (Fig. 325, 12, 13). The Clepsydroideae include Bertrand's genera *Ankyropteris* (Fig. 326) and *Clepsydropsis* (Fig. 325, 7, 8, 9). In all of them the pinna-trace departs as a closed ring, and the island extends along the whole length of the arms, except in *Clepsydropsis* where the arms appear to be undeveloped, as they are also in the "assumed primitive type" (6). The Dineuroideae may be divided into two series according to the relative extension of the parenchyma-islands. In the first of these the island extends the whole length of the arms, and the pinna-trace is given off as a long continuous band: as in Bertrand's genera *Metaclepsydropsis* and *Diplolabis*, and in *Z. primaria* (10, 11). In the second series the island hardly extends

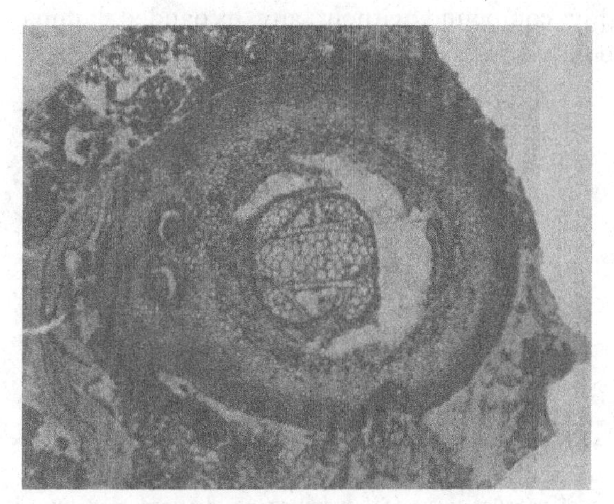

Fig. 327. *Etapteris Scotti.* Transverse section of rachis. In the main bundle, the sieve-tubes on either side of the middle band are preserved. On the left the two traces are entering the common base of the paired pinnae. On the right a pinna-bar is seen. (× 9.) S. Coll. 2009. (From a photograph by Mr W. Tams. After Scott.)

beyond the lateral termination of the cross-bar, and the pinna-trace is usually already double at or near to the point of origin. This includes Bertrand's genus *Etapteris* (12, 13, also Fig. 327). Lastly *Dineuron* is related to this group (14), but it shows a simplicity like that of *Clepsydropsis.* These genera last named may be held as representing a primitive type from which the more elaborate types were derived; and it is significant that that type exists in the ancient *Clepsydropsis antiqua* (Fig. 328), and with slight modification in the Devonian *Asteropteris* (Fig. 324). It thus appears that the Zygopterid leaves advanced far along special lines of development of their own, and that they culminated in the production of two rows of appendages on each side of the rachis. This last condition may well be accepted as the result of a very early dichotomy of each pinna of the single-rowed type.

Notwithstanding the good state of preservation of the rachis, and often of the axis of the Zygopterideae, the knowledge of the upper regions of their leaves is remarkably imperfect.

STAUROPTERIS

It is otherwise with *Stauropteris oldhamia* Binney, a fossil of frequent occurrence in nodules of the Coal Measures; another species (*S. burntislandica*) has been described by P. Bertrand from the Lower Carboniferous of Burntisland. Though the stem of *Stauropteris* is unknown, the structure of the petiole and of the upward ramifications of the erect frond has been followed out in detail, as well as the form and characters of the distal sporangia. The frond had a more nearly radial structure than any of the Zygopterideae, though it is constructed on a plan similar to that of the four-rowed types. The genus is held to be somewhat apart from that family, but the natural relationship to it can hardly be in doubt. The frond was highly compound without any expanded lamina. Its general habit was probably shrubby, with numerous upward-directed branchlets of

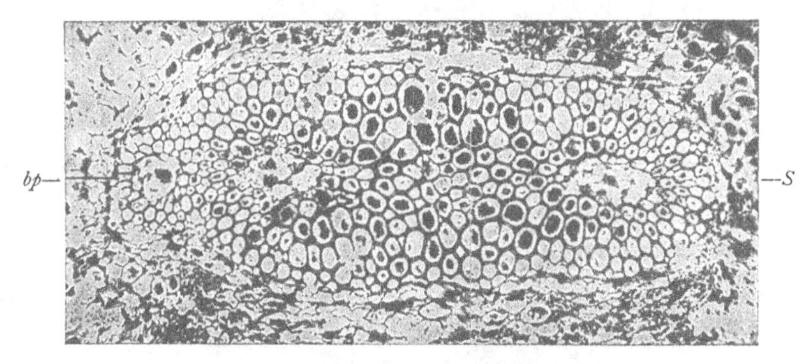

Fig. 328. *Clepsydropsis antiqua* Unger, var. *exigua*. The foliar trace of a primary petiole, with a ring of vascular tissue (*S*) separating off, destined to a secondary petiole. *bp*, peripheral loop. (× 70.) (After P. Bertrand.)

almost cylindrical form supported by a central rachis. The structure of this is shown in Fig. 329. Centrally the xylem is cruciform in section. It is composed of four wedges often united at the centre, but they may be more or less detached. A group of protoxylem lies peripherally near to the outer margin of each, whence the supply for the successive pairs of pinnae is derived alternately right and left. The xylem is surrounded by phloem which intrudes between the wedges with groups of large sieve-tubes. The whole vascular tract takes an almost quadrangular form in the upper regions of the frond (Fig. 330). It is surrounded by a broad band of thick-walled parenchyma limited by an epidermis with stomata; but the hypodermal tract differs from the rest in being a lacunar tissue, probably photosynthetic. It widens out in the upper regions, forming a considerable proportion of the bulk of the branchlets. Comparison has been made with the similar photosynthetic tissue in *Psilotum*.

The appendages appear in alternating pairs right and left of the very slightly bifacial upper region of the rachis. They branch again repeatedly, and the appearance they collectively presented must have been that of "a feathery plexus of delicate green branchlets devoid of lamina, some

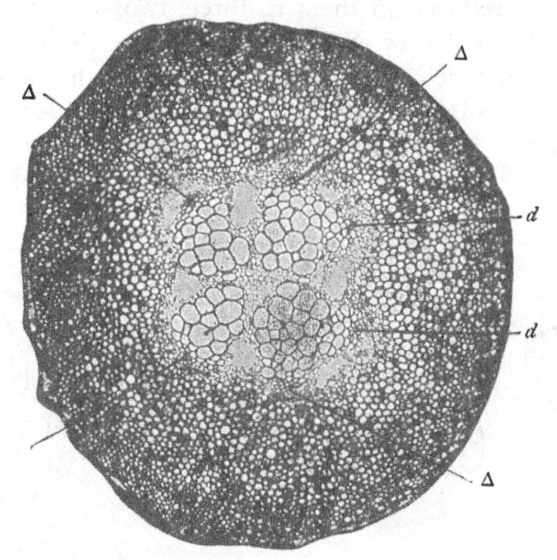

Fig. 329. *Stauropteris oldhamia* Binney. Transverse section of a petiole. Δ, Δ, Δ, Δ, the four fundamental poles; *d*, *d*, poles of protoxylem passing off. (After P. Bertrand.)

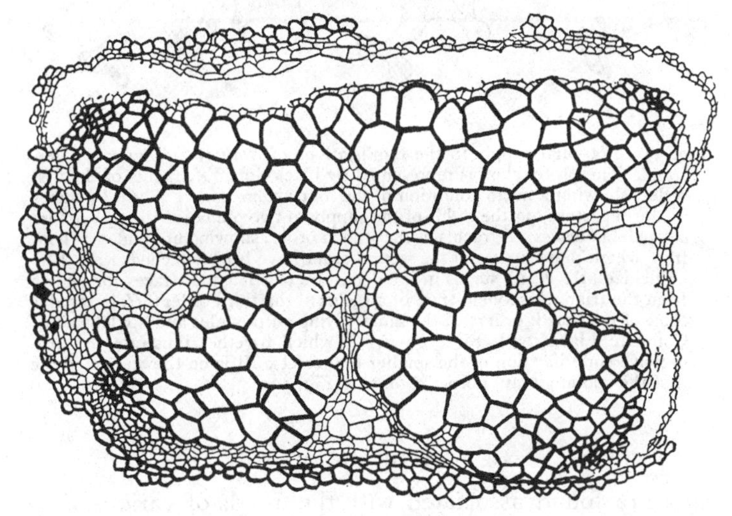

Fig. 330. *Stauropteris oldhamia.* Transverse section of vascular bundle of main rachis, showing the cruciform xylem with protoxylem near the four angles. Groups of large sieve-tubes are seen in the bays of the wood, and small-celled phloem between wood and cortex, and in the middle. (×60.) S. Coll. 2202. (From Tansley, *New Phytologist.* After Scott.)

of which bore terminal sporangia" (Seward). The mode of origin of those branches is shown in Fig. 331. The branching is evidently comparable to that of the Dineuroideae. Structurally their vascular tract becomes simpler in the upper branchlets (Fig. 331, *H*, *I*), the number of protoxylems being reduced in them to three, two, or lastly to one single central strand. Aphlebiae (*a*, Fig. 331) are borne right and left by the successive pairs of pinnae and pinnules. So long as the base of the plant remains unknown it is impossible to say what relation this strange structure bears to other parts. It may perhaps represent the chief organ of a plant in which the distinction of leaf and axis as constituents of the shoot was not existent.

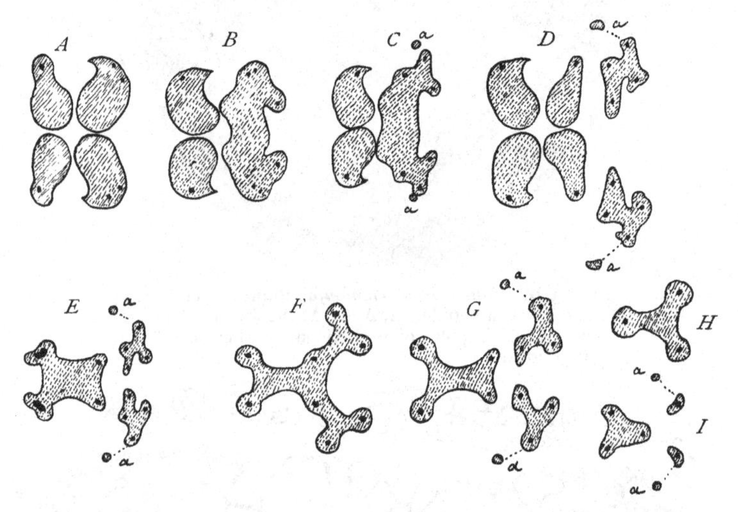

Fig. 331. Vascular supply to the ramifications of *Stauropteris*, after P. Bertrand. The protoxylem is represented by black dots. *A*, rachis of the first order, showing a mean condition of the four xylem-tracts; *B*, *C*, *D*, stages in the departure, on the right, of the supply to two pinnae of the first order; *a*, aphlebia-traces; *E*, rachis of the second order, showing undivided xylem, from which pinna-traces of the second order have been detached, with their aphlebiae (*a*); *F*, *G*, stages in the departure of triangular traces for pinnae from the triangular xylem-tract of a rachis of the third order; *H*, triangular trace of the fourth order; *I*, the same, giving off pinnules and aphlebiae (*a*). Compare Fig. *I* with Figs. *D*, *E*, *G*, which together illustrate the progressive simplification of the smaller branchlets. It is on the ends of these that the sporangia are borne. Compare Fig. 332.

SPORANGIA

Sporangia are found associated with the fronds of various Zygopterids: but it is only in a few instances that their connection has been proved with the parts that bore them. Three good examples are, however, fully established in their connection with the parent plants, viz. *Stauropteris oldhamia*, *Etapteris Lacattei*, and *Corynepteris*. In all of these the position of the sporangia was either terminal, or marginal. Their structure was massive, the wall relatively thick, and the spore-output from each was high, while the opening mechanism was either rudimentary or primitive. In some of them a vascular strand traversed the pedicel to a point closely below the spore-sac itself. These features all indicate a relation to certain living Eusporangiate Ferns: and it is with them that these plants are to be ranked both on the basis of their vegetative and of their propagative organs. On the other hand comparison may be made with the fossils of the Rhynie Chert.

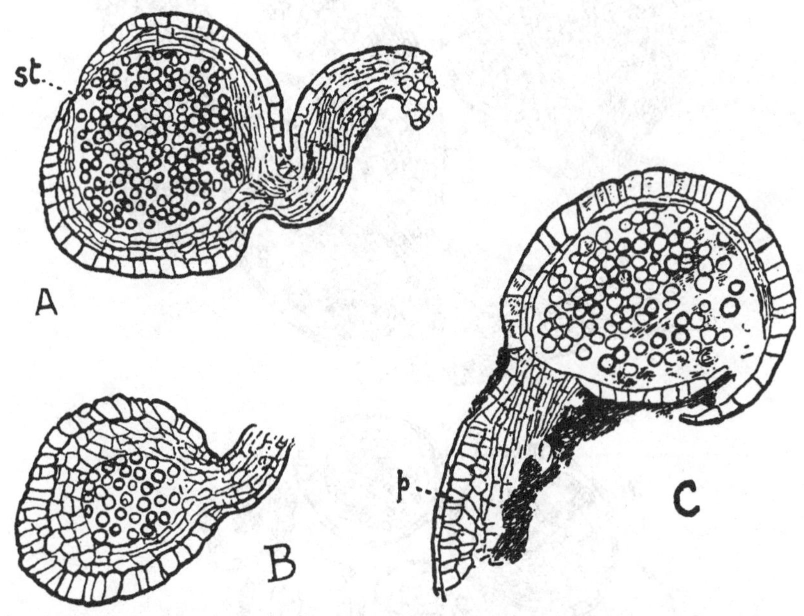

Fig. 332. *Stauropteris oldhamia* Binney. A = sporangium in nearly median section, attached terminally to an ultimate branchlet of the rachis: st = stomium. Scott. Coll. 2213. B = sporangium in tangential section attached to a short piece of the branchlet. Scott. Coll. 2207. C = sporangium with wall burst, attached as before: p = palisade tissue of branchlet. Scott. Coll. 2219. (All figures × about 50.) (From sketches by Mrs D. H. Scott.) (Specimens are from Shore, Littleborough, Lancs.)

The sporangia of *Stauropteris* are isolated on the ends of ultimate branchlets which show the anatomical characters of the plant (Fig. 332). Each is approximately spherical, being about 0·7 mm. in diameter. The sporangial wall is composed of several layers of cells of which the outermost has its cells enlarged and strengthened. There is no specialised annulus, but a well-marked stomium appears at the distal end. A thin cylindrical vascular strand traverses the stalk. The sporangia appear to vary in size, but in a large sporangium about 150 tetrahedral spores appear in a single vertical section, which indicates a spore-output of 1000 or more. The spores are all alike. The inference from these facts is that the sorus is monangial, and its type Eusporangiate, but that these terminal sporangia are of a more rudimentary kind than is seen in any living Fern.

The sections of *Etapteris Lacattei* bearing sporangia, originally described by Renault, have lately been re-examined by Dr P. Bertrand, who confirms the reference of the sporangia to the genus by the characteristic vascular structure of the pedicels. The sporangia were pear-shaped, and were borne

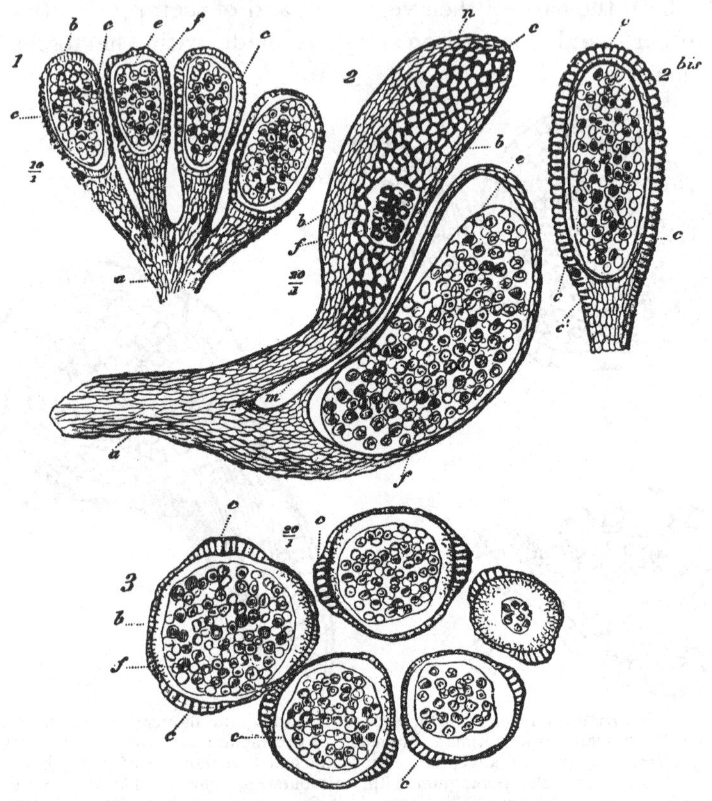

Fig. 333. *Etapteris Lacattei.* 1 = Group of four sporangia on a common pedicel (*a*). (× 10.) 2 = Two sporangia on pedicel: the upper shows the annulus (*c*) in surface view, with spores exposed at *f*; the lower is in section. (× 20.) 2 *bis* = Sporangium cut in plane passing through annulus. 3 = Group of sporangia in transverse section. (× 20.) Lettering common to the figures: *a*, common peduncle; *b*, sporangial wall; *c*, annulus; *e*, tapetum (?); *f*, spores; *m*, pedicel of individual sporangium; *n*, probable place of dehiscence. All after Renault. (From Scott's *Studies.*)

in groups upon a common stalk (Fig. 333), through which a vascular strand passed dividing into smaller strands traversing the several pedicels. The sporangial wall consisted of at least two layers of cells: the outer was formed of larger cells, and was differentiated so that several rows of indurated cells traversed each side of the sporangium vertically to the summit. They thus formed a mechanical loop not unlike that of *Angiopteris* (compare Fig. 404, Chapter XX). Cut transversely the loop appears as a widening of the cells on either side, with thinner regions between them. The orientation of the sporangia thus constructed is evidently round a central point, with one thinner region turned centrally the other peripherally. In fact they constitute a simple uniseriate sorus, notwithstanding that each sporangium is pedicellate. The spores are again numerous: in the larger of the variable sporangia they must have numbered about 1000.

If the sporangial stalks were shortened and the sporangia thus grouped were associated more closely together, the condition characteristic of *Corynepteris* would result (Fig. 334). This genus is believed to be closely allied to *Etapteris*. Hitherto it is known only in the form of impressions. The leaves were of Sphenopterid or Pecopterid type, and the narrow pinnules bore each a single sorus of five or ten sporangia grouped round a common centre.

Fig. 334. *A* = *Corynepteris Essenghi* Andrae, from the Carboniferous (Westphalian). Fragment of a fertile pinna. (× 6.) *B* = *C. coralloides* Gutbier, from the Westphalian. Fragment of a fertile pinna. (× 4.) *B′* = sorus of the same species seen laterally. (× 28.) (After Zeiller.)

The annulus is here again a broad peripheral band, consisting of several rows of cells. The annulus of each sporangium is in juxtaposition with that of its next neighbour. It is as though a group like that of *Etapteris* were condensed upon a common receptacle (compare Fig. 333, 3). It is in fact a radial sorus, essentially of the Marattiaceous type. The existence of a definite annulus to each sporangium makes it probable that the sporangia were not actually coalescent, though closely grouped, otherwise the annulus would be mechanically useless. These three examples suggest a possible origin of the uniseriate sorus, from isolated distal sporangia. If the pedicels bearing these were short and numerous, the sporangia would be disposed as in *Etapteris*. If so much abbreviated as to bring the sporangia compactly together, the receptacle would become a common one with the result as in *Corynepteris*. Thus it would be possible to regard the uniseriate sorus as a condensed and compacted tassel of branchlets each bearing a distal sporangium.

C. ANACHOROPTERIDEAE

The sori of a somewhat isolated type, *Anachoropteris*, from the Bohemian Middle Coal Measures, add a further step of structure (see Scott, *Fossil Botany*, 3rd. ed., Part I, p. 354). The stem is unknown. The leaves are characterised by a wide, strongly reflexed petiolar strand, having its convex curve adaxial, as in *Tubicaulis* (Fig. 153, *C*, Vol. I). The sori were borne on the incurved margins of small pinnules. Each consisted of four sporangia invested by a thick outer wall common to them all, so as to form a synangium. The sorus was borne on a receptacle having vascular supply. The whole structure suggests a small and simple, but marginal synangium of the type of *Marattia*.

The genus *Anachoropteris* Corda is held by Scott (*l.c.* p. 352) as the type of a family separate from the Zygopterideae and Botryopterideae. To this a provisional assent may readily be given.

Naturally nothing is known of the gametophyte of the Coenopteridaceae beyond the fact that the spores did germinate in *Stauropteris*, with the first results not unlike those seen in some modern Ferns.

Comparison

The Coenopteridaceae, comprising the three families now described, have features in common which justify their being grouped together under a comprehensive head, though as Scott remarks (*Studies*, p. 357), "we are at present without any clear indication of their supposed common ancestry." They were all plants of moderate size, some quite small (*Botryopteris*). Their shoot was frequently upright, and this was certainly so with the ancient Devonian *Asteropteris*: but others had clearly a prone habit sometimes with a greatly extended rhizome, as in *Metaclepsydropsis*. The similarity in structure of axis and leaf has been shown in *Botryopteris cylindrica* to be consistent with a theory of origin of leaf and axis from an indifferent dichotomous branch-system (p. 16). The form of the leaves is rarely well known, but the anatomy of the rachis indicates in some of them a form comparable with that of living Ferns: and even the circinate vernation of the young leaves has occasionally been preserved. Simple or curiously septate hairs are present, but no flattened scales, a fact which accords with the primitive position assigned to the family. Their anatomy is clearly based upon a primitive type such as that seen in *Botryopteris*. The stem is in all cases protostelic: but it is liable to elaborations which will presently be shown to be related to their actual size. Those elaborations consist on the one hand in differentiation of the xylem, on the other of modifications of its outline, and also of that of the stele itself. Their petioles were always monodesmic, which is a primitive state. But the meristele may be elaborately moulded: this is seen in less degree in the small plants of the Botryopterideae, but in sections of the larger petioles of the Zygopterideae the meristele shows very complex outlines, which however are all referable to amplifications of the primitive type. It will be shown that this also has its relation to size (see Vol. I, Chapter X). Such structure provides the leading anatomical features by which the Zygopterideae are characterised. The Anachoropterideae show a somewhat similar amplification, but along divergent lines. All these vegetative features accord with the early appearance of these Plants as fossils, and indicate that in them we are dealing with primitive and probably synthetic examples of the Filicales. Their most peculiar feature is the high degree of elaboration of the vascular system seen in some of them, though this is always closely related to the protostele and the monodesmic trace: and it demands some elucidation when found in families so primitive in type and early in occurrence.

If the illustrations of the Coenopteridaceae published in standard works be measured, and the results divided by the magnifying power, it is possible to obtain the approximate actual diameters of the parts together with the diameters of their conducting tracts. If these be tabulated in order, from those which are smallest to those which are largest, it is found that parallel with the increasing measurements there is an increase in complexity of the vascular system. The following table demonstrates the result for the stems of the Coenopteridaceae named, and in the last column the approximate ratio of the diameter of the axis to that of the stele is also given for some of them. The figures can only be approximate, since in certain examples the phloem and sheaths are imperfectly preserved, and the measurements have then been taken from the xylem alone:

TABLE I. *Stems of Coenopteridaceae*

Name	Stem diameter in mm.	Stele diameter in mm.	Remarks	Ratio
Botryopteris cylindrica ...	1·6	·8	solid	2–1
" " ...	2·0	·83	"	$2\frac{1}{5}$–1
" ramosa	2·7	·9	"	3–1
Diplolabis Römeri	—	3·2	differentiated xylem	—
Ankyropteris Grayi	—	4·2	diffd. and fluted	—
" "	17·5	3·75	" " "	5–1
Clepsydropsis australis ...	11·0	5·0	diffd. but not fluted	$2\frac{1}{5}$–1
Cladoxylon insigne	—	5·15	stellate	—
Asterochlaena laxa	35·0	14·4	deeply stellate, diffd.	2·5–1
" "	30·0	15·6	" " "	2–1
Asteropteris noveboracensis ...	26·6	16·0	deeply stellate: not diffd.	1·65–1

The inferences that may be drawn from Table I are these. First, it appears that where the size of the stele is small the protostele has a solid core of xylem (*Botryopteris*). Where the size is larger, as in *Diplolabis*, the xylem may be differentiated internally: where it is larger still (*Ankyropteris*) it is not only differentiated but its cylindrical form may be changed, giving the fluted character, though this is not always proportionate to actual size (*Clepsydropsis australis*). In still larger stems (*Cladoxylon*, *Asterochlaena*) the deeply stellate stele is reached, still with internal differentiation. But this is not always so, for in the most ancient of them all there is a deeply stellate outline of the xylem, but no internal differentiation (*Asteropteris*). As regards the ratio of the diameter of the stem to that of the stele it appears that in the largest the stele may expand in higher degree than the stem as a whole (*Asterochlaena*, *Asteropteris*): but there does not appear to be any constant ratio to absolute size (*Ankyropteris*). The table further brings prominently forward the fact that *Botryopteris*, which has commonly been regarded as a very primitive type, is far the smallest in the actual dimensions of stem and stele, a fact which accords well with the theoretical position.

Applying the same method of analysis to the petioles of the Coenopteridaceae a parallel series of results is apparent, and as the petioles are known in a larger number of examples the results are the more impressive and reliable:

TABLE II. *Coenopterid Petioles*

Name	Petiole diameter in mm.	Vasc. tract diameter in mm.	Remarks	Ratio (approximate)
Botryopteris ramosa ...	—	·65	meristele, simple oval	—
" antiqua ...	—	·7	" "	—
" hirsuta ...	—	1·0	" "	—
Dineuron pteroides ...	3·0	1·0	" "	3–1
Stauropteris oldhamia ...	3·0	1·1	small sized petiole: xylem 4-rayed	3–1
" " ...	—	1·08	4-rayed xylem	—
Asterochlaena laxa ...	—	1·2	meristele oval, with 2 loops	—
Clepsydropsis antiqua ...	—	1·2	" " " "	—
Etapteris Scotti	6·15	2·3	compact double anchor	3–1
Botryopteris forensis ...	—	1·8	three points well developed	—
" " ...	—	2·7	three points very deep	—
Ankyropteris corrugata ...	6·15	1·5	double anchor	4½–1
" "	—	2·4	wide double anchor	—
Asteropteris noveboracensis	—	2·3	Clepsydroid structure	—
Metaclepsydropsis duplex	9·0	2·0	very broad Clepsydroid	4½–1
Tubicaulis Berthieri ...	—	3·2	oval	—
Anachoropteris rotundata	—	4·6	broad inverse horse-shoe	—
Diplolabis Römeri ...	—	5·0	large Clepsydroid	—
Ankyropteris westphaliensis	11·5	6·3	large double anchor	1·8–1
" bibractensis	—	6·8	" " "	—

Putting aside certain anomalies, such as *Stauropteris* and *Asteropteris* which will be specially considered later, the increase in diameter of the petiole runs substantially parallel with the increasing complexity of its vascular tract. The measurements are here based chiefly upon the xylem, and must be held as approximate. The series starts as before from *Botryopteris* with its simple oval meristele of small size. With slightly larger dimensions the Clepsydroid type appears having a dumb-bell-shaped section. Where the dimensions are still larger this type is elaborated in various detail, with formation of the arms or "antennae," and finally they are seen to arch into an almost complete circle in the largest of all, viz. *Ankyropteris* with its striking double-anchor outline (Fig. 326). Meanwhile the ratio of size of petiole to size of the vascular tract gives no very distinct result till the last examples are reached, in which the trace is seen to have expanded in much higher ratio than the petiole that contains it. A word is due on the relatively late *Botryopteris forensis* of Permian origin: for its meristele is much larger than in other species, and it presents the characteristic three points which are specially deep in the large specimen figured by Bertrand. It is significant that this is the latest, the largest, and the most complex of the known species of *Botryopteris*. It would indicate that a phyletic increase in size bears with it increased complexity of vascular structure, and this is consistent with a like interpretation for the Zygopterids.

The ancient *Asteropteris* appears to be anomalous anatomically. It has one of the largest stellate xylem-masses (16 mm. in diameter), but the xylem was undifferentiated. The leaf-traces were, however, relatively small (2·3 mm.), and of the Clepsydroid type. This disproportion it shares with *Asterochlaena*, which has also a Clepsydroid leaf-trace. It seems probable that these two markedly stellate types show a structure consequent on enlargement of the axis and stele without any proportional enlargement of the leaves which it bears. In particular this seems probable for *Asterochlaena*, since there is in it no constant relation of number or position of the leaves to the flanges of the stele. Here stellation and phyllotaxis have not coincided. This discrepancy is readily intelligible if stellation be an independent consequence of increase in size. (See *Proc. R. S. E.* Vol. xliii, p. 117, 1923.) Other early Fern-like plants still imperfectly known, of which *Cladoxylon* is quoted in the table as a single example, suggest that in them also decentralisation and even disintegration of the vascular tract followed upon increase in size. (Solms-Laubach, *Abh. d. K. Preuss. Geol. Landesanstalt*, Heft 23, 1896; P. Bertrand, *Progressus*, IV, ii, pp. 249–254.)

Lastly, *Tubicaulis* and *Anachoropteris*, with their relatively large and simple meristeles in the petiole (3·2 mm. and 4·6 mm. diam.), appear to be also anomalous, other traces of similar dimensions being Clepsydroid. Their margins are revolute so that the convexity is directed adaxially (Fig. 315).

They represent a distinct line of development of the trace designated by P. Bertrand the "Inversicatenal" type. However different this type may be from the rest of the Zygopterids it has this in common, that the vascular tissue is decentralised, and is moulded into a curved tract which runs parallel to the outer surface of the petiole, while ventilating-connection between the outer cortex and the centrally-lying mass of parenchyma is maintained through the gap between the recurved margins. This state is cognate with, but complementary to, the **C**-shaped leaf-trace so usual in monodesmic Ferns of the present day (see Figs. 429, 432). In them, however, the converging margins point centrally. From the physiological point of view the result is the same as in *Anachoropteris*, though phyletically quite distinct. There is in fact an underlying principle related to increase in size of monodesmic petioles, which secures decentralisation while maintaining ventilation. This end has been gained in three different ways in primitive Ferns: (i) by the Zygopterids, (ii) by the Inversicatenal type, and (iii) by ordinary monodesmic living Ferns. Probably each is a reliable diagnostic character for its distinct phylum, though the three may all have originated from a common monodesmic source having a simple form of trace.

The upper leaf, though insufficiently known in most of the Coenopteridaceae, offers still more interesting characters. It must suffice to say that certain Coenopterids appear to have had pinnate leaves of an ordinary Fern-like type: but the interest centres round those Zygopterids which have their pinnae arranged in four rows or orthostichies. The problematical plant *Stauropteris* is naturally ranked with these (Figs. 325, 331). The base of this very common fossil of the Coal Measures is unknown, though the details of its leaf, with those of its solitary distal sporangia, have been fully made out (see above, p. 30). The question is then whether this strange plant is to be held as a very primitive type, perhaps stemless, with upward, radially constructed, branching shoot, or whether it is to be held as a specialised derivative from the Zygopterids. Till the base of the plant is known it seems useless to speculate: but it is significant in its bearing on the question that its sporangia are probably as primitive, in structure and in position, as any that have been referred to the Filicales.

The Zygopterideae have been divided into two groups according as the pinnae are biseriate (Clepsydroideae), or quadriseriate (Dineuroideae) Sahni. Scott points out that as regards antiquity there is nothing to choose between them, but that the early occurrence of *Stauropteris* shows that the origin of the quadriseriate type was itself very early. It has been suggested that the paired pinnae of the Dineuroideae resulted from a precocious dichotomy of primary pinnae. If that be true, the biseriate type would have preceded the quadriseriate. On the other hand, Lignier has suggested the derivation of the biseriate from the quadriseriate state by fusion. Kidston and Gwynne-

Vaughan point, however, to the great similarity of the traces of *Dineuron* (Fig. 325, 14), and of *Clepsydropsis* (Fig. 325, 7), and they suggest that their assumed ancestral form (Fig. 325, 6) was really the common starting-point for both groups. This appears to be a probable conclusion: both types are of early occurrence, *Clepsydropsis* from the Upper Devonian and *Dineuron* from the Culm. The difficulty that the more complex types of stem-structure are found among the biseriate Zygopterids need not be held as a serious obstacle. If as above suggested the stellate structure of the stele is a function of size, any Clepsydroid stem that increased in bulk might be expected to acquire it: and the more advanced stelar structure seen in either *Asteropteris* or *Asterochlaena* might be associated with a simple leaf-trace if the leaf did not increase in proportion. Accordingly it seems not improbable that from some simpler source two parallel sequences arose, the Clepsydroideae and the Dineuroideae. But what the actual relation of *Stauropteris* was to these must remain uncertain: though the relatively small size of its petiole and conducting tract, and the character of its sporangia, would indicate for it a relatively primitive position, related perhaps to *Dineuron*.

Those sporangia of the Coenopteridaceae which are sufficiently well known as regards structure and relation to the plants that bore them, present features interesting for comparison. They are all of the Eusporangiate type, with thick stalks and walls, a large spore-output of homosporous spores, and simple mechanical arrangements for dehiscence but not for ejection of the spores. They may be seriated according to their position and relations so as to illustrate steps which may well have occurred in the evolutionary origin of a simple uniseriate sorus, so prevalent among primitive Ferns. But in placing them in this sequence it must not be assumed that the examples actually formed a phyletic series: their seriation can only be used here for the purpose of illustrating a probable method of origin of a familar type of sorus.

A primitive state is that of *Stauropteris*, with solitary distal sporangia borne on the ultimate branchlets of a much-divided leaf. This position corresponds to that seen in the Psilophytales. If the pedicels were abbreviated and crowded together the condition seen in *Etapteris* would result, the position being still distal, with the individual sporangia facing inwards. Further compacting of such a group would lead to the sporangia being seated on a common pedicel, again facing inwards, as in *Corynepteris*: but here the position is no longer distal but lateral. A further step illustrated in *Anachoropteris* gives synangial fusion of the sorus, which was here borne on the incurved margin of the small pinnules. A "phyletic slide" of the sorus to the lower surface of a widening leaf-segment, such as is seen in many phyla of living Ferns (Chapter XII), would then give a soral state similar in structure and position to that seen in the Marattiaceae.

It thus appears that while the material facts are still insufficient for any trustworthy phyletic arrangement of the Coenopteridaceae among themselves, they nevertheless provide, as any synthetic group of organisms may well do, suggestions of value in relation to the principles used for that purpose. Their early occurrence coupled with the small size and simple structure of some of them; their dichotomous branching, the indeterminate distinction of axis and leaf; their simple hairs, not scales; and their primitive sporangia with numerous homosporous spores, collectively mark them out as generalised types. They ran into several lines of complexity of leaf-construction based upon dichotomy: they illustrate increasing vascular complexity in stem and leaf accompanying greater size of those parts: also advances are seen in the grouping and relations of sporangia. All of these being biologically probable advances may be recognised as valid, and probably true. They appear however to have been carried out separately and in various distinct types; and in some instances, such as the complex petiolar structure, not perpetuated in any more recent line of Ferns. Much may thus be learned that is of value in fern-morphology from these plants, which represent more nearly than any other known organisms the source from which the Class of the Filicales probably took its origin. It is significant that the genus *Botryopteris* is not only relatively small among the Coenopterids, but also very simple in construction, while the larger Zygopterids are more complex anatomically. It would appear probable that the former approach more nearly than the latter to some primitive stock. It has been suggested in Chapter XVII (Vol. I) that an archetype for Ferns may not improbably have shared the general characteristics of the Flora of the Rhynie Chert: and the problematical plant *Stauropteris* affords perhaps the most valid basis for comparison with those primitive fossils.

BIBLIOGRAPHY FOR CHAPTER XVIII

323. SEWARD. Fossil Plants. Vol. ii, p. 432. 1919.
324. SCOTT. Studies in Fossil Botany. 3rd Edn. 1920, p. 337.
325. RENAULT. Ann. d. Sci. Nat. Sér. vi. Vol. i. 1875.
326. RENAULT. Flore Foss. d'Autun et d'Épinac. Part ii, p. 33.
327. RENAULT. Cours de Bot. Foss. Vol. iii, Chap. viii.
328. WILLIAMSON. Plants of the Coal Measures. Part ix. Phil. Trans. 1878.
329. STENZEL. Die Gattung *Tubicaulis* Cotta. Cassel. 1889.
330. STOPES. *Tubicaulis Sutcliffii*. Mem. and Proc. Manchester Litt. and Phil. Soc. Vol. i. 1906.
331. KIDSTON. Trans. Roy. Soc. Edin. xlvi, Part ii. 1908.
332. SCOTT. Sporangia attributed to *Botryopteris antiqua* Kidston. Ann. of Bot. xxiv, p. 819. 1910.
333. BENSON. *Botryopteris antiqua*. Ann. of Bot. xxv, p. 1045. 1911.
334. CHAMBERS. Axillary strands of *Trichomanes*. Ann. of Bot. xxv, p. 1037.
335. BERTRAND. *Tubicaulis Berthieri*. Mém. de la Soc. d'Hist. Nat. d'Autun. xxiv. 1911.

336. BERTRAND. Progressus Rei Bot. Vol. iv. 1912. L'Étude anatomique des Fougères anciennes, p. 182.
337. BANCROFT. *Rachiopteris cylindrica.* Ann. of Bot. xxix, p. 531. 1915.
338. SOLMS-LAUBACH. Abh. d. K. Preuss. Geol. Landesanstalt. Heft 23. 1896.
339. GORDON. Structure and affinities of *Metaclepsydropsis duplex.* Trans. Roy. Soc. Edin. xlviii, Part I. 1909.
340. KIDSTON & GWYNNE-VAUGHAN. Fossil Osmundaceae. Part IV. Trans. Roy. Soc. Edin. xlvii, Part III. 1910.
341. GORDON. Structure and affinities of *Diplolabis Römeri.* Trans. Roy. Soc. Edin. xlvii, p. 711. 1911.
342. BERTRAND. Structure des stipes d'*Asterochlaena laxa.* Mém. de la Soc. Géol. du Nord. vii, I. Lille. 1911.
343. BERTRAND. Nouvelles remarques sur la Fronde des Zygoptéridées. Autun. 1911.
344. BERTRAND. Études sur la Fronde des Zygoptéridées. Lille. 1909.
345. TANSLEY. Lectures on the Evolution of the Filicinean Vasc.-Syst. New Phyt. Reprints. 2, p. 11. 1908.
346. JONES. *Lycopodium.* Trans. Linn. Soc. Vol. vii. 1905.
347. LIGNIER. *Stauropteris oldhamia.* Bull. Soc. Bot. de France. 1912.
348. SAHNI. On the branching of the Zygopteridean leaf. Ann. of Bot. xxxii, p. 369. 1918.
349. OSBORN. An Australian *Zygopteris.* Report Brit. Ass. 1915, p. 727.
350. SAHNI. An Australian *Clepsydropsis.* Ann. of Bot. xxxiii, p. 81. 1919.
351. BOWER. Proc. Roy. Soc. Edin. xliii, p. 117. 1923.

CHAPTER XIX

OPHIOGLOSSACEAE

THE preceding Chapter has dealt with organisms known only in the fossil state. This Chapter will treat of plants known as living objects. Scott remarks (*Studies*, Part i, 3rd Edn., p. 362) that "unfortunately nothing is known of the geological history of the Ophioglossaceae: we are therefore driven to compare directly a recent with a Palaeozoic group of plants." Nevertheless the points of agreement "seem sufficiently to justify the opinion that the Ophioglossaceae have more in common with the Botryopterideae than with any other known group of plants. The affinity must, however, be a somewhat remote one." That is the general result of recent comparisons, and it is well to state it in the terms quoted, so as to give at the outset some indication of the view to be developed here from the facts relating to the Ophioglossaceae, or Adder's Tongue Ferns[1].

The Ophioglossaceae include three genera of living plants: *Helminthostachys* with one species only, *Botrychium* with 34 species in Christ's Index, and *Ophioglossum* with 43 species. They all have well-marked characters in common, so that there is no doubt of their natural affinity. Their most distinctive feature is the fertile spike, a process which rises from the adaxial surface of the leaf, and serves as a basis for insertion of the sporangia: these are marginal in position upon it; they are of the Eusporangiate type, and are without any annulus. Their individual spore-output is very large. The plants of this cosmopolitan family are all perennials; they are mostly xerophytic and ground-growing, though some few are epiphytic. The method of their perennation is closely connected with their habit. The shoot is markedly megaphyllous. The stock is usually short, upright, and unbranched. It is buried in the soil, and is attached by thick radiating roots. The leaves, which are always stipulate, are borne in spiral sequence in the upright forms, and as a rule one large leaf is expanded in each season: rising above the soil, it carries out both photosynthesis and spore-production, and it dies down at the resting season, leaving the stock stored with nutrition for the next year. This is the habit for *Botrychium* and *Ophioglos-*

[1] Nevertheless, it may be well to bear in mind the existence of early fossils which have been referred to an Ophioglossaceous, or perhaps with equal claim to an Osmundaceous, affinity: such as *Phacopteris paniculifera* Stur. (E. and P. i, 4, Fig. 288). The reference rests rather on general habit than on details of structure, or on the propagative organs. Until these are better known, it seems wise to hold any decision in suspense.

sum, but in *Helminthostachys*, though its axis is vertical in the sporeling, there is a creeping rhizome in the adult, and the leaves are disposed in two rows on its upper surface. These facts point to an upright radial stock as the original form for them all, though it is departed from in the adult *Helminthostachys*, in which the shoot is dorsiventral.

The form of the sterile leaf-blade defines the three genera. In *Helminthostachys* it is usually ternate, each of the divisions being again divided: the venation is open (Fig. 335, *E*). In *Botrychium* the blade is variable in outline: in the smallest it may be simple and unbranched, in others it may be simply pinnate, as in *B. Lunaria*; or it may be doubly, trebly, or even four times pinnate, as in *B. virginianum*: but still the venation is always open (Fig. 335, *D*). This is in strong antithesis to the reticulate venation of *Ophioglossum*, in which the leathery blade is usually entire (Fig. 336, *C*), though in some species it may be irregularly lobed. Reticulation of the veins is general for the genus, and it is found even in the minute sterile blade of *O. Bergianum*. This is held to be a more advanced state, and being combined with other features it suggests that in *Ophioglossum* we see a later and derivative type of the Family. *Botrychium* and *Helminthostachys* will therefore be described first.

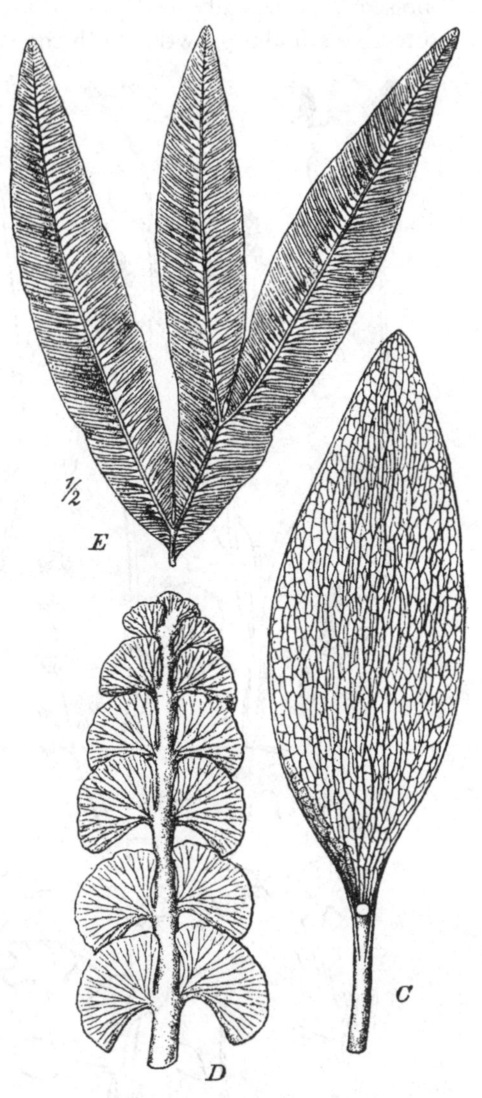

Fig. 335. *C*, *Ophioglossum vulgatum* L., the sterile blade, showing reticulate venation. *D*, *Botrychium Lunaria* Sw., sterile blade showing open dichotomous venation. *E*, *Helminthostachys zeylanica* Hook., a single lateral segment of the sterile blade, showing open dichotomous venation. (Natural size.) (After Diels, from Engler and Prantl.)

In the genus *Botrychium* the short upright stock is sheathed by the dry bases of the older leaves, and is usually unbranched. The plant is monophyllous as a rule, though occasionally two or more leaves may be expanded in one season, especially in the larger species. Each leaf bears at its base a protective stipular growth. Both the sterile blade and the fertile spike are

Fig. 336. *Botrychium simplex* Hitchc. Series of forms, *a–f*, *forma simplissima* Lasch.; *g–k*, *forma incisa* Milde; *l*, transitional to *forma subcomposita* Lasch.; *m*, has an enlarged fertile basal segment of the sterile blade; *o*, *r*, *forma composita* Lasch., *r* with four primary segments of the sterile part. From Luerssen, in *Rab. Krypt. Flora*: the drawings were from specimens in his herbarium. (Natural size.) Probably they illustrate progressive states of ontogenetic development.

highly variable within the genus. In the smallest forms both parts may be unbranched, as in *B. simplex* (Fig. 336, *a–d*): but the commonest condition is that where the sterile blade is simply pinnate, and the spike the same, or more highly branched, as it is sometimes in *B. Lunaria* (Fig. 337). In other species of larger size the sterile blade may be three or four times pinnate, and the fertile spike shows corresponding complexity, as in *B. daucifolium* or *virginianum*. The genus as a whole shows such gentle gradations of change from the simplest to the most elaborate that the unity of type is unmistakable. Various abnormal modifications of the leaf have been described for *Botrychium* involving the formation of accessory parts, such as doubling of the sterile blade, or increase in number of the fertile spikes: but no species is recognised in which these changes have become permanent. Abnormalities involving the distribution of the sporangia are the most important. All stages of the vegetative development of the fertile region have been found, even leading to the complete replacement of the sporangia by a structure like the normal sterile blade. But on the other hand it is not an uncommon thing for sporangia to appear upon the sterile region of the leaf (Fig. 338). Moreover not a part only, but even the whole

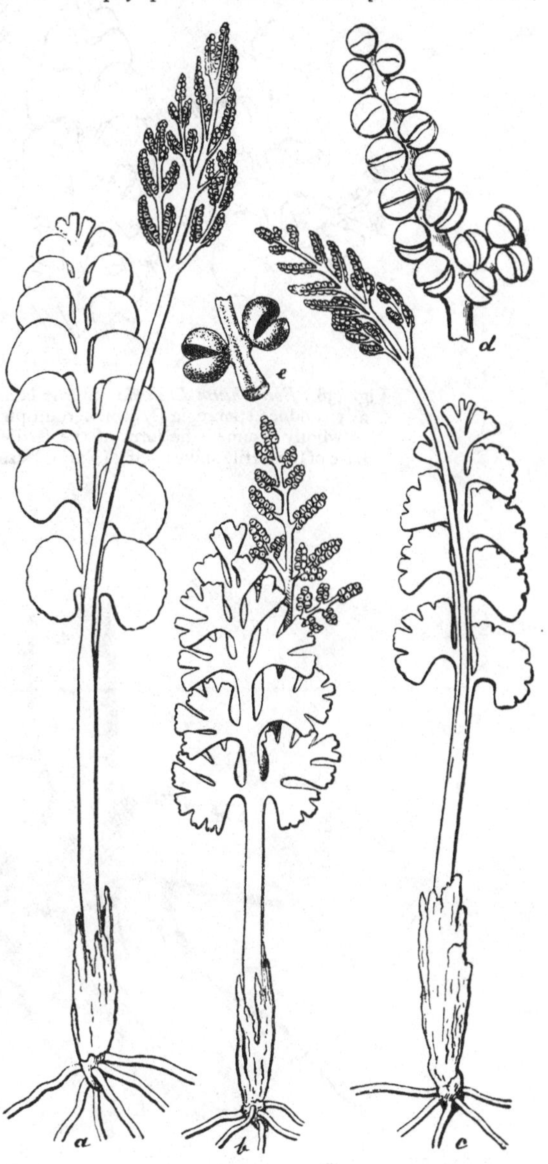

Fig. 337. *Botrychium Lunaria* Sw. *a*, *forma normalis* Roeper; *b*, var. *incisa* Milde; *c*, var. *subincisa* Roeper. (All natural size.) *d*, part of the fertile spike, with open sporangia, enlarged. *e*, two open sporangia somewhat bent asunder to show their attachment, enlarged. From Luerssen, in *Rab. Krypt. Flora*.

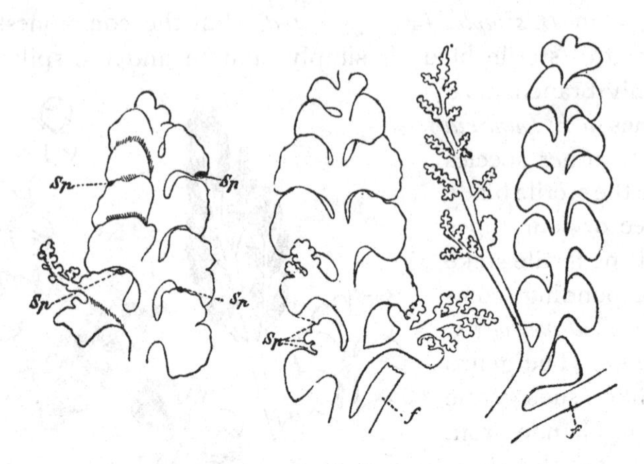

Fig. 338. *Botrychium Lunaria*. Sterile laminae, which occasionally produce sporangia (*sp.*) on certain pinnae, and have partly or wholly assumed the form of the fertile spike; *f*, shows the base of the fertile spike itself. (Natural size. After von Goebel.)

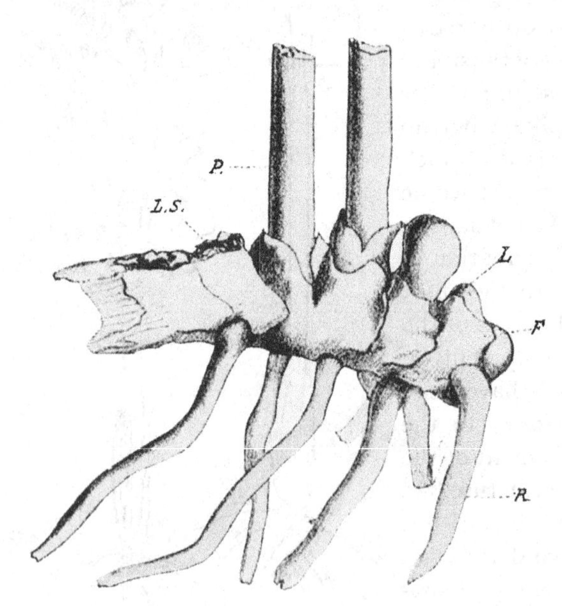

Fig. 339. *Helminthostachys zeylanica* Hook. Rhizome. (Natural size.) F = flap; R = root; L = leaf; P = petiole; $L.S.$ = leaf-scar. (After Farmer and Freeman.)

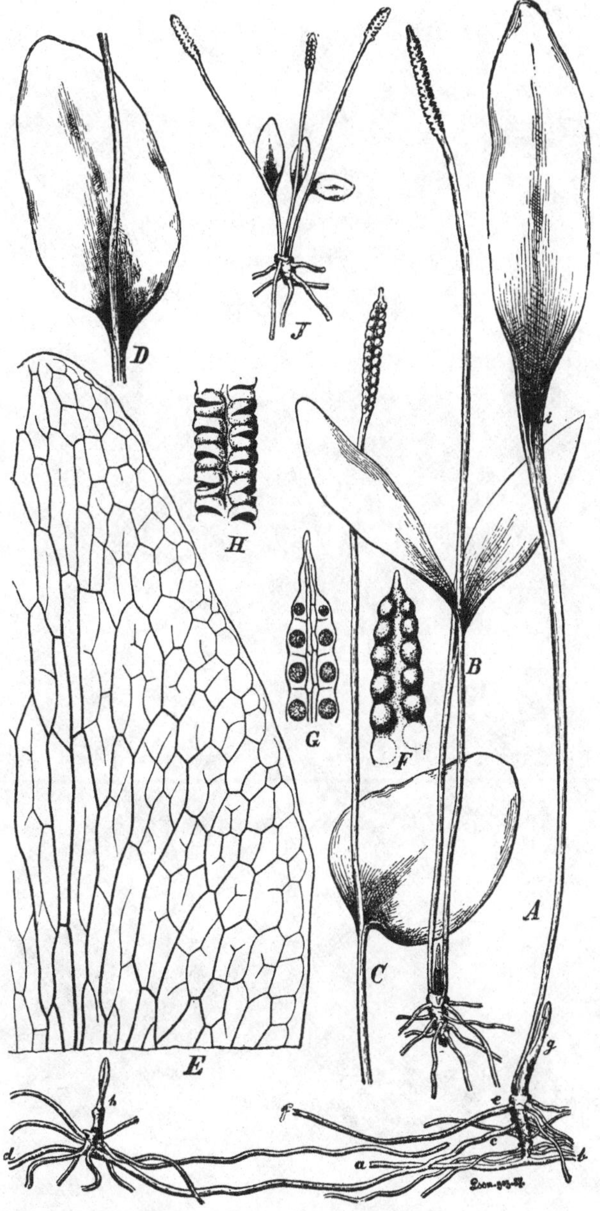

Fig. 340. *Ophioglossum vulgatum* L. *A*, old plant sprung as an adventitious bud from the root *a*, *b*: from its stem have sprung further roots (*c–d*, *e–f*), from one of which again an adventitious bud (*h*) has arisen; *g*, the leaf for the next vegetative period, still folded; *i*, an abortive spike attached to the expanded leaf. (After Stenzel.) *B*, an old plant with one sterile and one fertile leaf. *C* and *D* show form of leaf, with spike, and *E* the venation. *F*, *G*, *H*, details of spike. *J*, *Ophioglossum vulgatum* var. *polyphylla* R.Br. The figures *A–D* and *J* are half natural size. (From Luerssen, *Rab. Krypt. Flora.*)

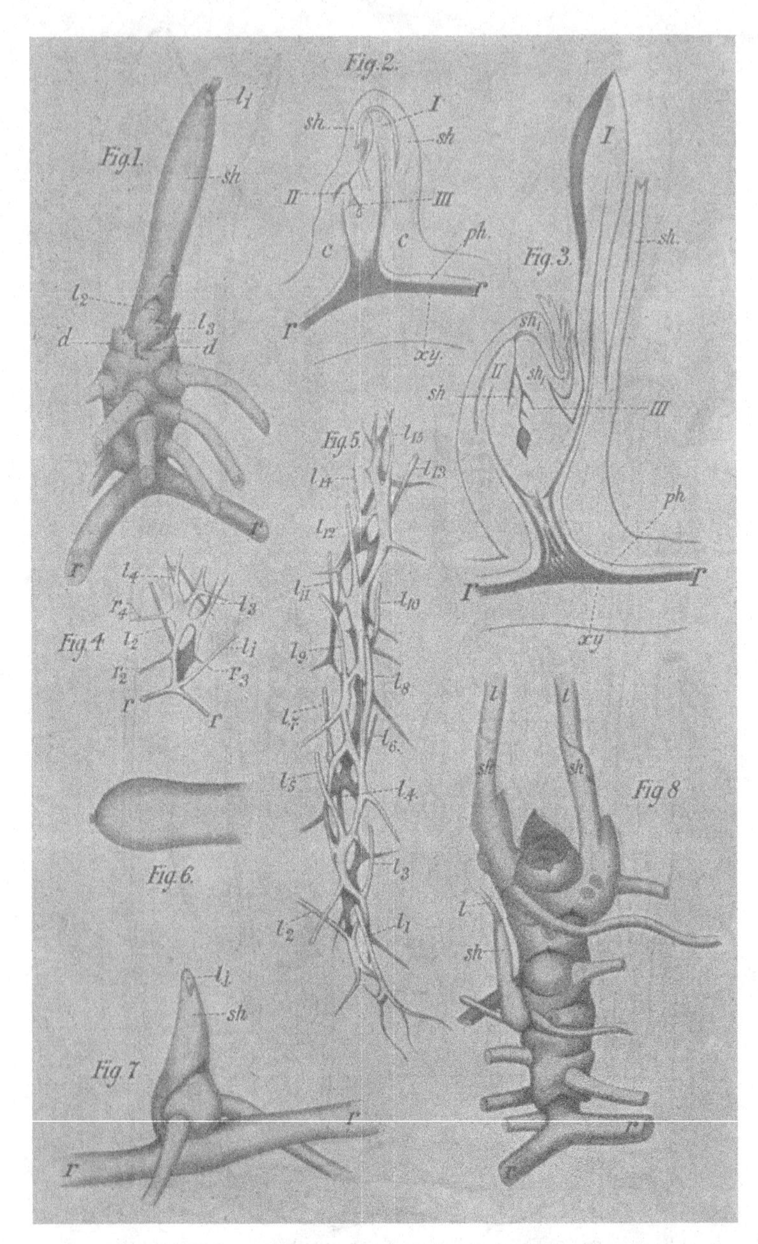

Fig. 341. *Ophioglossum vulgatum.* Fig. 1. Adult plant towards end of autumn; l_i, leaf of succeeding summer; *sh*, its sheath; l_2, second leaf; l_3, third leaf; *d*, debris of dead leaves; *r*, parent root. Fig. 2. Longitudinal section of a very young bud. *I*, *II*, *III*, leaves: *c*, cortex; *sh*, sheath; *xy*, xylem; *ph*, phloem. Fig. 3. Longitudinal section of an older bud, where the first leaf (*I*) is expanded; *sh*, sh_i, *sh*, sheaths of successive leaves. Fig. 4. Central cylinder of a very young bud prepared by maceration; *r*, parent root; l_i–l_4, traces of successive leaves; r_2–r_4, successive roots. Fig. 5. Central cylinder of an adult stem: l_1–l_{15}, the traces of successive leaves. Fig. 6. Enlarged apex of a root; the first appearance of a bud. Fig. 7. Bud slightly developed, where the first leaf has just pierced the sheath. Fig. 8. A false branching. (After Rostowzew.)

of the normally sterile blade may be involved. The importance of this from a theoretical point of view will be discussed later.

The adult *Helminthostachys* has a creeping rhizome, usually unbranched, bearing leaves alternately in two rows on its upper surface (Fig. 339), while roots which branch monopodially and are hairless spring from its flanks and undersurface. The stout petiole rising from a flap-like basal stipule bears the large ternate lamina, from the adaxial face of which springs the fertile spike. Borne right and left upon this are serried ranks of sporangiophores covering the margins, and each of them may carry a number of sporangia (Fig. 364, p. 68, *F*, *G*: also Fig. 372, p. 73).

The spike of *Helminthostachys* is often subject to accessory branchings, and these may be combined with correlative vegetative growth where sporangia are absent, as in *Botrychium*: the details show that a balance may subsist between the vegetative and the sporangial development (Fig. 344, *L–N*). Such changes are in line with those seen in other Ophioglossaceae.

The genus *Ophioglossum* is subject to considerable variety in detail of its numerous species, among which the Common Adder's Tongue, *O. vulgatum* L., occupies a middle position (Fig. 340). The plant consists of a short upright stock, covered externally by the scars of leaves of previous years: thick, hairless roots, which are commonly unbranched though occasionally showing dichotomy, radiate from it, one being inserted as a rule below the base of each scar; but this arrangement is not rigidly maintained here, and is departed from in other species, as well as in the other two genera. The stem terminates in a bud, and according to season the outermost leaf, or sometimes two of them, may be extended above ground: or the bud may be still enveloped by the ochrea-like stipule of the preceding leaf (*sh*, Fig. 341, 1–3). The apex of the axis is buried deep down among the successive leaves, and each of these is provided with a large stipular sheath covering all the younger leaves. There is no circinate vernation. Each leaf develops during three years; expanding in the fourth year it bursts the sheath of the preceding leaf, and forces its way upwards through the soil. The broadly ovate sterile blade finally unfolds as a fleshy expansion with reticulate venation (Fig. 340, *D*, *E*). From its upper surface, at the point of junction with the petiole, springs the fertile spike, a stalked body bearing along either margin of its upper part a dense row of sunken sporangia: but the tip of the spike is sterile (Fig. 340, *G*). Dichotomous branching of the shoot has been described by Poirault and by Petry (*Bot. Gaz.* 1915, p. 345), but it is rare: the deficiency is made up by the formation of buds which may appear either in relation to the axis as axillary buds (Fig. 341, 8), or more frequently upon the roots where they arise close to the apex (Figs. 340, *A*; 341, 7). The species is propagated freely by these root-buds.

The genus *Ophioglossum* is a variable one, and to obtain a conception of it as a whole other species than *O. vulgatum* must be examined. They are not all habitually monophyllous: several small species bear a plurality of leaves, for instance *O. Bergianum* (Fig. 342), where the spike is attached very low down on the narrow leaf, and the sporangia are few. In *O. crotalophoroides* Walt. four to six leaves may be simultaneously expanded, and this may go along with a bulbous distension of the storage stock. This condition is most frequent in small-leaved forms, and it may be held to connect the monophyllous habit with the polyphyllous strobiloid construction common in other Pteridophytes. But the genus also shows a capacity for amplification of the parts of the individual leaf by branching beyond the typical simplicity, as is seen especially in *O. pendulum* and *palmatum* (Figs. 343, 344): somewhat similar branchings are not uncommon in *O. vulgatum* itself, (Fig. 344, *J*, *K*) and other species. The large series of examples in Kew and the British Museum illustrate how frequently forking is related to the production of numerous spikes, or to the branching of the single spike: while the position which the spikes hold is usually not marginal, though this may occur occasionally. The majority of them are inserted upon the upper surface of the sterile blade, while the lowest of them is commonly nearest to the median line (compare Vol. I, Fig. 36, p. 30).

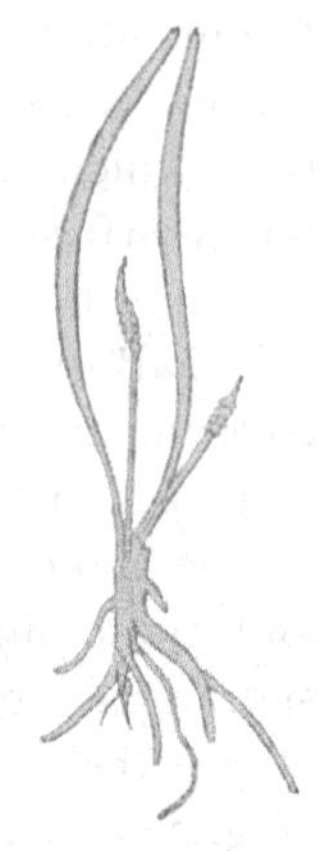

Fig. 342. *Ophioglossum Bergianum* Schlecht. (Whole plant, slightly reduced.)

There is a rough though not an exact parallelism between the number of the fertile spikes on a frond and the number of the lobes of the sterile blade. From 70 specimens of *O. palmatum* examined, ranging from those with a single sterile lobe to eleven, and from one fertile spike to seventeen, the totals come out as sterile lobes 328 and fertile spikes 373. This shows that there is a substantial parallelism, though it cannot be pursued into numerical detail. It is plain also that the leaves with the most lobes are those which are broadest: or speaking generally the number of fertile spikes increases with the assimilating area. The morphology of the ophioglossaceous spike will be discussed later (pp. 66, 87). Meanwhile it may be remarked that the facts for *Ophioglossum*, when a plurality of spikes is seen, do not appear consistent with any direct recognition of their pinna-nature. They point rather to some hypothesis of chorisis of a single original spike holding a median position; and they suggest the view that in *Ophioglossum* the fertile spike, whatever its morphological origin may have been, had become a morphological entity of the type seen in normal specimens of *O. vulgatum*, and that this was liable to fission and other morphological changes of form and of exact position. But these changes are no more suitable material for direct argument as to the nature of the spike than are the stamens or carpels of monstrous angiospermic flowers for the direct elucidation of the morphology of those parts.

Besides such amplifications within the genus there is also a line of simplification in *Ophioglossum*. It culminates in *O. simplex* Ridley (Fig. 345). This ground-growing mycorhizic plant has tall fertile spikes without any sterile blades. Anatomically as well as in form it resembles the epiphytic *O. pendulum*, but it is closer still to the rare *O. intermedium* Hook, which is also a ground-growing species. It appears probable that *O. simplex* forms the end of a series of reduction of the vegetative system consequent upon a mycorhizic habit and a shaded habitat. Here it would seem that the mycorhiza makes the nutrition of the large spike still possible in the dense wet forest in which the plant grows, notwithstanding that the usual assimilating organ is functionally absent. Reduction is, however, not apparent in the large spike itself: for provided nutrition be kept up from whatever source it would still retain its character, being essentially a spore-bearing organ.

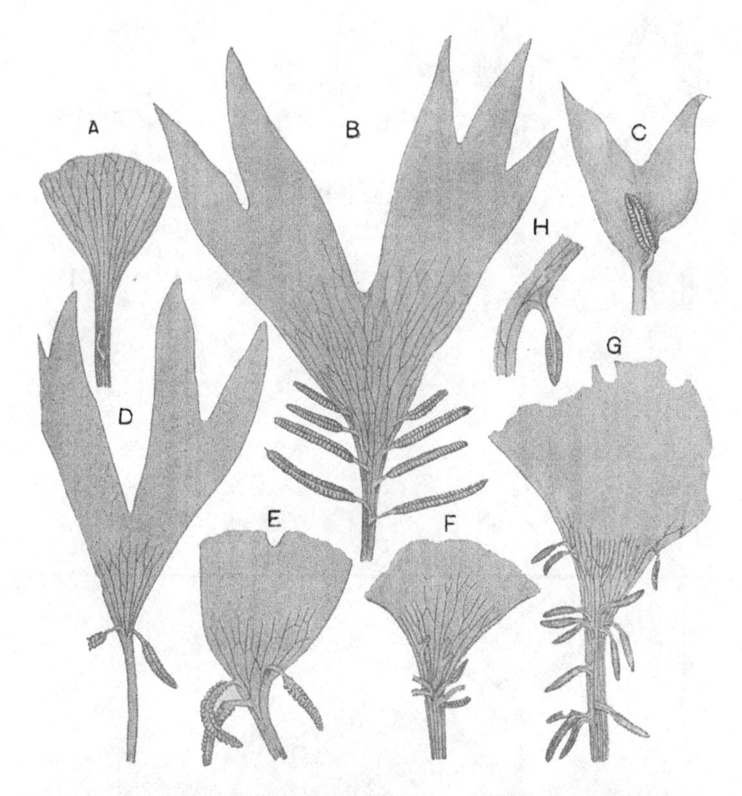

Fig. 343. *Ophioglossum palmatum* L. Drawings, slightly reduced, of specimens in the Kew Herbarium (except *B*, which is in Brit. Mus.), showing the various arrangements of fertile spikes, and their insertion as a rule intra-marginal.

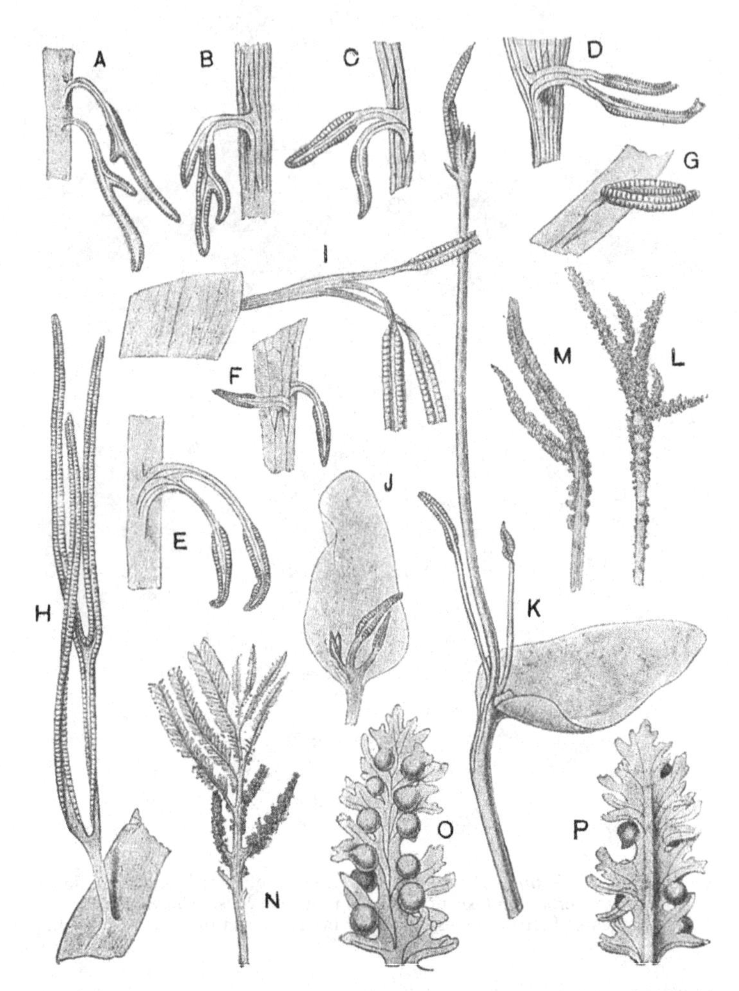

Fig. 344. *A–F*, various spikes of *Ophioglossum palmatum*, showing details of branching and insertion; *G*, *H*, *I*, spikes of *O. pendulum*; *J*, *K*, abnormalities of *O. vulgatum*; *L–P*, abnormalities of *Helminthostachys*; *O* and *P* are from drawings by Prof. von Goebel; *A–N* are one-half natural size.

In Lobb's specimens of *O. intermedium* (Hook, *Ic. Pl.* 995), as well as in Campbell's more recent collection (*Eusp. Ferns*, 1911, Pl. 4, *A*), the sterile lobe is still present though sometimes almost obsolete. The facts suggest that in certain derivatives from *O. pendulum* which have forsaken epiphytism for a soil-habit, mycorhiza present in them all has been advanced to be the main source of food, with reduction of the sterile blade as the result. Such departures as these from the usual type are best interpreted as secondary, superposed upon a type already standardised in the form held as normal for the genus. Taken together with the reticulate venation universal for the genus, they mark *Ophioglossum* not as the most primitive, but probably the most advanced of the three genera. Anatomy also supports this view.

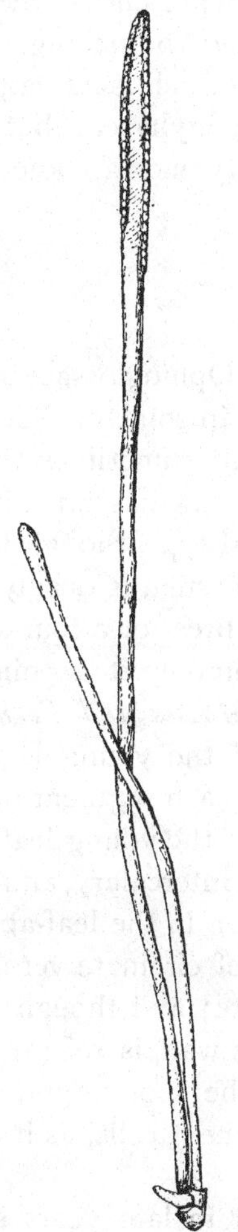

Fig. 345. *Ophioglossum simplex* Ridley, slightly reduced. Three leaves are seen inserted on a short stock, but the leaves appear to consist each of a fertile spike, with no sterile lamina.

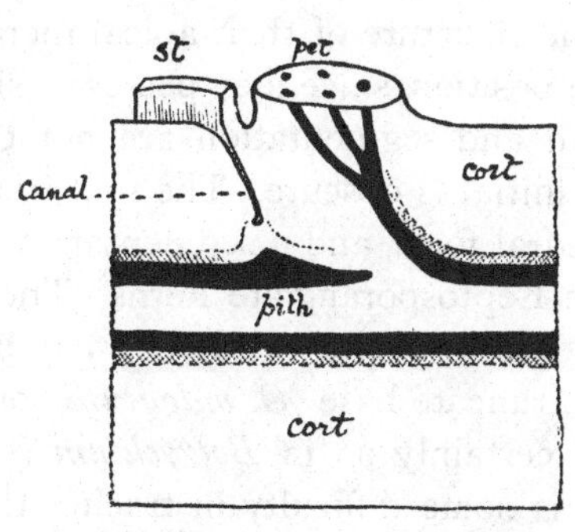

Fig. 346. Longitudinal section through a leaf-insertion of *Helminthostachys* showing the petiole (*pet.*) and stipule (*st.*), in the axil of which a canal arises leading obliquely down to a dormant bud. (After Gwynne-Vaughan.)

The formation of axillary buds has been proved in all the three genera. They were first seen by Gwynne-Vaughan in *Helminthostachys*, as minute groups of embryonic cells seated each at the base of a narrow axillary canal above an interruption of the endodermal cylinder (Fig. 346). There is one present in each leaf-axil. Commonly they are dormant, but may be stimulated into activity by injury of the main shoot. Similar axillary buds have been described by Bruchmann and by Lang as practically constant in *Botrychium Lunaria*, and they occasionally develop into actual branches. The facts for *Ophioglossum* have been less clearly made out, but in presence of the drawings of Rostowzew we cannot deny the existence of lateral branching, apparently of axillary origin, in that genus also (Fig. 341, 8). Such facts supply a basis for comparison not only with the living Hymenophyllaceae, but also with the ancient Zygopterideae, in both of which axillary buds are known to exist (Lang, *Studies*, iii, p. 47).

MERISTEMS

In the structure of their apical meristems the Ophioglossaceae occupy a middle position, since they possess a single initial in some of their parts, but its form and segmentation are not constant, and sometimes the identity of the initial is obscure. The root in all three genera has an initial cell of tetrahedral form, and a segmentation of the usual type, though less regular than in Leptosporangiate Ferns. The apex of the stem is deeply sunk, and a single initial is present, but it may be either a three- or a four-sided prism with a truncate base (*O. vulgatum*), or more commonly it is pointed below, and is certainly so in *Botrychium* (Campbell, *Mosses and Ferns*, p. 262). There is some difficulty in tracing the origin of the young leaves: but it seems probable that one of them corresponds to each segment of the initial cell, as in some Leptosporangiate Ferns. At first the young leaf also shows a single initial, but subsequent growth is largely intercalary, and there is no accurate segmentation comparable with that seen in the leaf-apex of most Ferns. This fact goes along with the absence of circinate vernation. The sporangia of the whole family are eusporangiate; and though the sporogenous tissue, together with part of its covering wall, is referable as a rule to a single initial with regular segmentation, the sporangium as a whole originates by outgrowth of a number of constituent cells, as it does in the Marattiaceae (Figs. 370, 371).

The detailed study of segmentation has lost in late years much of its glamour. We are still ignorant of the causes influencing it, and it will not resume its lost position till a better knowledge it attained. But it still keeps its comparative value, however little its causes may be understood. The general thesis has been developed in Vol. I, Chapter VI, that there has been a progression from a more robust type of cellular construction as seen in Eusporangiate Ferns, to a less robust as seen in Leptosporangiate Ferns, characterised by definite and regular segmentation. The meristematic characters of the Ophioglossaceae, though not very distinctive, rank most nearly with those of the Marattiaceae. In them, as will be shown later, the young and slender parts may often have a single initial, though its identity is lost in the adult. But the sporangial structure is quite distinctive of a primitive state: and what is seen in *Equisetum* shows that a typically eusporangiate plant may yet have a single initial with very regular segmentation in its stem and root; this is so also in the Ophioglossaceae. The general conclusion is that the Ophioglossaceae fall in with certain other Eusporangiatae in their apical constitution. On the other hand, the only direct basis for comparison with the Botryopterideae is through Kidston's specimen of the leaf-tip of *Zygopteris corrugata* (Vol. I, Fig. 104, p. 109), which shows an apical structure quite comparable with what might be seen in a modern *Botrychium*. This is in itself interesting, but the fact should not be accorded any undue importance.

Vascular Anatomy

The Ophioglossaceae have as a rule naked surfaces when mature, but sparse hairs may be found near the stem-apex, and especially in *O. palmatum*. They are mostly sappy, soft-tissued plants, and their ground tissue, being used for storage and other purposes, does not present features of comparative value. Accordingly the interest centres round the vascular system. Following the general view above stated, *Botrychium* and *Helminthostachys* will be considered first, and *Ophioglossum* will subsequently be compared with them.

The Root

The roots are hairless: their massive cortex, which is commonly used for storage, is delimited from the central stele by an endodermis of a relatively primitive type (Fig. 347). The stele is of a quite usual construction: in *Helminthostachys* it is frequently hexarch or even heptarch: in *B. virginianum* it may be tetrarch or triarch, but in most species of *Botrychium* and *Ophioglossum* it is diarch or even monarch. Most of the species of *Eu-Ophioglossum* have monarch roots, and this is conspicuously the case in *O. vulgatum*. In *O. Bergianum* the structure may be diarch or monarch, the latter having been observed in roots close to their base (Fig. 348). With the monarch structure goes terminal bifurcate branching, while monopodial branching is seen where the structure is more complex, as in *O. pendulum*, *Botrychium*, and *Helminthostachys*. Thus both dichotomous and monopodial branching are found in the same genus. Possibly dichotomy is restricted to monarch roots: this was suggested by Van Tieghem, who remarks that if the monarch root divides, we know beforehand that it will dichotomise (*Ann. Sci. Nat.*, v Série, T. xiii, p. 108). The larger polyarch roots of the family have a structure reminiscent of other Ferns, and particularly of the Marattiaceae. It is a tenable view that the simpler roots are reduced from

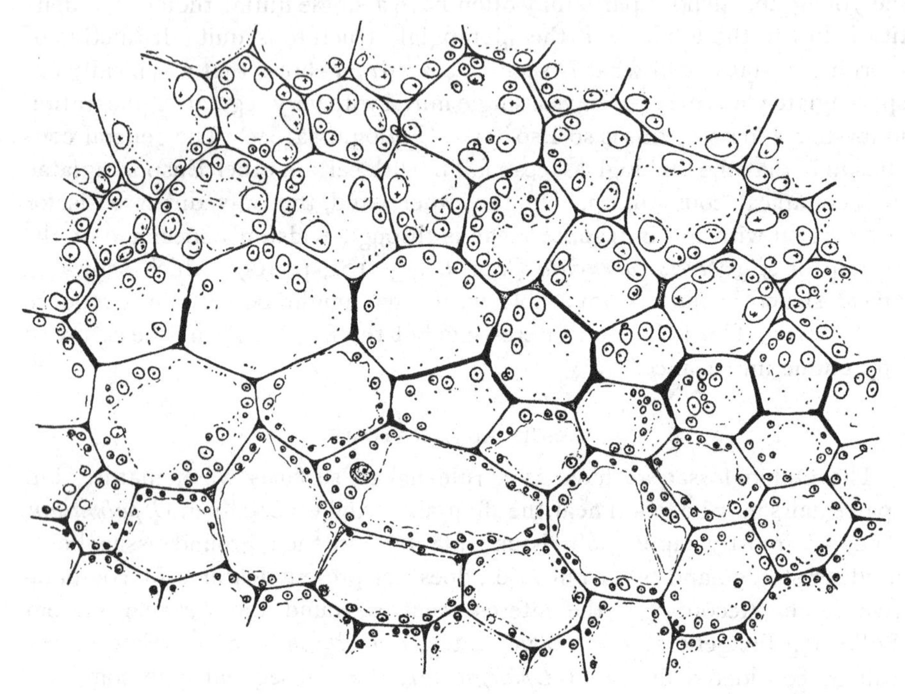

Fig. 347. *Helminthostachys zeylanica*: part of a transverse section of a root (Gwynne-Vaughan Coll. slide 589. ×66). The endodermis, recognised by the characteristic structure of its radial walls, marks a boundary between the outside cortex, with large starch grains (here above) and the inner conjunctive parenchyma (here below), with smaller starch grains. (Drawn by Dr J. M. Thompson.)

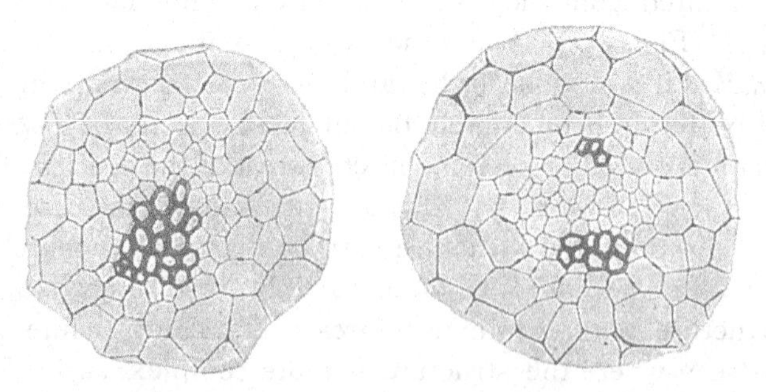

Fig. 348. *Ophioglossum Bergianum* Schlecht. Transverse sections of the stele of the root, the one showing two unequal groups of xylem, the other only one. (×200.)

that more complex type. But the fact that the monarch structure may appear at the very base of insertion both in *Botrychium* and in *Ophioglossum* presents a difficulty in accepting this.

The Stem

All the Ophioglossaceae have a definite stele in the basal region of the axis delimited by an endodermis, which is, however, of a rather rudimentary type. In the ontogeny of Pteridophytes a coherent body of tissue called the stele, partly made up of elements having a truly cauline origin, exists from the first, and it serves to connect up adjacent leaf-traces. Where the axis preponderates over the leaf this composite nature of the stele is readily seen at an early stage of the ontogeny, as in Lycopods. But where as in the

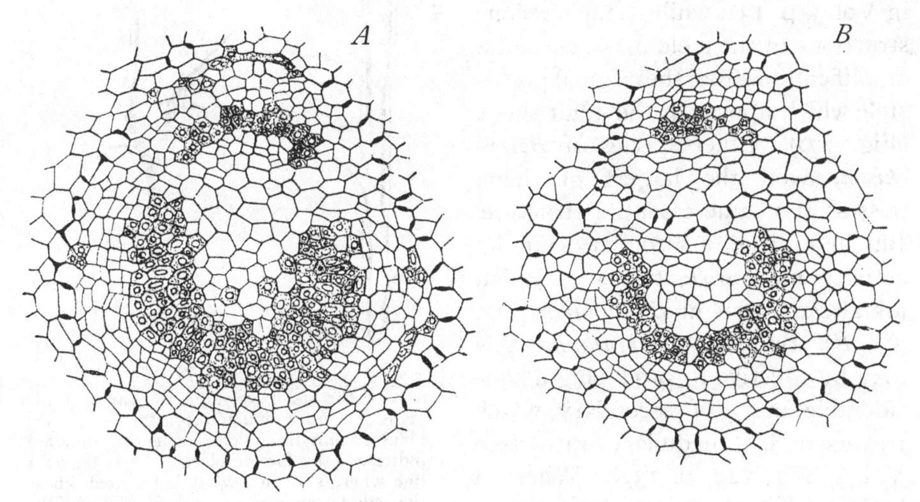

Fig. 349. *A*, *B*, transverse sections of the young stem of *Botrychium Lunaria*, showing medullation, the departure of the leaf-trace, and the first steps of cambial activity. Note isolated tracheides in the pith of *A*, and the complete endodermal investment of *B*. (× 125.)

Ferns the leaf preponderates over the axis in the young shoot, the stage when this composite nature can be recognised will appear later (see Vol. I, Chap. VII, p. 139). It happens that the Ophioglossaceae in their extreme types, and doubtless as a consequence of their monophyllous habit, present peculiar difficulties in this recognition. These are, however, less serious in *Botrychium* and *Helminthostachys*, in which the cauline factor is stronger, than in *Ophioglossum* where it is peculiarly weak. The whole question is closely related to that of the constitution of the shoot itself, and the relation of axis and leaf not only structurally but also in time of origin. The view adopted here is that the two constituents of the shoot co-exist from the first, as indeed embryology clearly demonstrates in all but extreme instances: but that the balance of their importance in the shoot may vary.

In *B. Lunaria* a protostele with solid xylem is found at the base of the young plant. Passing upwards it becomes medullated, the change being intra-stelar, apparently resulting from an absence of thickening in the innermost procambial cells: but not uncommonly isolated tracheides lie embedded in the pith thus explaining its origin (Fig. 349). The further steps seen in *Botrychium*, resulting in foliar gaps with a partial internal endodermis, have already been described in Vol. I, p. 131, while Lang's reconstructions of the stele make clear the amplifications from the original protostele which appear in the adult shoot (Fig. 350). Other species of *Botrychium*, even the largest of them, possess the same essential structure. But in addition a secondary cambial activity may arise, traces of which are already seen in *B. Lunaria* (Fig. 350, *B*). But in *B. virginianum* this may be effective in producing a considerable mass of secondary wood, traversed by medullary rays (see Vol. I, Fig. 129, p. 137). Where a leaf-trace is given off a sector of the xylem-ring becomes detached, and passes outwards, thus opening the xylic ring: but throughout the departure of the leaf-trace in the young plant, the stele and the trace are both enclosed by endodermis, and it is only in the adult that the foliar gap opens giving direct connection between the cortex and the pith. Similarly with the root-traces, which are given off irregularly in position and number, as will be seen from Lang's reconstructions.

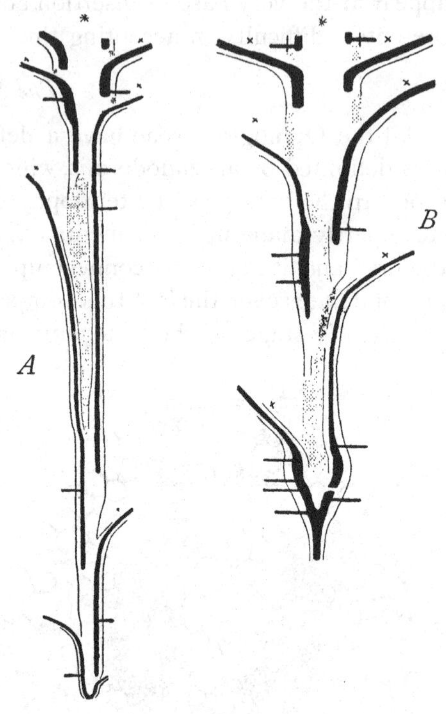

Fig. 350. *Botrychium Lunaria*. *A* = reconstruction of the stelar structure by Lang, of his plant *E*. *B* = a similar reconstruction of his plant *F*, only the xylem and endodermis are indicated, the former black; the latter as a line where seen in section, but dotted where the endodermis is seen in surface view, as if the stele were split in half. The leaves are really arranged spirally, but they are here represented as though they arose alternately. The level of origin of the root-traces is indicated. The proportions have been altered from those in nature so that the stele is represented as broader in proportion to its length than is actually the case: (×) represents axillary buds; (*) the apex. (After Lang.)

The massive creeping rhizome of *Helminthostachys*, bearing its leaves less crowded and only on its upper side, gives a better chance of elucidation of its stelar structure than do the short upright stocks of *Botrychium*, and it has been fully examined by Farmer and by Lang. In the juvenile stem the protostele has a solid xylem: a pith arises higher up by development of parenchyma in its centre. Different plants or regions exhibit successive stages in medullation: first there is an inner xylem with scattered parenchyma, then a mixed pith with parenchyma predominating centrally, and finally a small pith free from tracheides. Thus the pith is of intra-stelar origin. The stele is dilated upwards till in the adult stem it appears as a solenoxylic ring, bounded externally by endodermis and phloem. Finally in large rhizomes a second endodermis lines the xylem internally (Fig. 351). Lang has greatly extended the interpretation of this adult structure. The stele can be followed to the apex of the stem beyond the youngest leaves

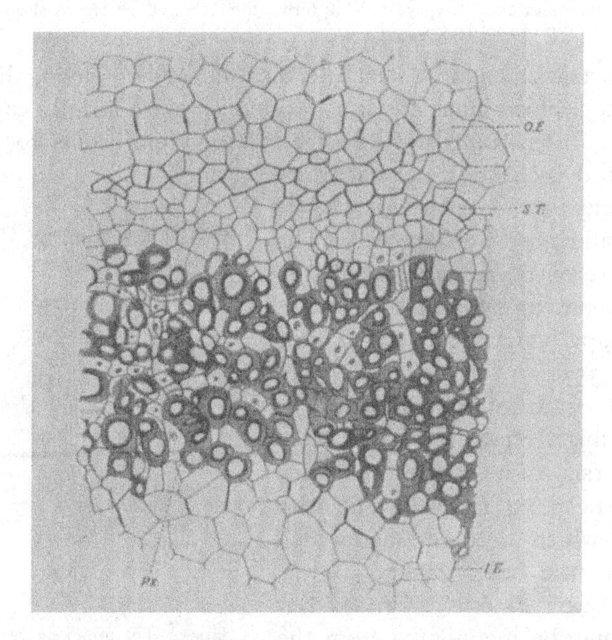

Fig. 351. *Helminthostachys zeylanica.* Transverse section of part of the stele and cortex of the rhizome: *O.E.*, outer endodermis; *I.E.*, inner endodermis; *S.T.*, sieve-tubes. (After Farmer and Freeman.)

and roots, and hence it is actually cauline. It gives off sectors of itself as leaf-traces obliquely right and left from its upper side (Fig. 352), and the upper region is thus common to stem and leaf. But the lower ventral portion does not give off leaf-traces at all, and it is cauline. The xylem-ring is mesarch: it is distinguished as consisting of an outer and an inner ring with the protoxylem elements between them. The leaf-trace which is always undivided shows variety in its mode of departure. In large rhizomes it passes off as a mesarch portion of the stelar tube (Fig. 352), but the inner xylem is scanty, and dies out as the trace passes outwards. In smaller rhizomes

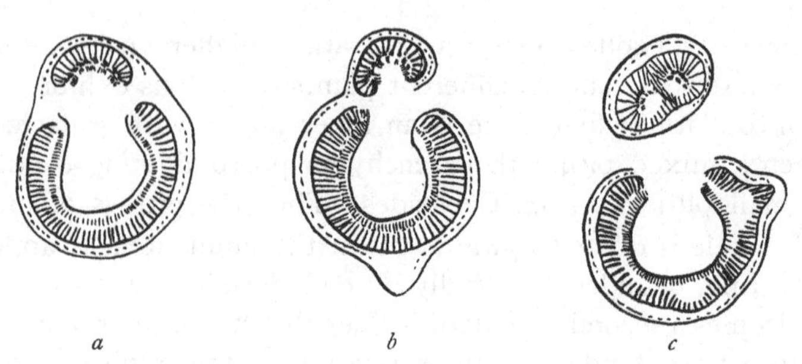

Fig. 352. Origin of the leaf-trace of *Helminthostachys*, after Lang: *a*, before the endodermis opens; *b*, the separation of the leaf-trace; *c*, the leaf-trace rounded off, and "Clepsydroid" in form; the stele not yet closed shows the origin of root-traces.

the trace departs with endarch structure, the inner xylem having disappeared from the trace-sector. Thus the inner xylem is more strictly cauline than the outer. In *Botrychium*, however, the inner xylem seems to have been wholly replaced by pith. Such facts are out of harmony with a phytonic theory such as that advanced by Campbell. The stelar structure of *Helminthostachys* is in effect a rudimentary type of solenostele, with foliar gaps which do not as a rule overlap (Fig. 353).

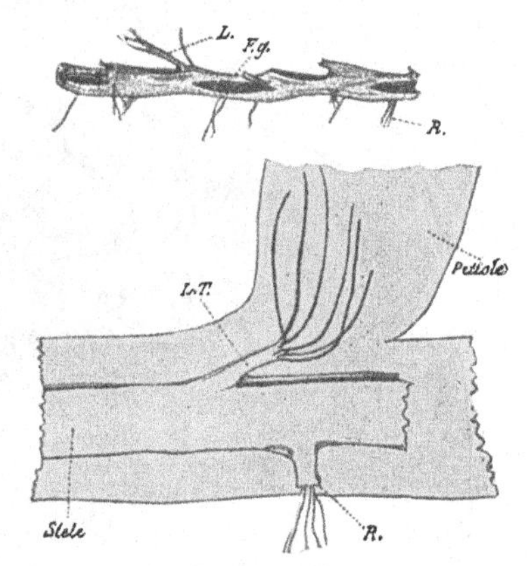

Fig. 353. *Helminthostachys zeylanica* Hook. The upper figure represents the vascular skeleton, dissected out. *L.* = leaf-trace; *R.* = root-strand; *F.g.* = foliar gap. The lower figure shows the rhizome-stele giving off a leaf-trace *L.T.*, which breaks up above into separate petiolar strands. *R.* = root-trace. (After Farmer and Freeman.)

The stelar state of *Ophioglossum* is variable, as might have been expected from the diversity of its species. In some the base of the plant is protostelic with solid xylem, which becomes medullated upwards as in the other genera (Bower, *Ann. of Bot.* xxv, Pl. xxv, Fig. 2). But in others it may be medullated from the first: and this is seen particularly at the base of the root-buds of *O. palmatum*. At the lower part of the plant an outer endodermis may be found in the smaller species (*O. Bergianum, capense, ellipticum* Poirault): but in most species it is absent. Passing upwards the medullated stele expands into a reticulum with leaf-gaps, well shown in Rostowzew's dissections (Fig. 341, 4, 5). The meshes are large and the leaves arranged in a compact spiral: as a rule a root-trace passes off below each leaf-insertion. The leaf-trace in *O. vulgatum*, and in *Eu-Ophioglossum* generally, is undivided as in the other genera. Transverse sections of the stock accordingly show an interrupted ring of vascular strands consisting chiefly of xylem, and without any endodermis. Near the base of the stock the vascular ring may appear more complete, but higher up the disrupted stele appears as a ring

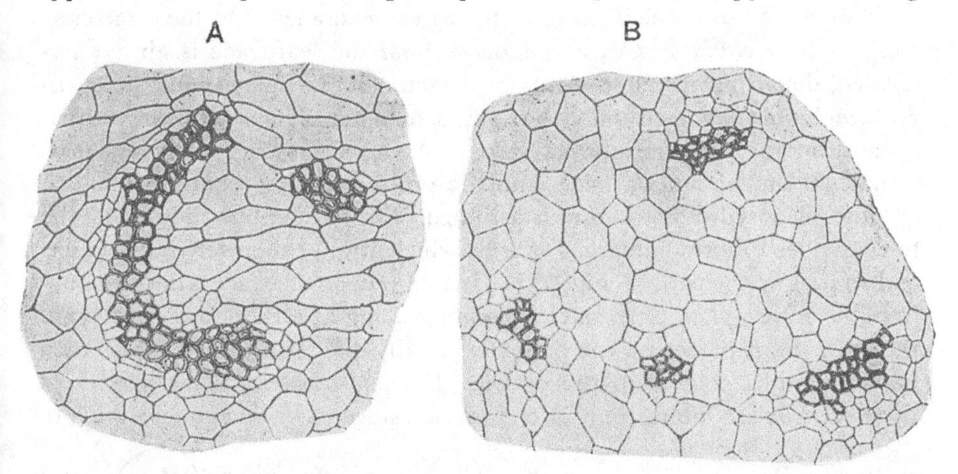

Fig. 354. *Ophioglossum Bergianum* Schlecht. *A*, transverse section of the stock, showing a semi-lunar stele, with a wide foliar gap into which a small leaf-trace is entering. *B*, another section showing probably the result of overlapping of the foliar gaps. No endodermis is seen. (× 200.)

of meristeles embedded in parenchyma (Fig. 354). It is, in fact, a rudimentary type of dictyostele. In some species the stock is distended for storage, as in *O. palmatum* (see Vol. I, Fig. 36). Here the vascular structure has been found to follow the same scheme as that described, but it is embedded in massive parenchyma (Fig. 355). In this species, however, and also in *O. pendulum* and *simplex*, the structure is further complicated by the fact that the leaf-trace is divided from the first, while endogenous roots may traverse both pith and cortex (Fig. 356). (See Bower, *Ann. of Bot.* 1911, p. 227.) These facts indicate that *Ophioglossum*, and in particular *O. palmatum*, *pendulum* and *simplex*, possess the most advanced type of stelar structure in the Family as judged by the usual criteria of comparison for Ferns. As bearing on the question of origin of the pith in these plants the record of strands

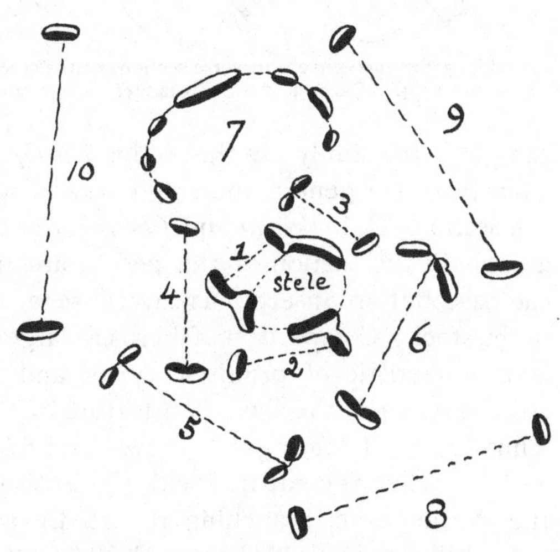

Fig. 355. Vascular system of the distended stock of *Ophioglossum palmatum*: the endogenous roots are omitted for clearness. The leaf-traces are numbered, and the marginal strands are connected by dotted lines. Only the marginal strands are represented in leaves 8, 9, 10. (Enlarged.) The leaf-traces come off as two separate strands, dividing up later, and the resulting strands arrange themselves in a circle in each petiole.

of xylem in the pith of *O. pendulum* has its interest in confirming the opinion of its intra-stelar origin (Petry, *Bot. Gaz.* March, 1914, p. 185).

Examination of the leaf-traces of the three genera leads to the same conclusion. In *Helminthostachys* and *Botrychium* the leaf-trace is always undivided, departing in adult plants as a sector of the solenoxylic stele. In *Helminthostachys* the strand is composed of outer xylem with or without some proportion of the internal xylem. As it departs the outer xylem is completed on the adaxial side, and the protoxylem segregating into two groups a Clepsydroid structure is assumed. But as it enters the petiole the trace divides to form numerous strands which arrange themselves in a circle, and so pass up the petiole (Fig. 353). In *Botrychium* the trace is derived only from the outer xylem of the stele, the inner being as a rule obsolete even in the axis. As it passes up the petiole it usually divides into two, which then take a parallel course up the petiole. The leaf-traces of some species of *Ophioglossum* are interesting for comparison since they form an exception to

Fig. 356. Successive transverse sections of the stock of *Ophioglossum pendulum*, showing the origin of each leaf-trace as numerous separate strands: this is an advanced mode of origin.

the rest of the family. In the section *Eu-Ophioglossum* the leaf-trace comes off as in the other genera, the single strand soon branching into three. But in the sections *Ophioderma* and *Cheiroglossa* the numerous strands arranged in a circle in the section of the petiole are not united into a single strand at the base, but are inserted as several separate strands upon the vascular stele of the stock. Comparison of Ferns at large shows that the concrete leaf-trace is characteristic of primitive types, and that its separation into distinct strands down to the base is a feature of those that are late and derivative. Thus by this feature §*Ophioderma* and §*Cheiroglossa* are shown to be late and derivative as compared with *Eu-Ophioglossum*. The strands pass on into the sterile blade, branching repeatedly as they go. The venation of the juvenile leaves in *Botrychium* and *Helminthostachys* is a scorpioid sympodium (Fig. 357), which is a derivative of equal dichotomy, as is shown by comparison of the young leaves of *Osmunda* (Fig. 416, p. 127): but the cotyledons of *Helminthostachys* and *Botrychium* step direct into a position that is only secondary in *Osmunda*. The venation of the adult blade of these genera is only a further and variously cut development of the plan thus initiated (Fig. 358). But in all the species of *Ophioglossum* the sterile blade shows a closed reticulum, certain veins of which lying near to the centre are thicker, and

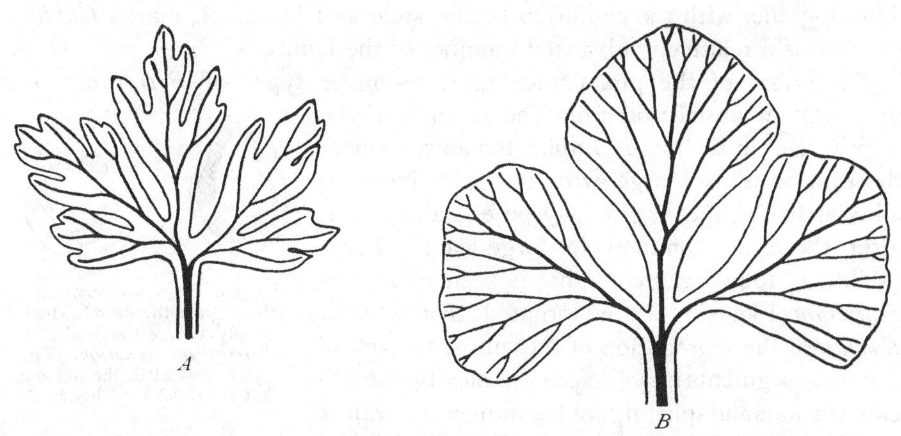

Fig. 357. *A* = cotyledon of *Botrychium virginianum* (× 4). *B* = juvenile leaf of *Helminthostachys*, probably an actual cotyledon: from the collection of Dr Lang (× 4). Here the ontogeny starts from a stage which appears relatively late in the series of *Osmunda* (Fig. 416), or of *Anemia*.

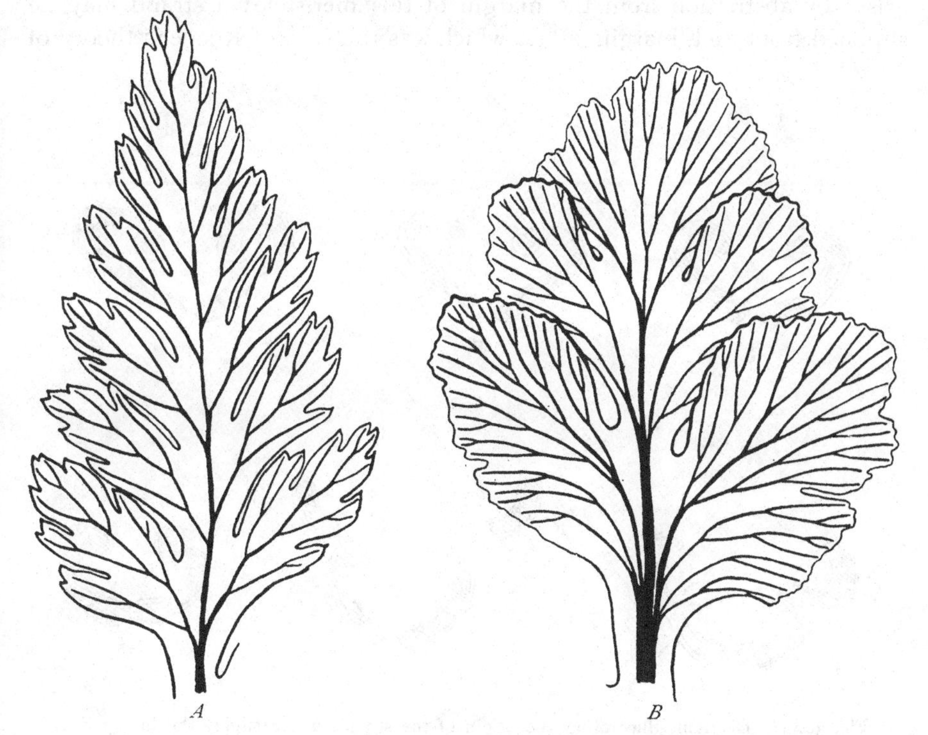

Fig. 358. *A* = apex of adult leaf of *Botrychium virginianum* (× 2). *B* = distal end of a leaf of *Botrychium jenmani* (× 4), showing a more fully webbed condition.

form a sort of midrib. Free vein-endings are seen both within the meshes of the network and also at the leaf-margin. Venation of this type originated only in the Mesozoic Period, and it is held to be secondary and derivative. This, together with the condition of the stele and leaf-trace, marks *Ophioglossum* as a relatively advanced member of the family.

The origin of the pinna-trace in the simpler types of *Botrychium* is by marginal abstriction from the strands of the midrib. But in the more complicated leaves, where those strands are large with strongly incurved adaxial hooks, the origin may be extra-marginal after the type common in large-leaved Ferns (Vol. I, p. 174, Fig. 170). This is seen in *B. virginianum* (Fig. 360). In the ternate leaf of *Helminthostachys* the segregation of the supply to each of the three segments may be accompanied by complicated fusions and splittings of the numerous strands.

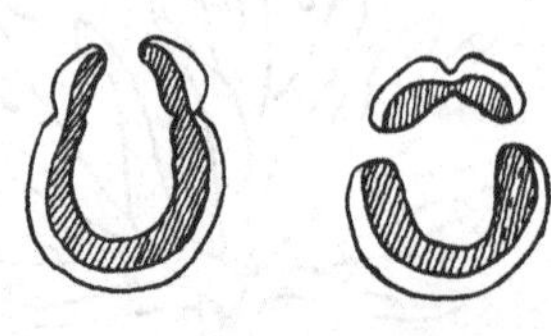

Fig. 359. Diagrams illustrating the origin of the strands which supply the fertile spike in *Botrychium ternatum*. The xylem is shaded, the phloem left clear. (After Chrysler.)

The vascular supply to the fertile spike originates in a manner similar to that of the pinna. In certain species of *Botrychium* and in *Ophioglossum*, it arises by abstriction from the margin of this meristele: a strand may be supplied from each margin, a fact which was the base of Roeper's theory of

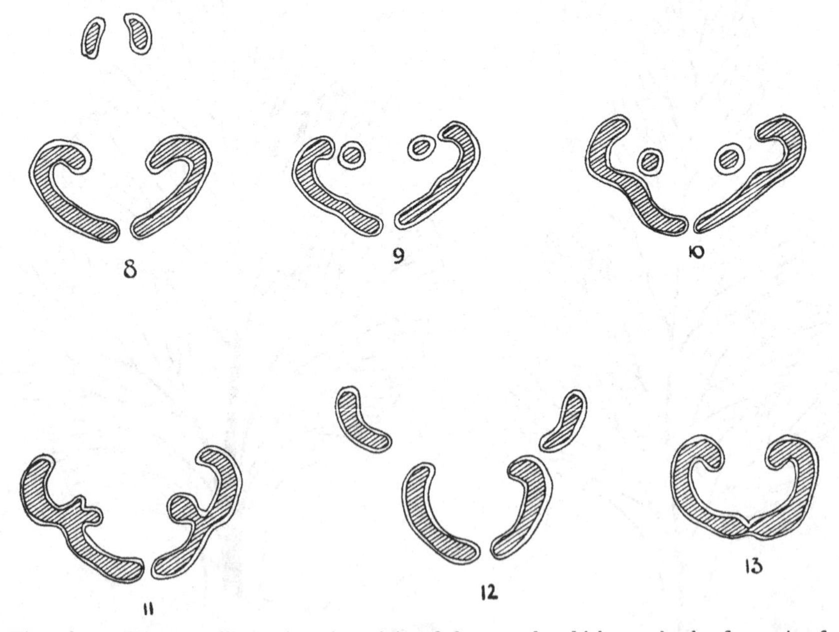

Fig. 360 *A*. Diagrams illustrating the origin of the strands which supply the first pair of pinnae of the sterile segment in *Botrychium virginianum*. Fig. 8 follows Fig. 7 in the sequence. (After Chrysler.)

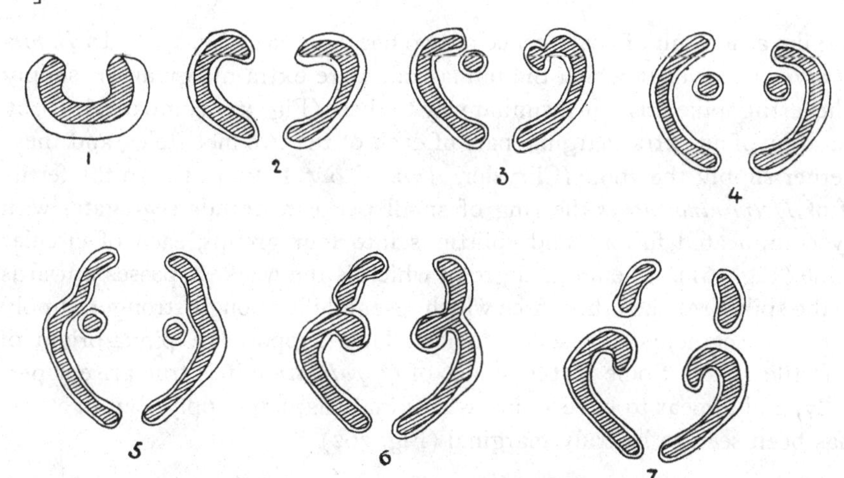

Fig. 360 *B*. Diagrams illustrating the origin of the strands which supply the fertile spike in *Botrychium virginianum*. Fig. 1 is from the lowest section: xylem is shaded, phloem left clear. Compare Figs. 359, 360 *A*. The adaxial side of the petiole is placed upward. (After Chrysler.)

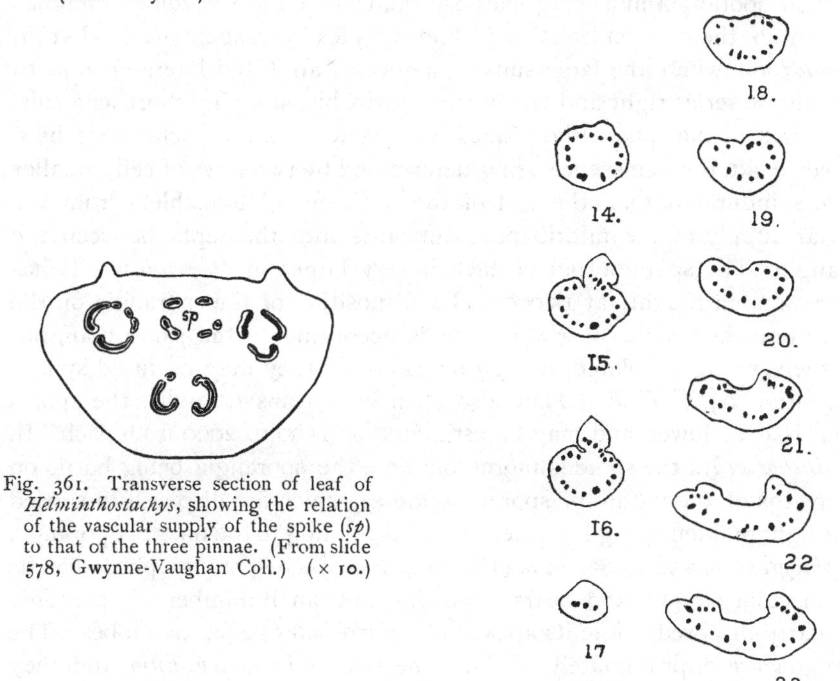

Fig. 361. Transverse section of leaf of *Helminthostachys*, showing the relation of the vascular supply of the spike (*sp*) to that of the three pinnae. (From slide 578, Gwynne-Vaughan Coll.) (× 10.)

Fig. 362. *Ophioglossum palmatum*. Successive transverse sections of the fertile leaf: 14–16 show the origin of the vascular supply to the lowest of its spikes; 17, transverse section of the stalk of that spike; 18–23, successively higher sections showing the origin of the supply to the second and third spikes. (× 4.)

the spike as a result of coalescence of two basal pinnae (Fig. 359). In *B. virginianum*, however, in which the pinna-traces are extra-marginal, the supply to the fertile spike arises in a similar way to these (Fig. 360, compare *A*, *B*), by separation of an extra-marginal part of each of the two meristeles, and these together supply the spike (Chrysler, *Ann. of Bot.* 1910, p. 1). In the fertile leaf of *Helminthostachys* the ring of small petiolar strands segregates with very complicated fusions and splittings into four groups, each of circular outline (Fig. 361). The adaxial group which is the weakest, passes upwards into the spike (*sp*), the other three which resemble it, though stronger, supply the sterile segments. These anatomical facts support the pinna-origin of the fertile spike. Some of the spikes of *O. palmatum* it is true arise superficially, and appear to have an intra-marginal vascular supply: but in others it has been seen to be truly marginal (Fig. 362).

Structure and Development of the Fertile Spike

It is a significant fact that all the fertile spikes of the Ophioglossaceae bear their sporangia in a marginal position, though there is some difference of detail in their exact relations. The simplest arrangement is that in *Ophioglossum*, where the large sunken sporangia are fused laterally so as to form a dense series right and left of the midrib, but stopping short, as a rule, of the apex of the spike (Fig. 364, *A*, *B*). When mature each opens by a transverse slit, the dehiscence being determined by two rows of cells smaller and less indurated than the rest of the wall. Small branchlets from the vascular supply of the midrib pass outwards into the septa between the sporangia. The spore-output of each is very large: in *O. pendulum* it has been estimated roughly at 15,000. The disposition of the sporangia on the usually branched spike of *Botrychium* is according to the same plan, but with the sporangia isolated, though occasionally they may be fused synangially (Fig. 364, *C*, *D*, *E*). They also open by a transverse slit: the spore-output is here lower, and may be estimated at 1500 to 2000 from each. In *Helminthostachys* the spike is more complex, the sporangia being borne on outgrowths of the nature of sporangiophores, which are disposed in serried ranks right and left along lines clearly corresponding to the lines of sporangia in *Ophioglossum* and *Botrychium* (Fig. 364, *F*). Each sporangiophore has its own vascular supply, and bears a varying but small number of sporangia irregularly disposed, while its apex ends in irregular vegetative lobes. The sporangia are approximately of the same size as in *Botrychium*, and they open by longitudinal slits (Fig. 364, *G*). It is clear that the spikes and sporangia are comparable in all the three genera, and may be held as variants of the same organs.

The development of the fertile spike has been traced in *Ophioglossum* from its first beginnings. The leaf itself originates very close to the initial cell of the deeply depressed apex of the axis. The sheathing stipule which envelopes the growing point as well as all the later leaves, is formed early: the spike appears above it in a median position on the adaxial face of the leaf, but near to its base (Fig. 363). The outgrowth soon becomes pointed and turns upwards. It consists of a rounded outgrowth with a four-sided pyramidal initial cell at the centre ("*x*" in Fig. 365, *A*, *B*, *C*, *E*), but its identity appears to be soon lost, and the construction passes over to that with four initials (Fig. 365, *F*, *G*). Consequently the spike comes to be composed of four quarters separated by walls at right angles, as seen in transverse section. As the spike grows older a special band of cells, which may be called the sporangiogenic band because it gives rise to the sporangia, is developed running along the lateral margins of the slightly flattened spike. It is derived from two regular rows of cells which form part of the two abaxial quarters of the spike: but usually the regularity of their arrangement is liable to interruptions. The sporangia formed from this band appear in continuous linear series: the details of their development are well seen in the large spike of *O. pendulum* (Figs. 366, 367). The cells of the sporangiogenic band dividing periclinally and anticlinally, certain cell-groups derived from the inner products soon begin to show more dense protoplasmic contents. These are recognised as sporogenous groups (Fig. 366, *A*, *B*, *C*). The inner product of the band is thus segregated into alternate blocks of sterile and fertile tissue. But the whole of the cells of the sporogenous groups do not form spores: a peripheral part of each group takes the character of a tapetum, and becomes disorganised as the development of the spores proceeds. The outline of the sporogenous group that remains is very variable: but when most regular it is as in Fig. 367, *C*, *D*, *E*. Meanwhile in the broad bands of sterile tissue between the sporangia vascular bundles make their appearance, connected as branch-bundles with the general system of the spike. The development of the spike in *O. vulgatum* and *reticulatum* has been found to be essentially the same, though on a less bulky scale.

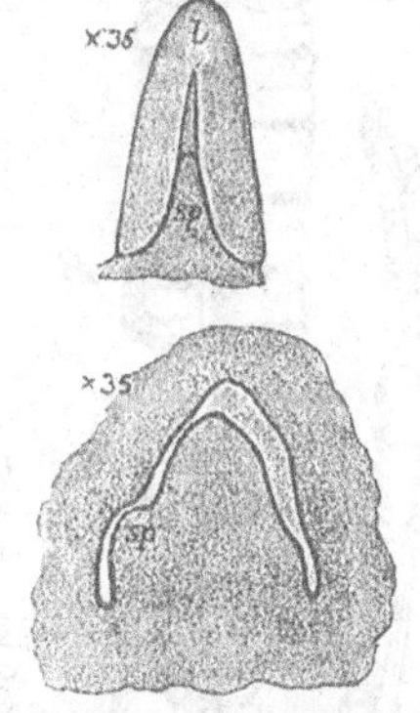

Fig. 363. *Ophioglossum vulgatum*. The lower drawing shows a longitudinal median section of a young leaf, with the spike (*sp*) arising about halfway up its adaxial face. The upper drawing shows a rather older leaf in frontal view. (× 35.)

The structure of the sporangium of *Ophioglossum* as it approaches the stage of separation of the spore-mother-cells and of the tetrad-division, is seen in Fig. 368. The tapetum, consisting of several ill-defined layers of cells, becomes disorganised, its protoplasts fusing into a continuous plasmodium, while the nuclei persist (Fig. 369). The sporogenous tissue breaks up into packets, or finally into isolated cells, while the tapetal plasmodium penetrates between them. Normally all the spore-mother-cells undergo tetrad division, and form spores, though occasional cells may become disorganised (*st.* Fig. 369, *A*). When ripe each sporangium bursts by a horizontal slit, already defined structurally in the tissue of the wall: it gapes as the tissues dry up, but there is no mechanical annulus.

The origin of the leaf in *Botrychium* and of the fertile spike have been described by Bruchmann for *B. Lunaria* (*Flora*, 1906, p. 213). He found the spike to originate nearer the apex of the leaf than in *Ophioglossum*, probably

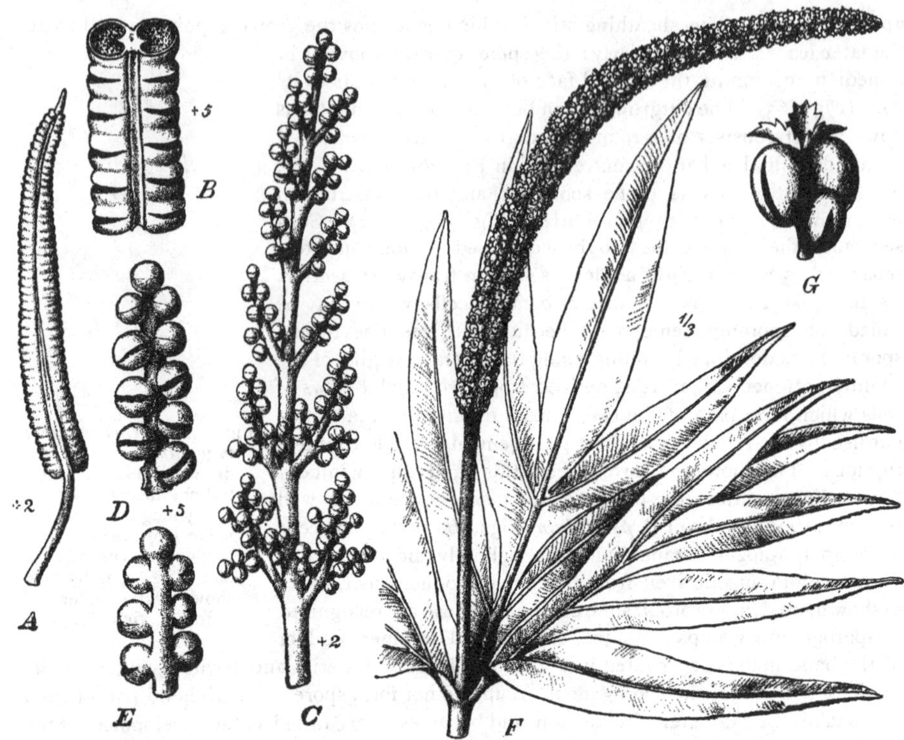

Fig. 364. *A*, *Ophioglossum palmatum* L., a single fertile spike with sporangia still closed; *B*, part of the same with sporangia ruptured; *C*, *Botrychium Lunaria* Sw., a fertile spike; *D*, a branch of the same with ruptured sporangia, seen from within; *E*, the same seen from without; *F*, *Helminthostachys zeylanica* Hook., sterile and fertile regions of the leaf; *G*, branch of the latter with a group of sporangia, and at the apex the irregular terminals of the sporangiophore. (*A*, *B*, *C*, *E*, after Bitter, in Engler and Prantl; *D*, after Luerssen; *F*, *G*, after Hooker.)

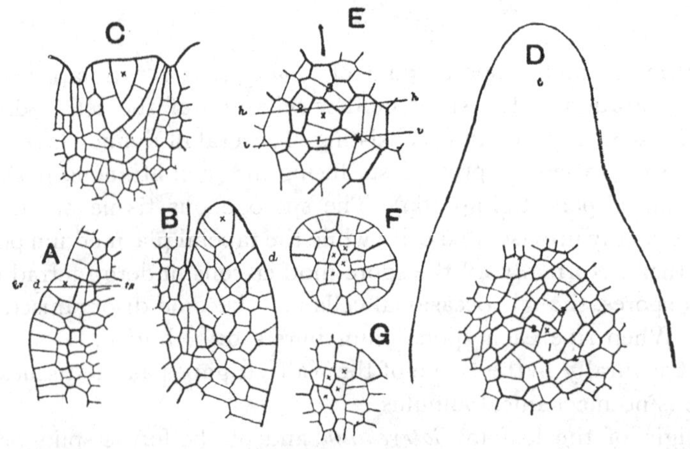

Fig. 365. *Ophioglossum vulgatum* L. *A*, median radial section through a very young spike showing an initial cell (*x*); *B*, similar section of an older spike; *C*, transverse section of a leaf, as along a line (*tr*) in *A*, traversing a young spike, *O. reticulatum* L.; *D*, tangential section of leaf (*l*) traversing a young fertile spike; *F*, *G*, transverse sections from the apex of a young spike of *O. vulgatum*, showing four initials (*x*). (× 100.)

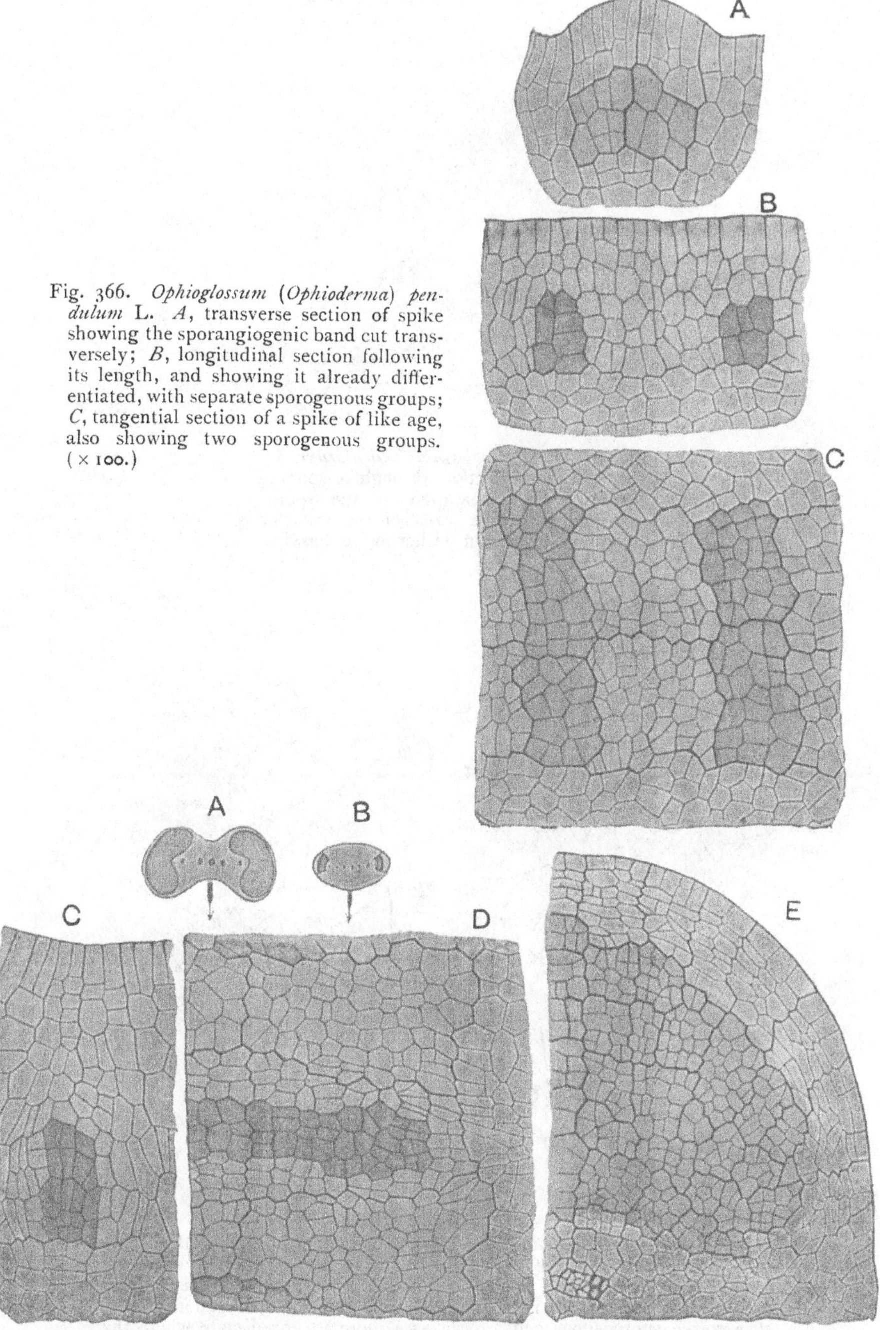

Fig. 366. *Ophioglossum* (*Ophioderma*) *pendulum* L. *A*, transverse section of spike showing the sporangiogenic band cut transversely; *B*, longitudinal section following its length, and showing it already differentiated, with separate sporogenous groups; *C*, tangential section of a spike of like age, also showing two sporogenous groups. (× 100.)

Fig. 367. *Ophioglossum* (*Ophioderma*) *pendulum* L. *A*, *B*, transverse sections of spikes of different ages to show sporangia and vascular bundles, slightly enlarged. *C*, a single sporangium older than in Fig. 366 *B*, in radial longitudinal section: the tapetum lightly shaded surrounds the darker sporogenous mass; *D*, tangential section of like age, showing one sporogenous mass darkly shaded: the smaller shaded groups will form the vascular strands; *E*, part of a transverse section of an older sporangium of *O. reticulatum*. (× 100.)

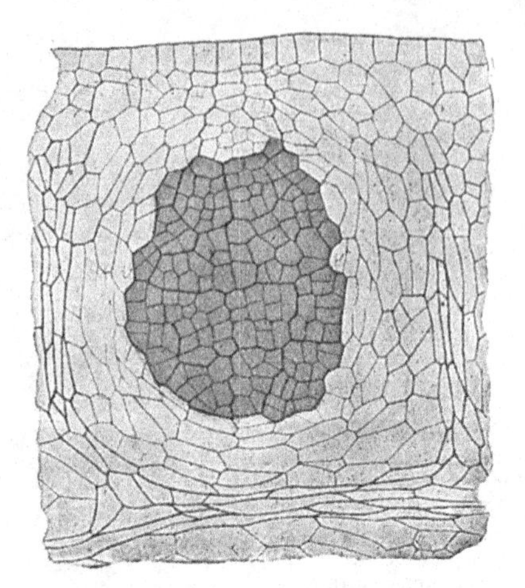

Fig. 368. *Ophioglossum reticulatum* L. Longitudinal section through a sporangium before separation of the spore-mother-cells: the walls of the vascular tissues are drawn rather more heavily. (× 100.)

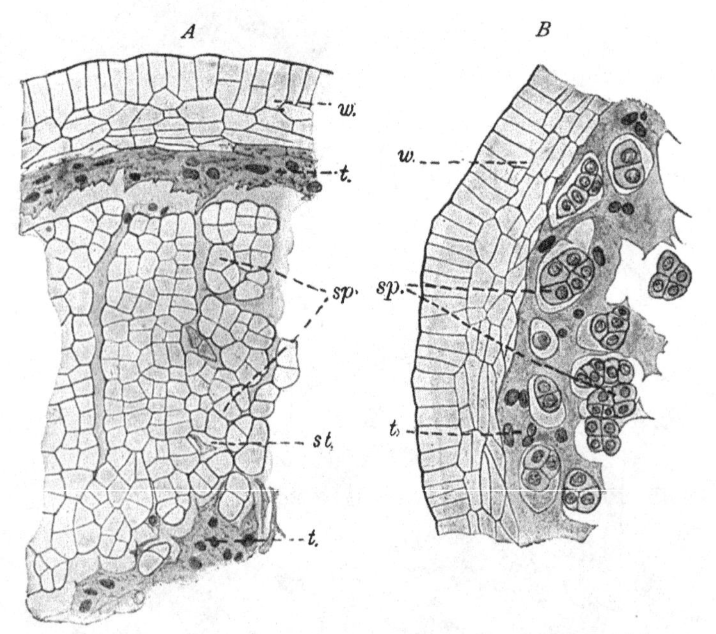

Fig. 369. *Ophioglossum vulgatum* L. Portions of sporangia showing the sporogenous tissue in progress of disintegration. In *A* the tapetum (*t.*), evidently derived from more than a single layer of cells, has formed a plasmodium with many nuclei, which is beginning to penetrate the sporogenous tissue, *st*, a sterile sporogenous cell. *B* shows a more advanced state where the sporogenous cells (*sp.*) appear in small clusters, or isolated, embedded in the tapetal plasmodium (*t.*): *w.* = sporangial wall. (× 100.)

from one of the latest adaxial segments of the leaf-initial (*l.c.* Fig 363). Both parts retain their active initial cells till about the time of origin of the lateral pinnae. He describes how the origin of these in *Botrychium* is by a process quite distinct from that of the fertile spike: the latter appears in a median position with a definite apical cell from the first: the pinnae arise in acropetal order by marginal growth (*l.c.* p. 218). Steps in development of the sporangia of *Botrychium*, normally separate from one another, are shown in Fig. 370. The sporangium is eusporangiate, but the sporogenous tissue originates from a parent cell that is the inner product of a single superficial cell which divides periclinally. The stalk is thick: the wall may consist of several layers of cells, and the spore-output of the sporangium is large. A peculiarity is that

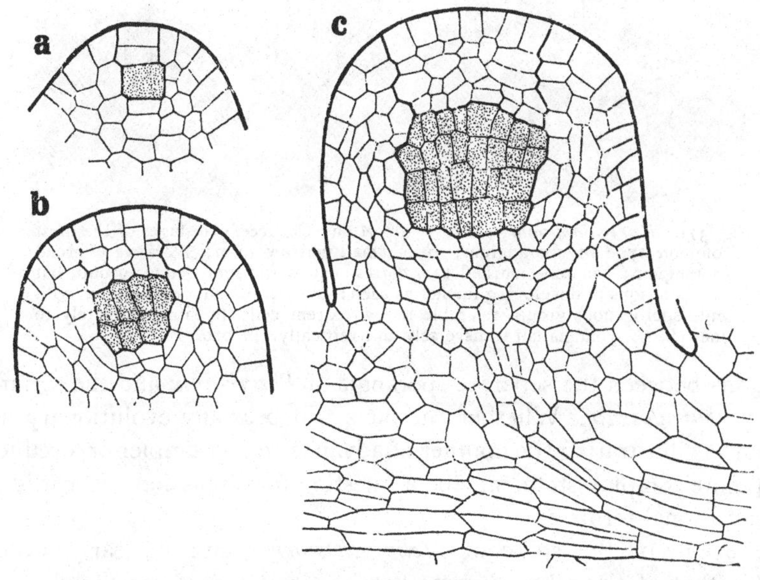

Fig. 370. *a*, *b*, *c*, successive stages of the development of the sporangium in *Botrychium daucifolium*. (×250.) Note the very massive stalk.

each sporangium may have its own vascular supply by a strand which terminates immediately below the capsule. This is an exceptional condition in living Ferns, but it has its parallel in the fossil *Stauropteris*. A further point is that the identity of the sporangia is not always maintained. In *B. Lunaria* sporangia may be found coherent together, and this condition has been worked out developmentally in the larger *B. daucifolium*, in which such cohesions are more frequent (Fig. 371). The drawings *A—C* show sporangia of the normal type; but others (*D*, *E*) have a broader form, leading to a doubling of the sporogenous group: it is only a slight step from *E* to *F* or *G*, that is, to a state where two distinct sporogenous groups are present, with a rounded contour for each. Such examples illustrate gradual

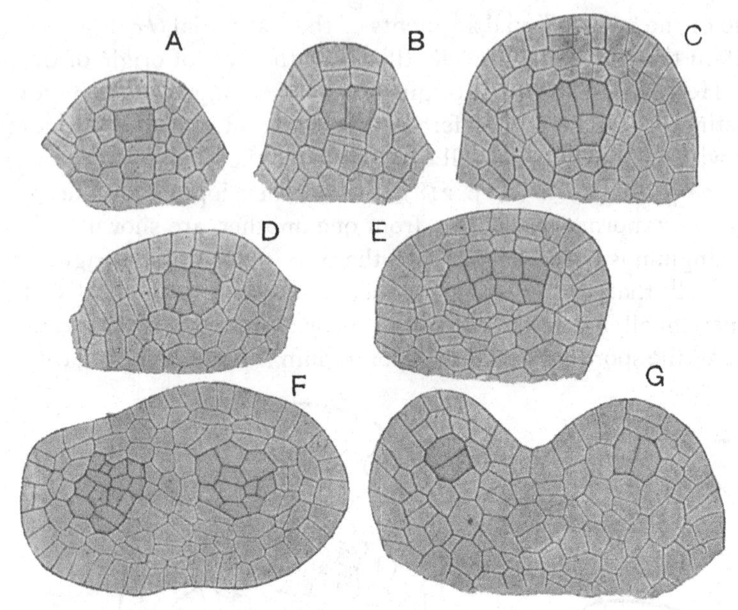

Fig. 371. *Botrychium daucifolium* Wall. *A* and *C*, successive stages of the most common type of sporangium; *B*, a small narrow form; *D*, a very broad sporangium: the cells marked (×) correspond to the sporogenous group, but show no signs of developing further as such; *E*, a still broader sporangium with wide sporogenous group, referable to two parent cells, possibly ultimately to one; *F*, *G*, synangia cut transversely and vertically. (× 200.)

transitions between the separate sporangia of *Botrychium* and the synangial state of *Ophioglossum*. Whether the facts will bear any evolutionary interpretation, as illustrating the manner of advance from a simpler or a reduction from a more complex state, may be a question for discussion. But it is clear that transitional forms exist.

The origin of the spike of *Helminthostachys* and its early structure resemble that of the other genera: but its tip is curved over while young, and is protected by the segments of the sterile frond, which are themselves invested at first by the stipular sheath of the next older leaf. The young spike seen from without appears as in Fig. 372. The young sporangiophores, very variable in size and arrangement, are densely clustered along the margins, corresponding thus roughly in position to the marginal rows of sporangia in the other genera. Transverse sections through the lateral regions corresponding to the sporangiogenic band of *Ophioglossum* disclose a fan-like tracery of the cell-walls with rather deep superficial cells (Fig. 373, *a*). It is from these that the sporangiophores originate, as outgrowths very irregular in size and arrangement (Figs. 373 *c*, 374 *a*). There seems to be no fixed type of segmentation of the cells which form them. As they increase in size their apex may be occupied by a wedge-shaped cell; more frequently no

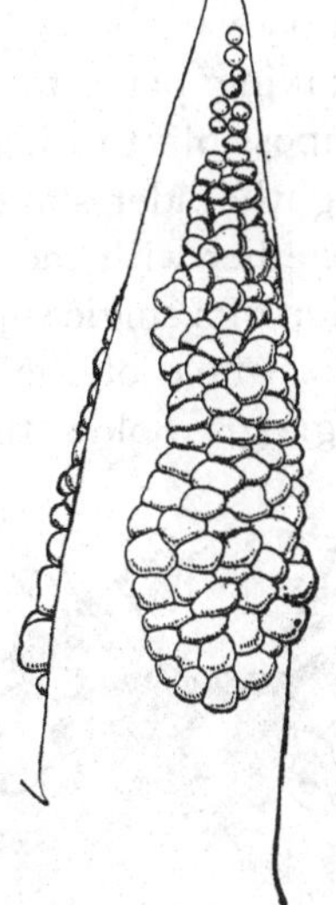

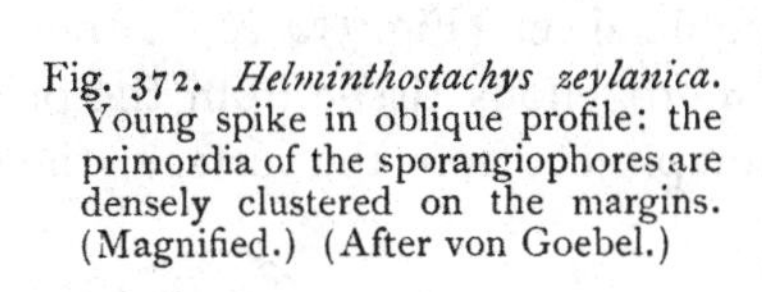

Fig. 372. *Helminthostachys zeylanica.* Young spike in oblique profile: the primordia of the sporangiophores are densely clustered on the margins. (Magnified.) (After von Goebel.)

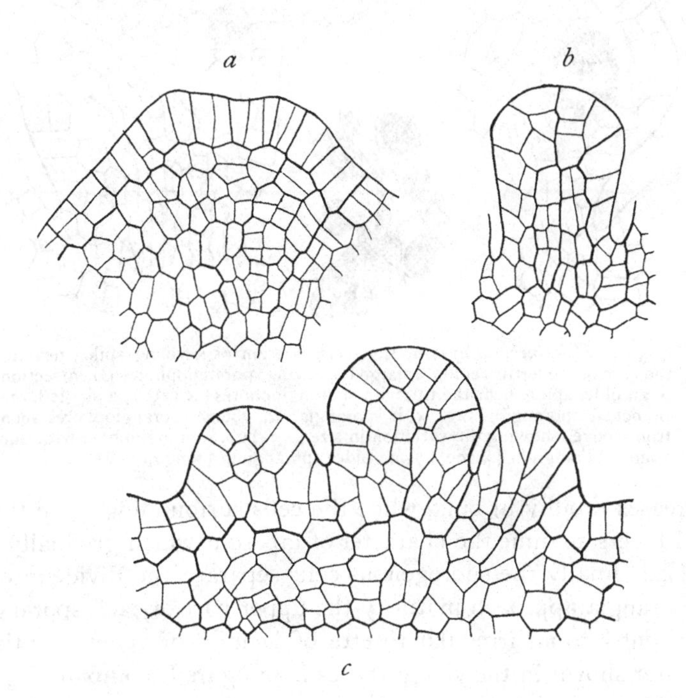

Fig. 373. *Helminthostachys.* *a*, transverse section of the fertile spike, through the region where sporangiophores will be formed; *b*, young sporangiophore; *c*, three young sporangiophores, showing their mode of origin, and differences in size. (× 200.)

such cell can be found (Fig. 373, *b*, *c*). Fig. 373, *b*, shows the most regular type of these very variable organs. Already in the centre of it longitudinal divisions are taking place to form the vascular strand, but in such early stages it is impossible to distinguish the cells which will give rise to the sporangia. Slightly older stages show that the sporogenous group of each sporangium together with the superficial cells that cover it are referable to the segmentation of a single superficial cell (Fig. 374, *e*, *f*): the first periclinal wall defines the whole of the sporogenous tissue from the protective wall. As the sporangia grow older they project from the surface: the sporogenous

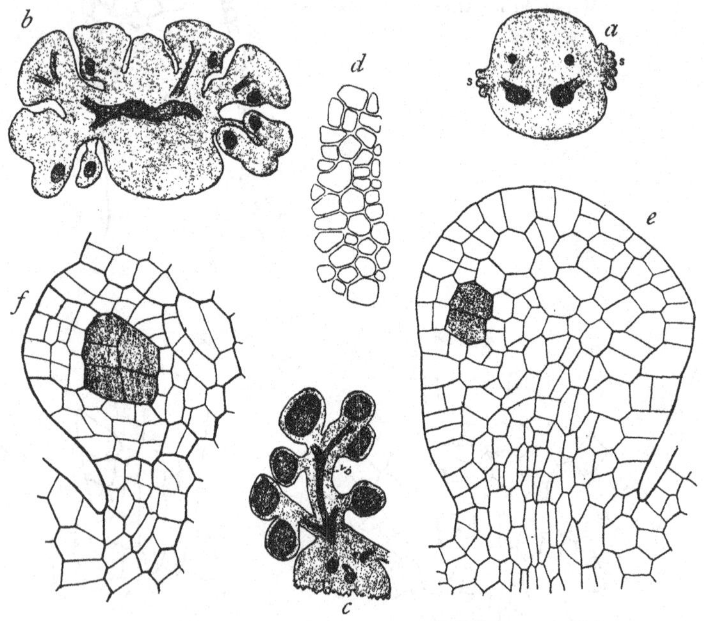

Fig. 374. *Helminthostachys*: *a*, transverse section of a young spike, near to the base of the fertile region, enlarged, showing sporangiophores (*s*); *b*, section of an older spike, near the tip, with sporangiophores (× 12); *c*, a single large branched sporangiophore, with sporangia; *d*, young sporangiophores seen from above, showing the variation in size (× 24); *e*, sporangiophore with one young sporangium (× 200); *f*, an older sporangium (× 200).

mass increases rapidly in bulk, while the cells surrounding it, to the extent of several layers, assume the character of tapetum, which gradually becomes disorganised: finally the sporogenous cells separate and divide into tetrads. As the sporangia approach maturity the upper part of each sporangiophore may grow out into an irregular rosette of laciniae of vegetative tissue, but these are not shown in the young stages here figured (compare Fig. 364, *G*).

It would be possible to arrange the genera of Ophioglossaceae in sequence from those with relatively large and indefinite sporangia to those having them smaller and more definite. A parallel sequence emerges also from a comparison of their fertile spikes. The simplest type as regards form is *Ophioglossum*, with its two series of sunken sporangia. In *Helminthostachys* the external form as well as the development may suggest that the sunken sporangia of *Ophioglossum* are here replaced by the sporangiophores, with their separate and smaller sporangia. In *Botrychium* the spike itself is habitually branched, while the numerous separate and smaller sporangia continue to hold the same relative position as in the unbranched spikes of *Ophioglossum*. As in almost all organic sequences, we might regard this series either as an upward one of progressive elaboration or as a downward one of reduction, or thirdly, as marking a divergence from some central point. A decision on this point will have a special bearing on the probable phyletic relations of the family. But this will not be discussed till after the gametophyte generation and the embryology have been examined, and meanwhile opinion must be held in suspense. (See p. 87.)

THE GAMETOPHYTE.

The prothallus in all of the Ophioglossaceae is saprophytic, and it grows normally as a mycorhizic, colourless body underground. Campbell (*Eusp. Ferns*, p. 6) has traced the germination in several distinct instances, and found that after the first few cell-divisions in the germinating spores growth stops, unless there is infection by the mycorhizic fungus. This appears to give the necessary stimulus, and thereafter development proceeds. Chlorophyll in small quantity has been seen in some early stages, but not in all. Mettenius stated for *Ophioglossum pedunculosum* that chlorophyll may be developed if the prothallus appears above ground, and Bruchmann also found this for *Ophioglossum vulgatum*. But for all practical purposes the gametophyte of the Ophioglossaceae is saprophytic.

The adult prothallus of *Botrychium virginianum* (Fig. 375) grows about

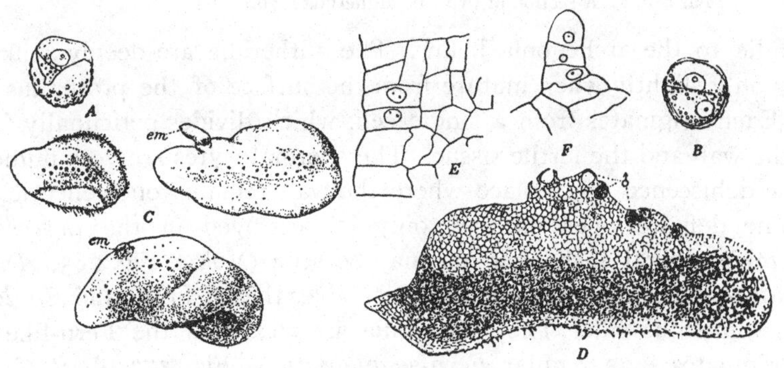

Fig. 375. Prothallus of *Botrychium virginianum* (after Campbell). *A*, *B*, germinating spore; *C*, three gametophytes, *em* = embryo (× 3); *D*, section of gametophyte (× 9): the shaded region is that occupied by the endophyte; ♂, antheridia; *E*, apical region of gametophyte (× 120); *F*, short multicellular hair, or paraphysis.

10 cm. below the surface of the soil, and is a flattened tuberous body as much as 20 mm. in length: it possesses sluggish apical growth, and may be lobed at the tip. It is pale to brownish in colour, and bears transient hairs, which offer entrance to the symbiotic fungus. Sections show that the infected region comprises the greater part of the central tissue leaving the periphery free. On the uninfected tissue, which is thicker at the upper flattened surface and includes the meristematic region, the sexual organs are borne. The antheridia are formed first near to the median line, while the archegonia appear later usually upon the flanks of the convex median ridge. It has been thought that this form and the arrangement of the sexual organs are convenient biologically, in presenting an upward face to rain soaking into the soil, and thus washing the sperms from the higher-lying

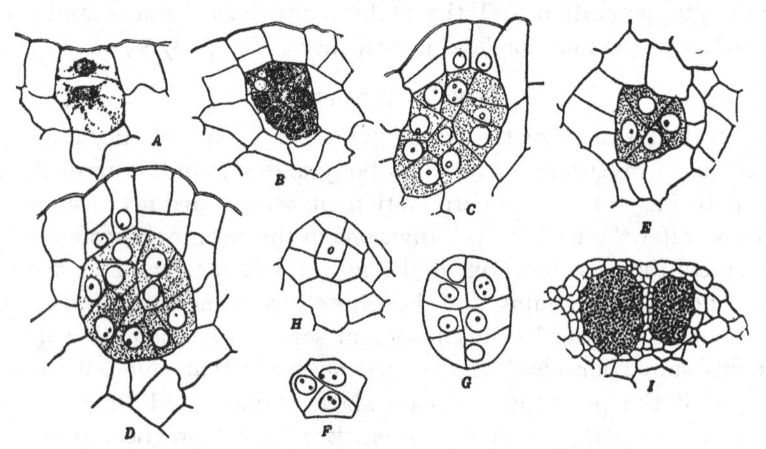

Fig. 376. Development of the antheridium of *Botrychium virginianum* (after Campbell). *A–D*, longitudinal sections of young antheridia (×240); *E–H*, transverse sections; *F*, *G*, show only the young spermatocytes; *H*, surface view, opercular cell, *o*; *I*, section of two ripe antheridia (×60).

antheridia to the archegonia below. The antheridia are deeply sunk, projecting only slightly when mature from the surface of the prothallus (Fig. 376). Each originates from a single cell, which divides periclinally (*A*) to form the wall and the fertile tissue. The spermatocytes are very numerous, and the dehiscence takes place where the wall remains one cell thick, and thus one definite opercular cell only is destroyed in the process. In *B. Lunaria* several such opercula may be seen (Vol. I, Fig. 283, *A*). The spermatozoid has about 1½ coils, and resembles that of Ferns and *Equisetum*, having numerous cilia. The archegonia are clearly of the Fern-like type: each originates by a regular Fern-segmentation (Fig. 377, *A*, *B*, *C*). The subsequent divisions in the central cell and in the cover-cell agree very closely with those in *Ophioglossum*; but the neck-cells are more numerous, and at maturity the neck projects much more strongly. There may be seven

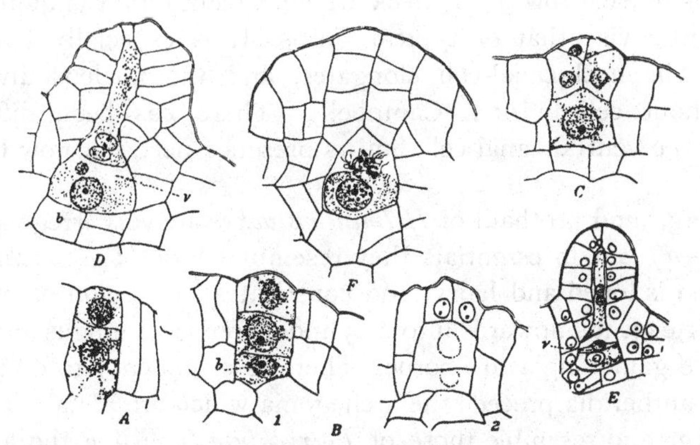

Fig. 377. Development of the archegonium in *Botrychium virginianum* (after Campbell). *A–D*, longitudinal sections (× 240); *E*, ripe archegonium, showing ventral canal-cell, *v* (× 125) (after Jeffrey); *F*, recently fertilised archegonium, showing spermatozoids within venter (× 240).

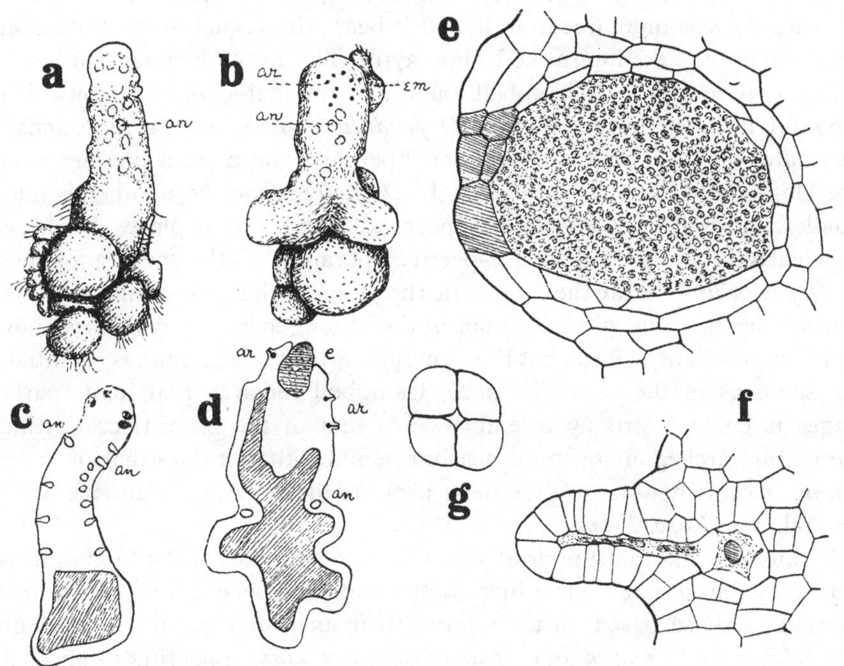

Fig. 378. Prothalli and sexual organs of *Helminthostachys zeylanica* (after Lang). *a*, *b*, prothalli seen from without; *c*, *d*, in section, with mycorhizic regions shaded; *e*, antheridium in longitudinal section; *f*, *g*, archegonia. (*a*, *b*, *c*, *d*, ×7; *e*, *f*, ×200). *an* = antheridium, *ar* = archegonium, *e* or *em* = embryo.

or eight cells in each row of the neck, which, except that it is quite straight, resembles otherwise that of typical Ferns. It is especially like that of *Osmunda*. The neck-canal-cell elongates, and the nucleus divides, but usually without cell-division (Campbell). There has been difficulty in recognising the ventral-canal-cell, but its presence need not now be held in doubt.

The underground prothalli of *Helminthostachys* are very irregular in form (Fig. 378, *a–d*), but in essentials they resemble those of *Botrychium*. The lower portion is lobed and hairy, and constitutes the region of mycorhizic nutrition. It extends upwards into a cylinder free from fungus and hairless, on which the gametangia are borne. There is a tendency to dioecism, but in any case antheridia precede the archegonia which are distal. The antheridia are large and resemble those of *Botrychium* (*e*), while the archegonia have, like *Botrychium*, projecting necks (*f*, *g*).

The prothallus of *Ophioglossum* is irregularly cylindrical. In *O. moluccanum* it may be 5–10 mm. in length (Campbell), but in *O. vulgatum* it may attain a length of 6 cm., and it is occasionally branched. There is often an enlarged base, and it is here that the mycorhizic fungus is present. The upward-growing apex is more free from it, and it bears the sexual organs, which are very numerous and intermixed (Fig. 379). The antheridia are sunken as before, and very large. Campbell found 250 spermatocytes and upwards in the section of one antheridium of *O. pendulum*, giving several thousands of spermatozoids for each antheridium, "perhaps more than in any other Pteridophyte" (*Eusp. Ferns*, p. 23). In *O. moluccanum* the number is much smaller, and that in *O. vulgatum* appears to take a middle place. In this we see clearly a parallel with the respective sporangia of the species, for those of *O. pendulum* are far the largest in the genus. The archegonia of *Ophioglossum* are more deeply sunk than in the other genera. *O. pendulum* shows this in a pronounced form, but the structure of the archegonium is essentially the same as in the rest (Fig. 380). Campbell remarks that in its earlier stages it bears a striking resemblance to that of the Marattiaceae, which the mature archegonium more nearly resembles than it does that of *Botrychium*. Even when mature the neck projects little above the surface of the prothallium (*Eusp. Ferns*, p. 29).

It thus appears that the prothallia and sexual organs of the Ophioglossaceae may be seriated according to the details above given, along lines parallel to those based on their form, their anatomy, and their sporangia, *Botrychium virginianum* and *Ophioglossum pendulum* appearing as specially divergent types.

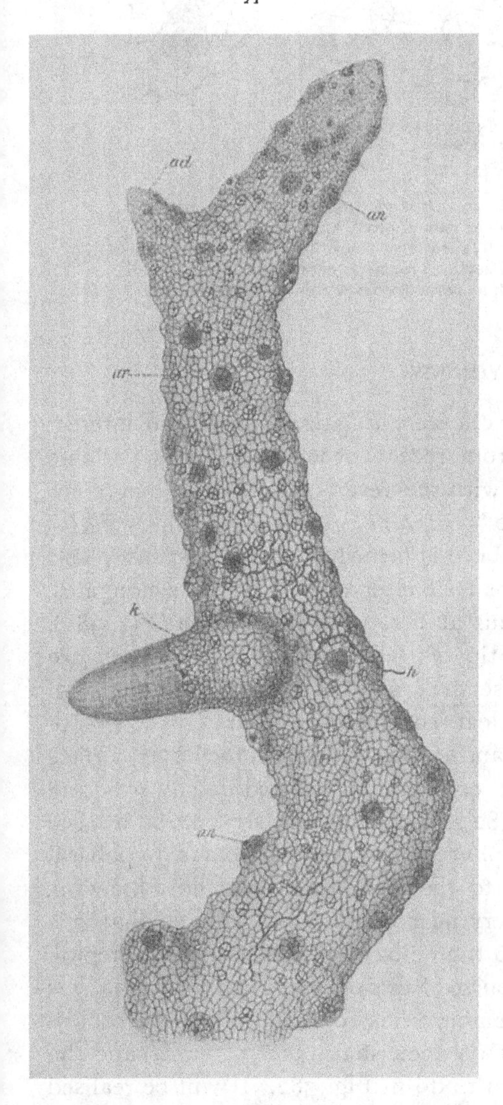

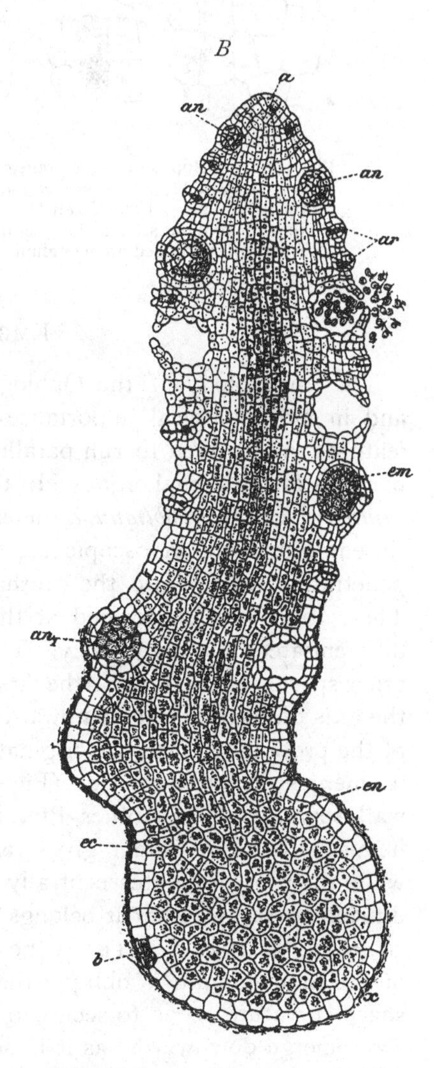

Fig. 379. *Ophioglossum vulgatum*. *A*, a prothallus seen from without (× 30). Numerous antheridia (*an*) and archegonia (*ar*) are present. (*k*) is a young sporophyte with a strong primary root; (*ad*) is an adventitious branch; (*h*) the dark brown hyphae of a fungus, branches of which penetrate the prothallus. *B*, longitudinal section of a prothallus: (*a*) apex, (*b*) base; at the apex developing antheridia (*an*), and archegonia (*ar*); below are empty sex-organs: (*em*) a young embryo, (*en*) the endophytic fungus occupying the lower part of the thallus. (After Bruchmann.)

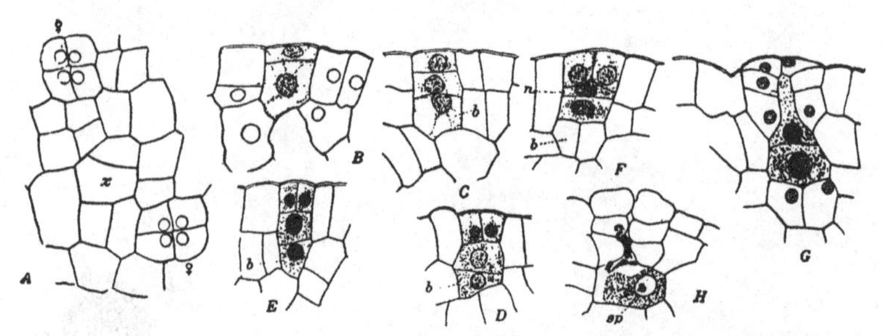

Fig. 380. Development of the archegonium in *Ophioglossum pendulum* (after Campbell) (× 135). *A*, transverse section of the gametophyte apex, showing two young archegonia ♀, and apical cell (*x*). *B–G*, successive stages in the development of the archegonium, seen in longitudinal section: *n*, neck canal cell; *b*, basal cell. *H*, recently fertilised archegonium: *sp*, a spermatozoid within the egg-nucleus.

EMBRYOLOGY.

The embryology of the Ophioglossaceae has gained greatly in interest and in morphological importance from recent observations. Its variable features will be seen to run parallel with the results of the foregoing comparisons of the sexual organs. In the "*ternatum*" group of the genus *Botrychium*, as seen in *B. obliquum*, the embryo is furnished with a suspensor, and the embryogeny is endoscopic (see Vol. I, Chapter XV). The zygote elongates, penetrating deeply into the prothallus at first without division (Fig. 381). The embryo itself is formed at the tip of this suspensor, and its parts are differentiated relatively early. These are essentially similar to those of other species of the genus: the first leaf (cotyledon) appears on the side of the axis directed obliquely upwards, and it breaks through the upper surface of the prothallus: the root originates on the side directed downwards, and it emerges on its under side (Fig. 382). Campbell has shown that a basal wall separates, as in other Pteridophytes, an epibasal from a hypobasal hemisphere. The former gives rise to the stem-apex and the cotyledon, while the root originates centrally very near to the basal wall, so that it is difficult to say whether it belongs to the epibasal or the hypobasal region (*Ann. of Bot.* 1921, p. 141). In the course of this development, since the axis of the archegonium is oblique the embryo has to execute a more or less sharp curvature so as to secure that its apex shall point upwards, and the root emerge downwards, as it is seen to do in Fig. 382. It will be realised that this is a consequence of the presence of the suspensor and of the endoscopic orientation.

The embryogeny of *Eu-Botrychium* (*B. Lunaria* and *simplex*) differs from this in the fact that there is no suspensor, and the embryo shows no curvature, being exoscopic in orientation, and the archegonium pointing usually upwards. The octant-walls appear directly in the zygote itself, and they are

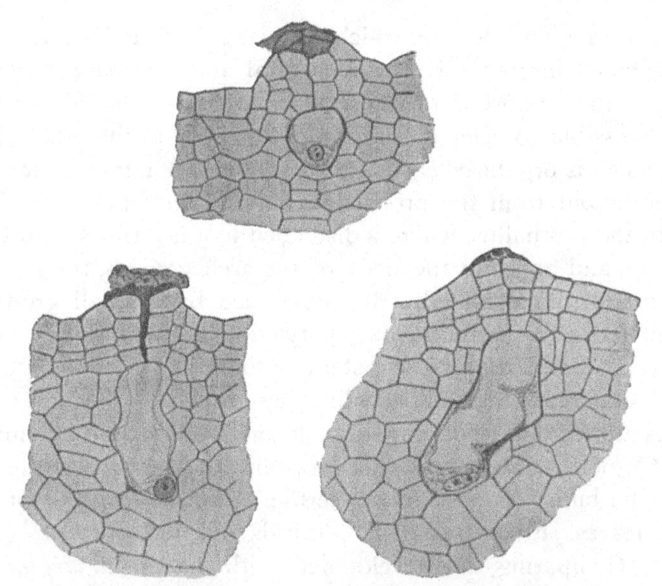

Fig. 381. *Botrychium* (*Sceptridium*) *obliquum*. First stages in the embryogeny. Before the first segmentation the zygote grows into an elongated tube, cut off later as the suspensor, which burrows its way irregularly into the tissue of the prothallus (× 150). (From sections lent by H. L. Lyon.)

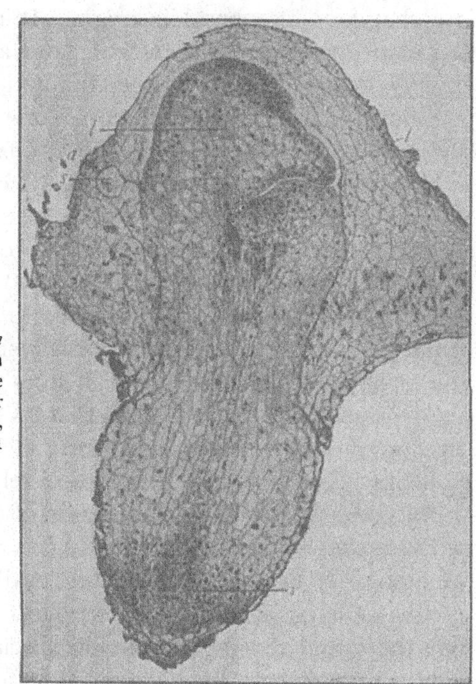

Fig. 382. *Botrychium* (*Sceptridium*) *obliquum* Muhl. Photo-micrograph of a section through a gametophyte and young sporophyte. The root has already protruded from the under side of the gametophyte. *a* = archegonium, *s* = suspensor, *t* = stem-tip, *l* = first leaf, *r* = root (× 60). (After H. L. Lyon.)

followed by less regular divisions which disguise them in the resulting ellipsoid body. The exact limits between the epibasal and hypobasal parts are lost, and owing to the late origin of the several parts of the embryo Bruchmann found it impossible to refer them to one source or to the other (Fig. 383). The root, which is organised early, grows first in a horizontal direction, and bursts laterally out from the prothallus, but the remainder of the embryo rests within the prothallus, where a distended foot is formed. On this ovoid cellular body, and opposite the neck of the archegonium, the apical cell of the stem arises: it is immediately overarched by a small growth which Bruchmann takes for a rudimentary cotyledon, and the embryo is now as in Fig. 384, *A*. Even at this early stage it contains the endophytic fungus. Successive roots may then follow, while the growth of the bud remains in this species almost quiescent, though it forms a succession of small leaves (Fig. 384, *B*): of these about the eighth appears above ground, the rest serving to protect the bud. A rudimentary fertile spike may appear on some of these scale leaves. From this point onwards the development is as in the adult plant. Comparing this development with that in *B. virginianum* the relative position of the parts is essentially the same: the chief differences are in their proportion. The root and foot are larger, and the axis later in definition in *B. Lunaria*: also in *B. virginianum* the first leaf is itself expanded above ground, while the same difficulty exists in defining whether the root is epibasal or hypobasal in origin. It may be held that what is seen in *B. Lunaria* and *virginianum* is a later and derivative state compared with *B. obliquum*, resulting from the elimination of the suspensor, which liberates the embryo from the endoscopic orientation, and makes the awkward curvature unnecessary (Jeffrey, *Proc. Canad. Inst.* 1898).

Helminthostachys shares with the "*ternatum*" group the more primitive endoscopic embryogeny with a suspensor. Lang (*Ann. of Bot.* 1914, p. 21) has shown how on the roughly cylindrical prothallus the axis of the archegonium is oblique or even horizontal (Fig. 378). Consequently a like curvature to that in *B. obliquum* will be necessary; it is seen to occur, and the relation of the parts in the sporeling is closely similar also. At first the axis of the young sporeling thus produced is upright (Fig. 385, *a*, *b*). It is only as the plant passes to the adult state that its creeping habit is assumed. This is the only member of the family that shows this prone position, and it may be held to be secondarily acquired in relation to its heavy foliage.

In *Ophioglossum* the first division of the zygote is transverse to the axis of the archegonium, and as in *B. Lunaria* there is no suspensor, and the orientation is exoscopic. The first wall separates the epibasal from the hypobasal hemisphere; but reference of the parts to any definite relation with the initial cleavages is made specially uncertain by the fact that the embryo attains considerable size before any differentiation occurs (Fig 386).

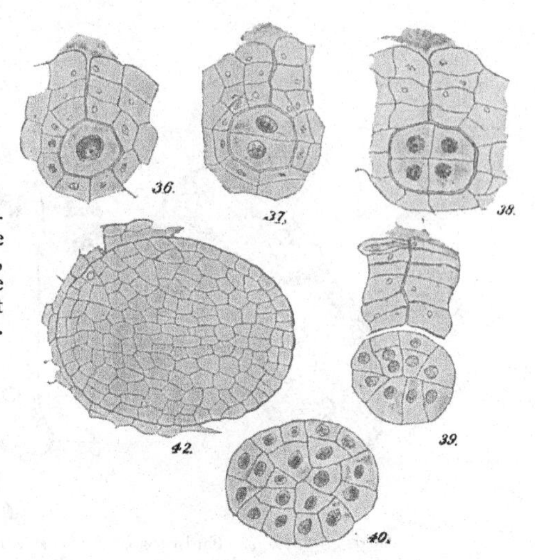

Fig. 383. *Botrychium Lunaria* L. 36 = a fertilised archegonium, 37 = zygote showing the first segmentation, 38 = embryo of four cells, 39, 40 = embryos cut in direction of the archegonium, 42 = an embryo breaking out of the prothallus. 36–40 (× 225). 42 (× 150). (After Bruchmann.)

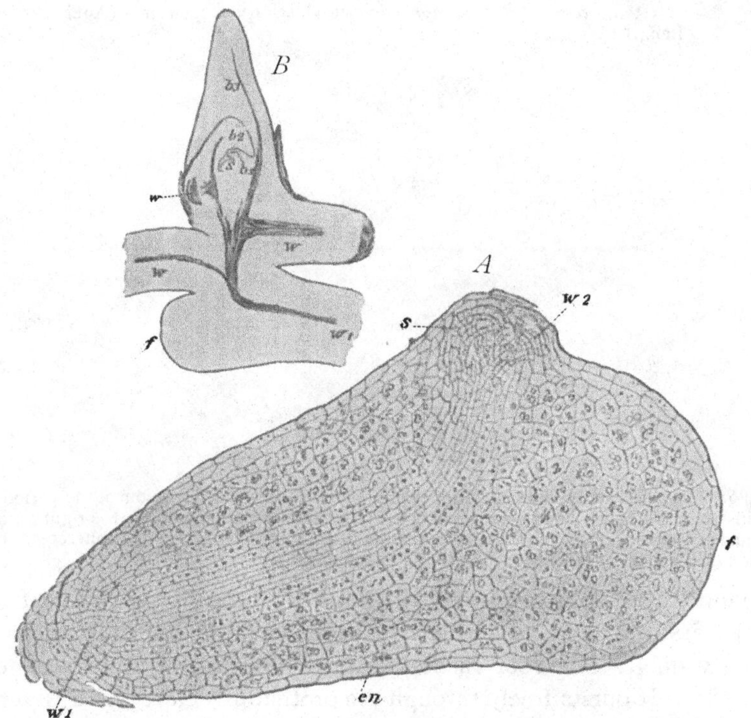

Fig. 384. *Botrychium Lunaria* L. The lower figure (*A*) represents an old embryo with well-developed foot (*f*): w_1 = apex of first root, *s* = apex of the rhizome with the second root, w_2. The endophyte (*en*) is already in the cells (× 52). The upper figure (*B*) is a diagrammatic section of a seedling with six to eight roots, of which three are in the plane of section. *f* = foot, w_1 = first root, *w* = other roots, *s* = apex of rhizome, b_1–b_3 = developing leaves (× 6). (After Bruchmann.)

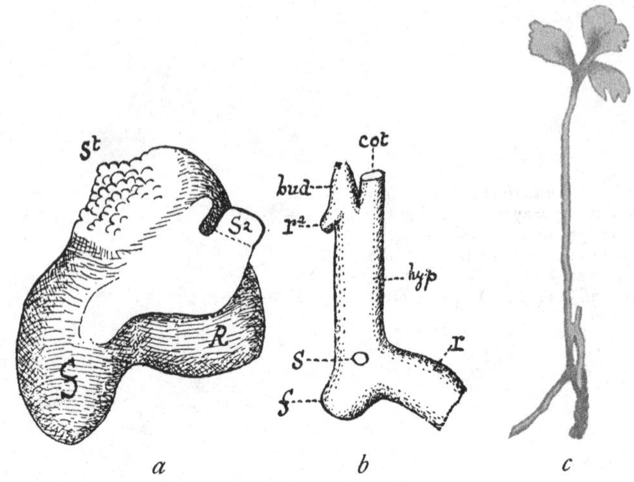

Fig. 385 *a*, *b*, *c*. Embryos of *Helminthostachys*. *a*, shows a young state; *b*, is more advanced; *c*, is still older with upright axis, and shown natural size, while *a*, *b* are enlarged. *f*=foot, *R*, *r*, and r_2=roots, *st*=stem, *cot*=cotyledon, *hyp*=hypocotyl, *s*, s_2=suspensor. (After Lang.)

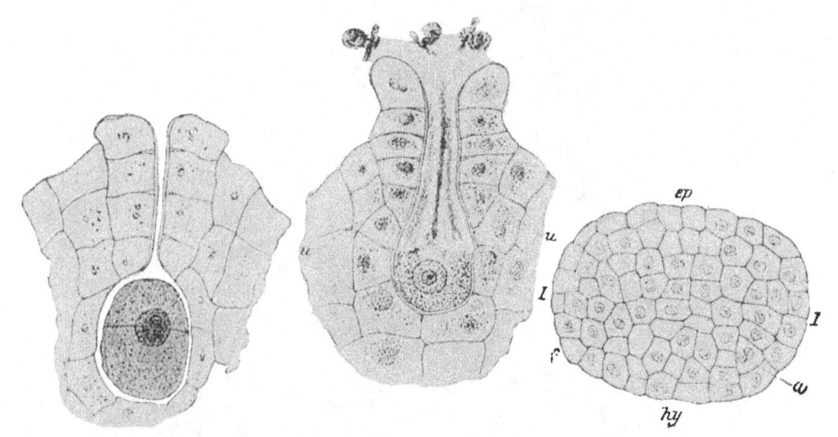

Fig. 386. *Ophioglossum vulgatum*. The central figure shows an archegonium at the period of fertilisation. The left-hand figure shows the first division of the zygote. To the right a more advanced embryo. *I*, *I*, the basal wall; *ep*, epibasal; *hy*, hypobasal hemispheres; *f*, the region of the foot; *w*, root (× 225). (After Bruchmann.)

Bruchmann states, however, for *O. vulgatum* that the hypobasal half gives rise to the first root and foot: the latter is never large, but only appears as a slight swelling. The root rushes forward forming its apical cell early (*w*, Fig. 387). It bursts freely through the prothallus before there is yet any definite trace of the axis or cotyledon (Figs. 379, 387). Up to this time the embryo is stored with nutritive substances, but there is no endophytic fungus, and the development of *O. vulgatum* up to this stage may occupy several vegetative seasons. The differentiation of the long-delayed shoot is accompanied by the origin of the second root, which appears endogenously close to the proximal end of the vascular strand of the first. Immediately above this, and opposite the neck of the archegonium, the shoot appears, the cotyledon being on the same side of the axis as the first root. But the cotyledon

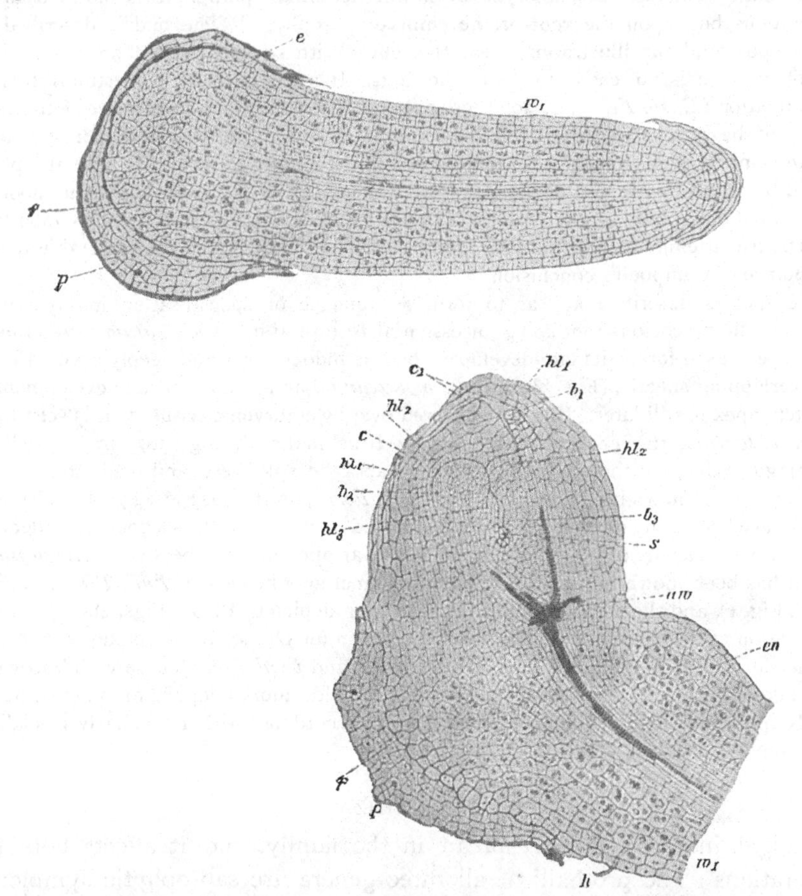

Fig. 387. *Ophioglossum vulgatum.* The upper figure shows a young sporeling in longitudinal section: *w*, first root with evident apical cell; *f*, the foot only slightly projecting; *e*, the epibasal region; *p*, the remains of the prothallus. The lower figure shows part of a large sporeling in longitudinal section: *p*, prothallus; *w*, first root; *h*, entering fungal filament; *en*, endophytic fungus; *aw*, insertion of second and third roots; *s*, apex of rhizome; p_1, p_2, p_3, leaves; *c*, canal; *hl*, *hl*, sheaths of leaves. (After Bruchmann (× 35).)

remains rudimentary, and is followed by a second leaf which may develope as a small sterile leaf expanded above ground. The third leaf may under favourable circumstances bear a fertile spike. The further development then follows as in the adult plant.

Campbell, having examined several tropical species of *Ophioglossum*, recognises three types of embryogeny in the genus: that of *O. vulgatum* above described; that of

O. moluccanum described by Mettenius and by himself; and that of *O. pendulum*, on which he has added largely to the earlier observations of Lang. The first type is characterised by early formation of the root and late development of the axis and leaf; in the second the leaf and root only appear to be developed at first; in the third roots only; and he states that in both *O. moluccanum* and *pendulum* the definitive sporophyte is "formed as an adventitious bud upon the root of the embryo sporophyte." The bud is described as endogenous, and the illustrations bear this out. With so experienced an observer as Campbell there is no reason to doubt the facts. It appears to be a question rather of interpretation (*Eusp. Ferns*, p. 40, Figs. 22–24). There can be no question of the near affinity of the various species of *Ophioglossum*. It would appear therefore more probable that the embryogeny should follow one plan subject to modifications, than that the plans should be so distinct in the different species as that one should form its main shoot by "first intention," while in other species it should be by adventitious origin of so important a part. An attempt to find another interpretation of the facts seems preferable to the acceptance of Campbell's conclusion.

The species described appear to form a sequence of specialisation in mycorhizic nutrition, the precocious root being an essential feature in it. *O. vulgatum* shows this in a less specialised form: its late-developed shoot is hidden away and deeply covered in by the overlapping sheaths (Fig. 387). In *O. moluccanum* and *pendulum* the development of the stem-apex is still later. If it were covered over by embryonic tissue, as it is seen to be in *O. vulgatum*, the channel described by Bruchmann (*c*, Fig. 387) might well be indistinguishable, but the originally organised apex be still there, and ready to awake to activity, as it is shown by Campbell to do (*Eusp. Ferns*, pp. 41, 42, Figs. 24, 25). The apex would have been lying dormant, and not be formed *de novo*. There is a good precedent for such complete closure of a similar channel above an apex in the tubers of *Phylloglossum*: here it has been shown how completely that channel may be closed (*Phil. Trans.* II, 1885, Pl. 71, Fig. 1), and the steps leading to it have been depicted (Pl. 72, Figs. 28–33). A like obliteration of the channel described by Bruchmann for *O. vulgatum* would give the conditions described by Campbell for *O. moluccanum* and *pendulum*. These embryos would then take their place not as the primitive basis for wide morphological arguments, but as highly specialised modifications of embryology in accordance with a peculiarly specialised nutrition.

MYCORHIZA.

Fungal infection is widespread in the family, and it affects both the generations. The prothalli of all three genera are saprophytic symbionts, and the existence of chlorophyll in them is exceptional. There is some exactness in the segregation of the infected areas, the hyphae being present in the basal regions of the prothalli, but they are excluded from the upper and more particularly from the superficial parts, in which the apical growth is located, and the sexual organs are borne (Figs. 375, 378, 379). This distribution has its analogies with what is seen in symbiotic Orchids. The distribution of mycorhiza in the sporophyte has been fully stated in *The Origin of a Land Flora* (p. 477), where the literature is quoted up to that date. In the Ophioglossaceae it appears to be inconstant: it is seen in the adult plant of *O. vulgatum, pendulum* and *simplex*, and in the latter it is certainly associated with marked morphological peculiarities. In *Helminthostachys* it is present in the roots of the young but absent in the adult plant. In *Botrychium* it has been seen in twelve species, but in varying abundance, and especially in the young. It is thus inconstant in its occurrence in the family, but it is seen to prevail in those species of *Ophioglossum* which are regarded as the most specialised. The mycorhizic condition does not necessarily entail a marked state of reduction, though in extreme cases this may be seen: but it appears to be commonly associated with a leathery type of foliage, such as that of most of the Ophioglossaceae.

COMPARATIVE TREATMENT OF THE OPHIOGLOSSACEAE.

This very distinctive and circumscribed family has been related systematically with the Filicales by most writers, a position which has been definitely confirmed by a large body of facts recently acquired. An alternative relation with the Lycopodiales was suggested by Mettenius (*Bot. Zeit.* 1867, p. 98); it was seriously considered by Strasburger (*Bot. Zeit.* 1873, p. 5), and by Celakovsky (*Pringsh. Jahrb.* 1884, p. 291), while Graf Solms Laubach (*Buit. Ann.* Vol. IV, p. 186) withheld a definite opinion in view of the incompleteness of the facts. The case for a Lycopodinous affinity was further developed on the basis of many new facts in 1896 (Bower, *Morph. of Spore-Producing Organs.* Dulau, London, 1896), and fully argued in 1908 (*Land Flora*, pp. 430–494). The question turns upon the morphological nature of that characteristic organ the "fertile spike." Is this an up-grade part illustrating a progressive evolution from a part similar in nature to the Lycopodinous sporangium, or is it a foliar part showing modification, mostly of the nature of reduction? In the former case the affinity would be with the Lycopodiales: in the latter with the Filicales. Naturally the facts of form, anatomy, development, and of physiological probability must all be brought into the final decision. Unfortunately Palaeontology is silent, except in so far as it supplies important comparisons with early fossils not recognisable as belonging themselves to the Ophioglossaceae. The first point will then be to decide the morphological character of that most distinctive part, the fertile spike.

Taking first the Lycopodinous alternative, the adaxial position and the spore-producing function are the same for the sporangium and for the spike. By comparison of the simplest spikes known it is possible to construct an up-grade series leading by successive steps of amplification, septation, and branching from the Lycopods and Psilotales to the most elaborate of Ophioglossaceous spikes. In such a series *Ophioglossum* would take its place as relatively primitive, and the extreme would be seen in *Botrychium virginianum*. As a purely morphological *tour-de-force* this is possible; but physiologically is it probable that it should happen in a family marked strongly by mycorhizic nutrition? Further, how does this view accord with the anatomical facts? Taking first the vascular supply of the spike of the

Ophioglossaceae, it has been seen in many instances that it arises either by marginal abstriction, or by extra-marginal segregation from the supply leading to the sterile blade (Figs. 359, 360, *B*). These are precisely the alternative methods of supply to the pinnae of Ferns (compare Vol. I, pp. 172–174), and incidentally in *B. virginianum* the supply to its own sterile pinnae arises in the same way (Fig. 360, *A*). From this the conclusion seems probable that the spike is of pinna-nature. But since the vascular supply habitually comes off equally from both sides, it was suggested by Roeper (*Bot. Zeit.* 1859, p. 241) that the spike really represents two lateral pinnae fused laterally, and *Anemia* with its two erect fertile pinnae was naturally cited as evidence. Though this suggestion lacks support from external morphology (unless the fissions of the spike in *Ophioglossum* be held as such) it may be held as a possible view, though nothing would appear more natural than that, as a laterally originated pinna moved to a median position, a vascular connection should be established secondarily with the other side of the leaf-supply. The fusion-theory may therefore be held as still "not proven." A consideration that strengthens the Fern-pinna theory as against comparisons with Lycopods and Psilotales is that in *Tmesipteris*, *Cheirostrobus*, and *Sphenophyllum*, in all of which adaxial spore-producing members exist, the origin of the vascular supply is median, and adaxial, not lateral as in the supply to the Ophioglossaceous spike. This seems to be a real and valid anatomical distinction, which ranges the Ophioglossaceae with the Filicales rather than with such plants as those named. Further, the positive and close analogy between the stelar morphology and the venation of the Ophioglossaceae on the one hand, and that of relatively primitive Ferns on the other, gives cogency to the conclusion that the spike is of pinna-nature, and that consequently the Ophioglossaceae are Filical in their alliance. Finally, this view accords with the fact that in *Botrychium* the sterile blade may not infrequently be in part or in whole fertile (Fig. 338, p. 46). The balance of evidence thus appears to be clearly in favour of ranking the Ophioglossaceae as primitive Ferns, and the spike as of pinna-nature.

In comparing the Ophioglossaceae among themselves it is necessary to be clear which are the more primitive types, and which the more specialised. In forming an opinion use must be made of the criteria established in Vol. I. A very obvious character is the venation. In *Botrychium* and *Helminthostachys* the venation is always open, in *Ophioglossum* it is as constantly closed: and not only is it closed, but it forms an elaborate small-meshed reticulum, such as appeared only in the Secondary Period, while the venation of the other genera is Palaeozoic in type. This at once marks off *Ophioglossum* as probably derivative[1]. Passing to stelar structure, *Botrychium* and

[1] Among the many disabilities consequent on the Great War has been the difficulty of interchange of scientific literature. To this I ascribe the omission of Prof. von Goebel to notice my memoir on leaf-architecture (*Trans. Roy. Soc. Edin.* 1916, Vol. XLI, p. 657) in his valuable study *Gesetzmässigkeiten*

Helminthostachys with their definitely circumscribed steles, their early protostely and subsequent medullation, and their uniformly undivided, and occasionally Clepsydroid leaf-trace, show clear anatomical relations with other early Fern-types, such as the Botryopterideae and Fossil Osmundaceae. More especially the slight secondary thickening of *Botrychium* finds its analogue in the Palaeozoic *Botrychioxylon*, and occasionally also in the Zygopterideae. On the other hand, the laxer stele of *Ophioglossum*, with its early loss of the endodermis and its expansion into an open dictyostele of unsheathed strands, presents an advanced state that reaches a climax in the distended sappy stem of *O. palmatum* (Fig. 385). Further, the divided leaf-trace of *O. palmatum*, *pendulum*, and *simplex* points out those species as anatomically advanced (Figs. 355, 356). On these grounds it may be held that *Botrychium* and *Helminthostachys* are relatively primitive members, while *Ophioglossum* (and especially § *Ophioderma* and § *Cheiroglossa*) will be regarded as the most advanced genus of the family.

How then do the facts relating to the fertile spike accord with this conclusion? How shall the branching and multiplication of spikes seen in *Ophioglossum* be harmonised with the idea of its being a relatively late type of the family? It has been the habit to regard *Ophioglossum* as primitive chiefly because its spike appears to be the simplest in form. Is this conclusion justified? It may be tested by comparison with known Palaeozoic fossils. In the Coenopteridaceae we see Ferns having sporangia of a type not far removed from those of *Botrychium* and *Helminthostachys*. The way in which those sporangia are borne, marginally or distally, accords with experience among the fossils. It requires no great strain of imagination to see in the unique spike of *Helminthostachys* a condensed form of such a type of sporophyll as that of *Stauropteris*: and this comparison is supported by the fact that a vascular strand passes to the base of each sporangium. On the other hand, the sporangium of *Ophioglossum* itself is not of a type known among fossils, any more than is the spike which bears it.

The suggestion that *Ophioglossum* is relatively late and derivative implies a progressive sinking of the sporangium into the tissue of the part that bears it, accompanied by an enlargement of the sporogenous tissue, and of the spore-output. No doubt it will be remarked that this is contrary to the general evolutionary progression in Ferns, so clearly brought out by the spore-enumerations, and by the comparisons of sporangial structure described in Vol. I, Chapter XIII. These demonstrate for Ferns at large the progressive decrease in size of the sporangium, and fall in the spore-output. This is

im Blattaufbau, Jena, 1922. The views which he advances appear to accord with the general arguments of this volume, and in particular with the opinion here expressed as to the venation in the Ophioglossaceae; but they will find more detailed application in the treatment of the more advanced Filicales to be given in Vol. III.

true: but the existence of a general trend in the evolution of a Class as a whole cannot be held to preclude the recognition of an opposite trend in individual families. Nature is not built on lines of logic, and is not throughout consistent. Opportunism is the ruling influence. In the present instance the biological conditions met by *Ophioglossum* as a whole, and by §*Ophioderma* and §*Cheiroglossa* in particular, are different from those of most Ferns. They are of xerophytic structure, with a strong tendency to mycorhizic nutrition. In the ordinary ground-species the habitat is as a rule exposed. In *Ophioderma* and *Cheiroglossa* the habit is even epiphytic and partially saprophytic, conditions which may well be expected to produce that leathery texture which their leaves show; also a deeply sunk state of their sporangia, these being otherwise without any protection. On the other hand, the very large spore-output may be recognised as a biological offset against the risks involved in establishing a progeny that is both partially saprophytic and also epiphytic.

A very special biological interest attaches to that curious degradation illustrated by *O. intermedium* and *simplex*. These species are clearly related to *O. pendulum*, of which they are probably ground-growing derivatives. The two last-named species show their advanced position anatomically by their divided leaf-traces. They are strongly mycorhizic. It has been suggested that the saprophytic nutrition in the wet and dark forest in which they grow has superseded their photosynthetic nutrition, and that abortion of the sterile blade has been the result. If this be true then these simplest-leaved Ophioglossaceae will be phyletically the most advanced.

The more complicated arrangements of spikes in the genus *Ophioglossum*, shown in Figs. 343, 344, have elsewhere been the subject of theoretical comparison with a view to the morphological elucidation of the spike. But it may be questioned whether such comparisons are enlightening, or even permissible. It cannot properly be assumed that these specimens have retrograded along the lines of their previous evolution. If *Ophioglossum* is really a late and derivative type of the family it would appear more probable that its spike, condensed and standardised, would behave as a morphological unit, subject rather to new fissions and interpolations than to normal retrogression. In fact that the spike may, like the stamen of an Angiospermic flower, show pleiomery. But such pleiomery cannot be read as elucidating the general morphological history. This is the interpretation which appears jnstifiable in such examples as are shown in Figs. 343 and 344. They may be compared with the condition of the sporangiophores of the Sphenophylleae as regards their position, and varying number. It is significant that these anomalies in *Ophioglossum* are characteristic of those sections of the genus which have the most advanced vascular structure, and in particular the disintegrated leaf-trace.

The gametophyte does not greatly help in the morphological comparison of the Ophioglossaceae. Nevertheless there are some significant features which though slight in themselves acquire additional weight because they point to the same conclusions as follow from the study of the sporophyte. *Botrychium* stands alone in having a flattened form of its prothallus: and though its sexual organs are borne upon the upper surface this is not unknown in the flattened green prothalli of Ferns, to which it is the nearest approach seen in the family. The sexual organs themselves are highly standardised for the family: but Campbell specially remarks upon the size of the antheridia of *O. pendulum*, and the number of its spermatocytes, "perhaps more than in any Pteridophyte." This runs parallel with the sporangia of the same species, which show a specially large spore-output. On the other hand, the archegonia of *Ophioglossum* are more deeply sunk than in the other genera, and this appears in the most pronounced form in *O. pendulum*. In fact the genera may be seriated roughly by their gametophytic characters with results corresponding to those based upon their anatomy, their sporangia, and their external form. Lastly, their embryology leads to a like result. *Botrychium obliquum* and *Helminthostachys* have endoscopic orientation of the embryo with a suspensor, and their embryos are subject in consequence to awkward curvatures during development. The sections §*Eu-Botrychium* and §*Osmundopteris* on the other hand have no suspensor, and their orientation is exoscopic, while their embryos have no curvature: and this is so for all the species of *Ophioglossum*. When finally we see in *O. moluccanum* and *pendulum* an extreme delay in the development of the stem-apex, accompanied by those peculiarities described in detail by Campbell, the cumulative conclusion seems justified that these illustrate not a primitive state, but probably the most advanced and specialised embryological condition that the family has produced.

Phyletic Arrangement of the Ophioglossaceae.

It is concluded then that the Ophioglossaceae are a family of primitive Ferns, allied especially to the Coenopteridaceae, but with some degree of affinity also with the Osmundaceae and the Marattiaceae. Within the family *Botrychium* and *Helminthostachys* appear to be relatively primitive genera, and *Ophioglossum* more highly specialised. This follows from the details of external form, venation, anatomy, and the characters of the gametophyte, sexual organs, and embryogeny above detailed. But it is difficult to assign to either of these more primitive genera the prior place.

Helminthostachys is represented only by a single species, of the limited Indo-Malayan region, a fact which may be held to indicate antiquity. The form and venation of the leaf, the stelar structure, and the undivided leaf-trace accord with this, while the construction of the fertile spike, unique among living plants, may find its nearest correlative in the fossils of a Zygopterid type. Among living plants it may have given rise by condensation to the spike of *Ophioglossum*. Though clearly a member of the family it stands as an isolated type.

The genus *Botrychium*, characterised by its branched spike, and its seriated but separate and projecting sporangia, may be divided according to the current system into three sections, of which that designated *Sceptridium* by Lyon (1905) is probably the most primitive. This section includes the *ternatum*-group of species as recognised by Prantl. Their primitive character is shown in the first instance by the endoscopic embryology, while the foliage takes a middle position as regards the triangular form of its lamina, its cutting and venation. The spike is inserted low down. The section *Eu-Botrychium* includes the species grouped with *Lunaria*, having smooth leaves, oblong or deltoid, and never more than doubly pinnate. They have exoscopic embryogeny. The section also includes reduced types such as *B. simplex*. The section *Osmundopteris* comprises the *virginianum* types, with relatively large leaves, some being five times pinnate, and hairy, and often of thin texture. The stem has an active cambium. The embryogeny is exoscopic. These characters are held to indicate a relatively advanced state.

The genus *Ophioglossum* is commonly regarded as typical of the family, and accordingly placed first (Christensen's *Index*, p. lviii; Hooker's *Synopsis Filicum*, p. 444; Engler & Prantl, *Naturl. Pflanzenfam.* I, 4, p. 465; Christ's *Farnkraüter*, p. 362). But on grounds explained above it should be placed last, as being derivative. It has been subdivided into three sections, of which that designated *Eu-Ophioglossum* includes *O. vulgatum*, and the large proportion of species characterised by a ground-growing habit, the sterile blade and spike as a rule simple, and the sporangia sunken and large. The stele is dilated and dictyostelic, usually without endodermis, and the leaf-trace is undivided at departure. The section *Ophioderma* includes the epiphytic *O. pendulum* together with the ground-growing species *intermedium* and *simplex*. The first is characterised by occasional irregular branching of the sterile blade and the spike, the large size of the latter and of its sporangia, and particularly by the subdivision of its leaf-trace at departure. Its embryogeny is exoscopic. Its mycorhizic habit leads in the allied, ground-growing species to the reduced structure seen in *O. intermedium* and *simplex*. Lastly, the section *Cheiroglossa* includes only *O. palmatum* which is epiphytic on rotting trees, and is characterised by frequent branching of its sterile blade and fertile spikes, thus leading to a plurality of spikes. The stem is distended, and the leaf-trace divided at departure. These features suggest a state of advance upon the *vulgatum* type. Accordingly the phyletic disposition of the family, as nearly as it can be shown by a linear sequence, would be thus:

OPHIOGLOSSACEAE.

(1) *Helminthostachys* (Kaulfuss, 1822). 1 species.
(2) *Botrychium* (Swartz, 1801). 34 species.
 § i. *Sceptridium* (Lyon, 1905).
 § ii. *Eu-Botrychium.*
 § iii. *Osmundopteris* (Milde, 1837).
(3) *Ophioglossum* (Linnaeus, 1753). 43 species.
 § i. *Eu-Ophioglossum.*
 § ii. *Ophioderma* (Blume, 1828).
 § iii. *Cheiroglossa* (Presl., 1845).

The primitive stock thus disposed in roughly phyletic sequence is not known to have given rise to any more advanced derivatives. The Ophioglossaceae appear to have terminated as a blind evolutionary series, and they stand to-day as an imperfectly modernised relic of the Palaeozoic Flora.

BIBLIOGRAPHY FOR CHAPTER XIX.

352. RUSSOW. Die Histologie d. Leitb. Kryptogamen. St Petersbourg. 1872.
353. HOLLE. Bau. u. Entw. d. Veg. d. Ophioglossaceen. Bot. Zeit. 1875, p. 241.
354. ROSTOWZEW. Réch. sur l'*Oph. vulgatum.* L. Kjobenhavn. 1891.
355. ROSTOWZEW. Beitr. z. Kenntniss d. Ophioglosseen. Mockba. 1892.
356. POIRAULT. Rech. Anat. sur les Crypt. Vasc. Ann. Sci. Nat. T. xviii. 1893.
357. BOWER. Studies. II. Ophioglossaceae. Dulau and Co. London. 1896.
358. JEFFREY. Gametophyte of *B. virginianum.* Trans. Can. Inst. 1896–7.
359. FARMER AND FREEMAN. *Helminthostachys zeylanica.* Ann. of Bot. xiii, p. 421. 1899.
360. ENGLER AND PRANTL. Natürl. Pflanzenfam. I. 4, p. 499. Here the earlier literature is very fully cited.
361. BOWER. *Ophioglossum simplex.* Ann. of Bot. 1904, p. 205.
362. BRUCHMANN. Prothallium von *O. vulgatum.* Bot. Zeit. 1904, p. 227.
363. BRUCHMANN. Prothallium von *B. Lunaria.* Flora. 1906, p. 203.
364. LYON. A new genus of Ophioglossaceae. Bot. Gaz. Dec. 1905, p. 455.
365. LANG. Prothalli of *Ophioglossum* and *Helminthostachys.* Ann. of Bot. 1902, p. 23.
366. LANG. Studies in the Morph. and Anat. of the Ophioglossaceae.
 I. Branching of *B. Lunaria.* Ann. of Bot. xxvii, p. 203. 1913.
 II. Embryo of *Helminthostachys.* Ann. of Bot. xxviii, p. 19. 1914.
 III. Rhizome of *Helm. zeylanica.* Ann. of Bot. xxix, p. 1. 1915.
367. CHRYSLER. Fertile spike of Ophioglossaceae. Ann. of Bot. 1910, p. 1.
368. CHRYSLER. Is *O. palmatum* anomalous? Bot. Gaz. lii, p. 151. 1911.
369. PETRY. Branching of Ophioglossaceae. Bot. Gaz. lix, p. 345. 1915.
370. PETRY. Anatomy of *O. pendulum.* Bot. Gaz. 1914, p. 171.
371. VON GOEBEL. Organographie. II Aufl., Teil. ii, Heft 2. 1918.

372. CAMPBELL. Mosses and Ferns. 3rd. edn. 1918. Here the literature is very fully cited.
373. CAMPBELL. Eusporangiate Ferns. Washington. 1911.
374. CAMPBELL. Gametophyte of *B. obliquum*. Ann. of Bot. 1921, p. 141.
375. CAMPBELL. Gametophyte of *B. simplex*. Ann. of Bot. 1922, p. 441.
376. BOWER. Origin of a Land Flora. 1908, pp. 430–494.
377. BOWER. Medullation in the Ophioglossaceae. Ann. of Bot. 1911, p. 537.
378. BOWER. Morph. of *O. palmatum*. Ann. of Bot. 1911, p. 227.
379. BOWER. The primitive Spindle. Proc. Roy. Soc. Edin. xliii, p. 1. 1922.

CHAPTER XX

MARATTIACEAE

THIS family is represented by seven genera of living Ferns. The characters of all the genera are well known, so that they form a sound basis for comparison with the related fossils. A number of these, dating back to the Palaeozoic Period, show strong similarity to the modern forms both as regards anatomy and the character of their sori. Their existence shows that the Marattiaceous type has been a very ancient one. The natural course will be first to consider the living Marattiaceae: we shall then proceed to compare them with their fossil correlatives.

EXTERNAL CHARACTERS.

The living genera are named, *Angiopteris*, *Macroglossum*, *Archangiopteris*, *Marattia*, *Protomarattia*, *Danaea*, and *Christensenia* Maxon (= *Kaulfussia* Blume). They are all intertropical, *Angiopteris* from the eastern, *Danaea* from the western, and *Marattia* from both hemispheres. The other four are from the Malayan region. The erect stock of *Angiopteris*, *Marattia*, and *Archangiopteris* is relatively short, massive, and unbranched: it is of a radial type, and is entirely covered by the persistent bases of the crowded leaves (Fig. 388). It continues directly the radial symmetry initiated in the embryo,

Fig. 388. *Angiopteris Teysmanniana* de Vriese. *A*, habit of a small plant, reduced to $\frac{1}{20}$. *B*, part of a pinna, natural size. From Bitter (Engler and Prantl).

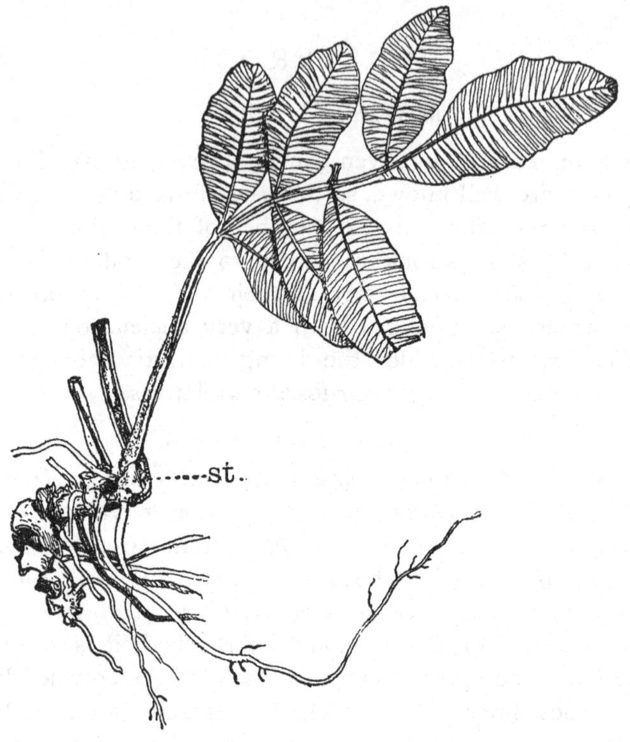

Fig. 389. A small plant of *Danaea alata* (× $\frac{1}{2}$). *st.* = stipules. (After Campbell, see Text.)

and this is probably a primitive condition. Some species of *Danaea* have also an erect radial shoot (*D. simplicifolia* and *elliptica*): others show at first an erect position and radial construction, but this passes over in the adult to an oblique position with a distichous arrangement of the leaves (*D. alata*, Fig. 389). *Protomarattia* and *Christensenia* are, however, strongly dorsiventral, with their leaves borne obliquely alternate on the upper surface (Fig. 390). The natural interpretation of these facts is that in the last-named Ferns the primitive radial and erect type has been relinquished in favour of the creeping habit, which goes along with a smaller bulk and greater elongation of the stem. In this they resemble *Helminthostachys* among the Ophioglossaceae.

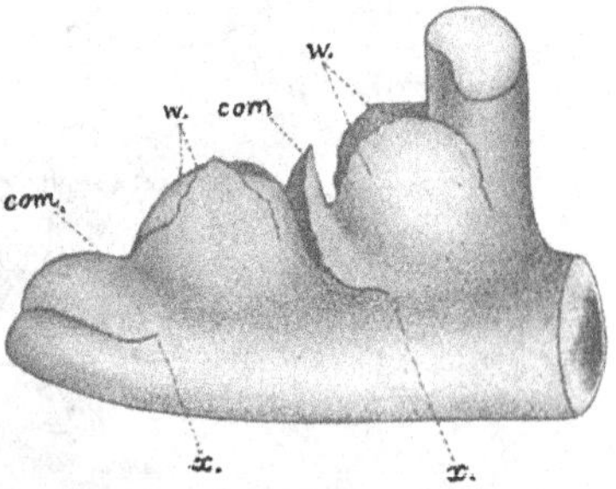

Fig. 390. Diagrammatic representation of the end of a rhizome of *Christensenia*. *w.* = wings of the stipule; *com.* = transverse commissure; x, x = points at which arrested apices may be found. (After Gwynne-Vaughan.)

The leaves are very diverse in outline, though conforming to a common type. The leaf-base in all of them bears stipular enlargements laterally, which are connected across the adaxial face of the petiole by a transverse commissure (Fig. 390; see also Vol. I, Fig. 96). Though these are characteristic for all the adult Marattiaceae, they are absent from the first and often from the second leaf of the sporeling. After the upper leaf decays they remain persistent, in close relation to the smooth scar which marks its attachment.

Scott (*Ann. of Bot.* Vol. xxvi, 1912, p. 60) has discussed the aphlebiae seen in *Zygopteris Grayi*, regarding them as modified pinnae of the leaf which have spread downwards on to the stem, but they were always in definite relation by their vascular strands with the leaf-traces. Their function was the protection of the growing points and young leaves. Basal growths laterally on the petioles are widely spread among living Ferns of primitive type, such as the Marattiaceae. It may be a question whether these are truly homologous with the basal aphlebiae of the Coenopterids; but in any case they correspond in position and in function. Such modern basal growths as those of the Ophioglossaceae, Marattiaceae, and Osmundaceae have been described as stipules, and they appear highly variable both in bulk and in the presence of a commissure, and may provisionally be held as being parts of the same nature as the basal aphlebiae. Buds, often described as adventitious, are formed on the stipules of the Marattiaceae, and propagate the plants if the stipules bearing them are cut away. These buds are stated not to be really adventitious, but to be laid down early, remaining dormant as are the axillary buds in *Helminthostachys* (Van Leewin, *Buit. Ann.* Vol. x, p. 202).

The upper leaf of the living genera varies considerably. It may be simply ovate, with marked midrib and acuminate apex, as in *Macroglossum* or *Danaea simplicifolia*: or it may be simply pinnate, as in *D. alata* (Fig. 389), or in *Archangiopteris* and *Protomarattia*: or the pinnation may be repeated, as in *Marattia* or *Angiopteris* (Fig. 388). In large plants the leaf may attain a high complexity of branching, while its length may be as much as 15 feet. In *Christensenia* the outline of the leaf differs from all the rest: the long petiole bears five palmately disposed lobes of broadly lanceolate form, with a general similarity of outline to the leaf of the Horse Chestnut (Fig. 392, *D*). The venation is simple and dichotomous in all the other genera, with open endings to the veins; but it is reticulate in *Christensenia*, approaching that of the *Drynaria*-type. The base of the leaf-stalk, and often the bases of the pinnae also, bear fleshy swellings or pulvini: here the stalk is liable to break on decay, leaving a clean scar. The texture of the leaf is usually leathery, but in *Danaea trichomanoides* there is a thin almost filmy character of the foliage, in obvious adaptation to its moist habitat.

The young surfaces of *Marattia* and *Angiopteris* are covered with hairs which are short and simple, while their terminal cells contain tannin and stain deeply. But in *Christensenia* and especially in *Danaea* conspicuous scales of a peculiar peltate form are abundant (Campbell, *Eusp. Ferns*, p. 150). Elongated scales of considerable size are present on the leaf-stalks of *Protomarattia* (Hayata, *Bot. Gaz.* lxvii, Pl. I, Fig. 4. 1919). A like condition was also described by Christ and Giesenhagen for *Archangiopteris* (*Flora*, lxxxvi, p. 72, Fig. 1. 7, 1899). It has been shown for Ferns at large that simple hairs are primitive and scales an indication of advance (Vol. I, Chapter XI). Thus *Christensenia*, *Danaea*, *Protomarattia*, and *Archangiopteris* stand out from the rest in respect of this minor feature as relatively advanced types: the three first named all have synangial sori, extended superficially far from the margin.

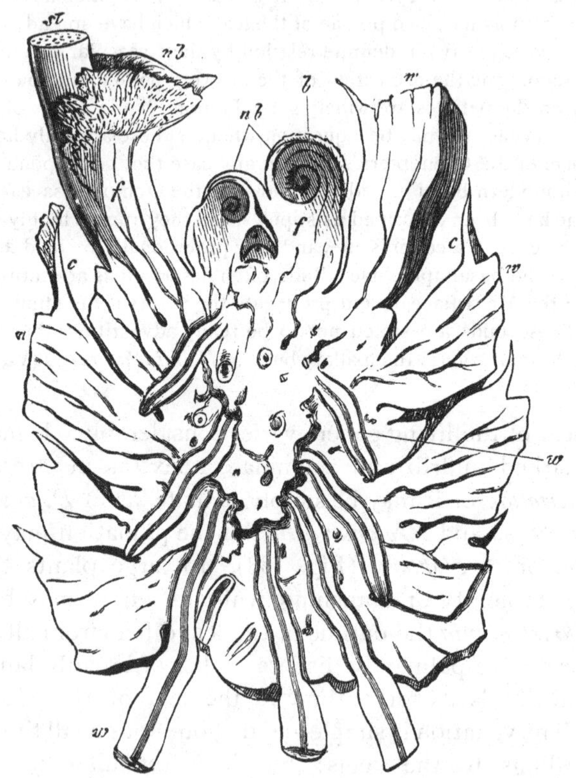

Fig. 391. Vertical section of the stem of a young plant of *Angiopteris evecta*. *b* = the youngest leaves still quite covered up by the stipules (*nb*); *st* = stalk of an unfolded leaf, with its stipules; *n* = the leaf-scars where leaves have fallen away; *c*, *c* = commissures of the stipules in longitudinal section; *w*, *w* = roots. (Natural size.) (After Sachs.)

The roots originate internally beneath the growing point of the stem (Fig. 391). In small plants there may be one root to each leaf, but in strong-growing plants the roots are more numerous. They take a course obliquely downwards through the tissue of the stock, finally issuing as robust roots which branch monopodially.

The sori in this Family are always intra-marginal on the lower surface of the leaf: they are distinct from one another, seated each upon a vein. As a rule they are disposed in a single series on either side of the midrib (Fig. 392). Probably in all the genera the sori had originally a close relation to

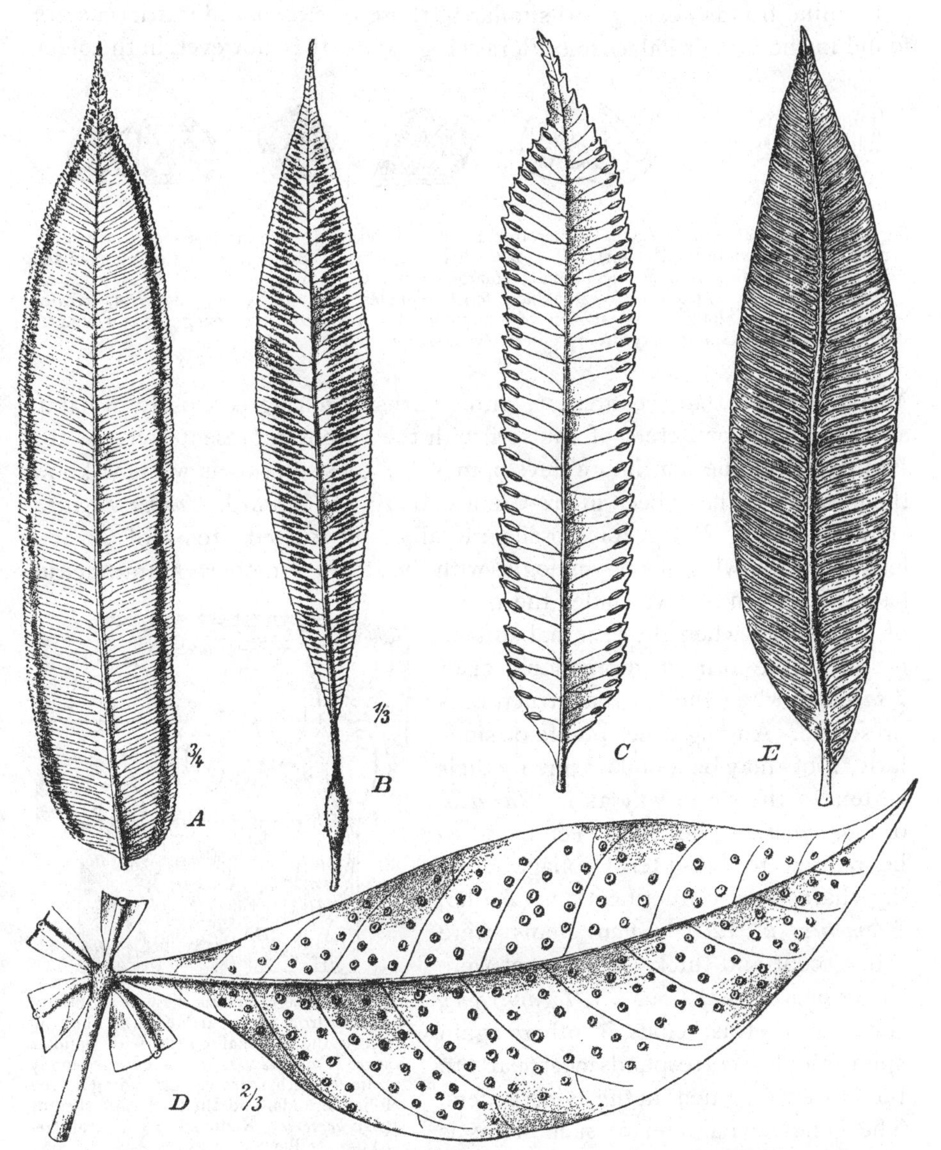

Fig. 392. Pinnae of five genera of the Marattiaceae. A = *Angiopteris crassipes* Wall.; B = *Archangiopteris Henryi* Christ and Giesen.; C = *Marattia faxinea* Sm.; D = *Christensenia* (*Kaulfussia*) *aesculifolia* Bl.; E = *Danaea elliptica* Sm. (A, C, D, E after Bitter; B, after Christ and Giesenhagen.) (From Engler and Prantl, *Nat. Pflanzenfam.*)

the margin, after the manner of *Angiopteris* and *Marattia*: and this position they actually have in the early fossil, *Corynepteris* (p. 33, Fig. 334). In all of them the sporangia of the single sorus are developed simultaneously, which is a primitive state (Vol. I, Chapter XIII).

Fernlike fronds bearing sori similar to those of existing Marattiaceae are found in the Upper Palaeozoic Floras (Fig. 393). It is, however, in the older

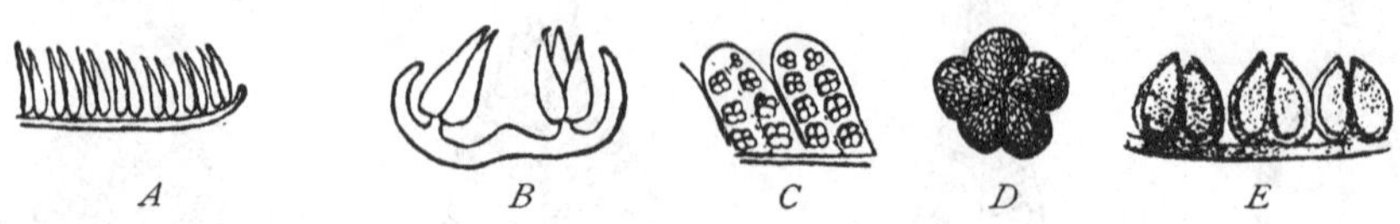

Fig. 393. *A*, *B*, *Scolecopteris elegans* Zenker, from the Lower Permian. *A*, transverse section of a fertile pinnule, enlarged (after Zenker). *B*, *Scolecopteris polymorpha* Brongn. from the Stephanian: longitudinal section of a fertile pinnule, enlarged (after Grand'Eury). *C*, *D*, *E*, *Asterotheca*. *C*, *A. Miltoni* Artis, from the Westphalian: fertile pinnules (×2). *D*, synangium of *Asterotheca* (×about 6). *E*, longitudinal section of a pinnule of *Asterotheca*, traversing three synangia, enlarged. (After Grand'Eury, from Zeiller's *Palaeobotanique*.)

Mesozoic rocks that we first encounter Ferns which agree closely in habit, as also in the characters of the sori with the recent representatives of the Family, while the maximum development of these Ferns as we now know them, seems to have been in pre-Cretaceous times (Seward, *Hooker Lecture*, 1922). Among Palaeozoic fossils radially constructed stems of greater length, but showing strong analogies with the stems of modern Marattiaceae, have long been known under the name of *Caulopteris* when the external surface is seen in the form of impressions: or of *Psaronius* when the internal structure is preserved. Among other points of similarity roots may be found traversing their cortex in the same way as in *Marattia* or *Angiopteris*. They have been found to be related to Pecopterid foliage bearing characteristic fructifications (Zeiller, *Elements*, p. 120). Such stems were arborescent and thick. The leaves were sometimes distichous (*Megaphyton*), sometimes tetrastichous, in others again spiral: this last corresponds most nearly to the leaf-arrangement in the living genera. The general character of such stems is suggested by Fig. 394. The conclusion seems clear that Ferns of arborescent character, but with leaves of different outline from the living Marattiaceae, existed in Palaeozoic times.

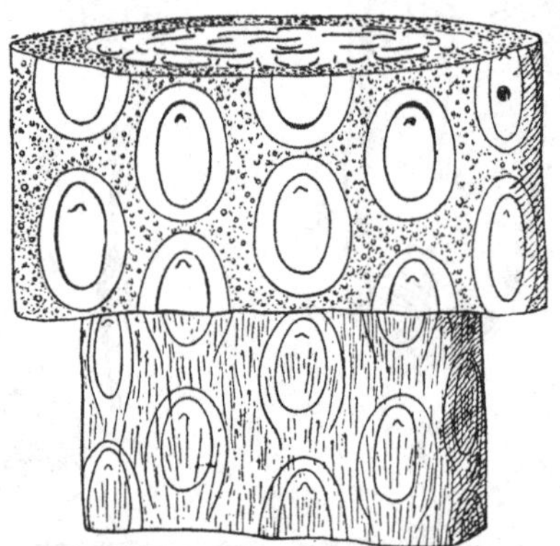

Fig. 394. Diagrammatic view of a trunk of a Fern from the Coal Measures, showing above the external cortex with petiolar scars (*Caulopteris*), and below the woody cylinder with scars corresponding to the foliar strands, and their sclerotic sheaths (*Ptychopteris*). Reduced to ¼ natural size. (After Zeiller.)

The Marattiaceae are relatively massive in the construction of their parts, and this appears especially in the adult state. It has been shown by Charles

(*Bot. Gaz.* li, p. 81. 1911) and by West (*Ann. of Bot.* xxxi, p. 361) that the stems and roots of the embryos may segment from a single initial cell: but this is only a temporary state. The stem and root of the adult show a bulky type of meristem not referable to a single initial. Schwendener (*Sitz. K. Preuss. Akad.* 1882) demonstrated this for the root, where he found four equal initials. A similar structure though less regular has been found at the stem-apex of *Angiopteris* and *Marattia* (Fig. 395). But the most convincing evidence of the robust structure of the vegetative system lies in the massive wings of the leaf-blade, which do not show segmentation of a single marginal series of cells, as in all Leptosporangiate Ferns, but a meristem referable possibly to two rows of initials (Fig. 396). All this harmonises with the eusporangiate state of these plants and demonstrates the relatively bulky construction of their whole sporophyte.

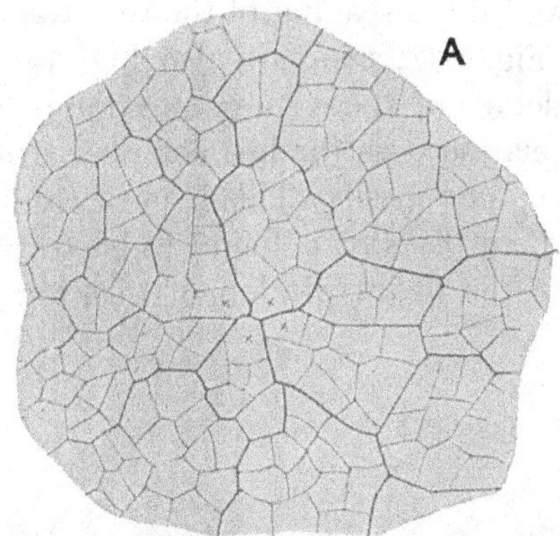

Fig. 395. *A* = apex of stem of *Angiopteris evecta*, seen from above. Apparently there are four initials ×, ×, ×, ×. (× 83.)

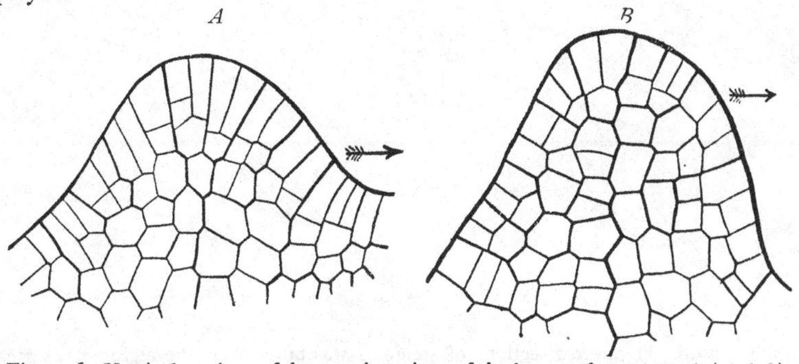

Fig. 396. Vertical sections of the massive wings of the leaves of *Angiopteris* (*A*, left), and of *Todea* (*B*, right), showing segmentation of a complex type.

Anatomy

The Marattiaceae are sappy Ferns, characterised by an absence of sclerenchyma, while their tissues are traversed by lysigenous mucilage-canals. Tannin cells also are widely distributed, either isolated or in groups, and they are present even in the hairs and scales (West, *Ann. of Bot.* 1915, p. 409). The vascular system of these Ferns is unusually complicated: in some of them it is undoubtedly the most complex of all living Peridophyta. The complication consists in the extreme sub-division of the vascular tracts both of axis and leaf, the strands being disposed in concentric circles, as though resulting from the disintegration of a polycyclic system. There is also frequent anastomosis of the separate strands, and a further source of confusion lies in the formation of roots in relation with the internal meristeles, and the threading of the root-traces through the rest on their way outwards (Fig. 397; compare also Fig. 391). It will be unnecessary to describe the details of these complex arrangements: it must suffice to bring them into relation with the less complex structure of other Ferns, and with that of the cognate fossils. This is most readily done by reference to the sporeling: for there the axis is found to be traversed by a coherent monostele: and it is to the disintegration of this, together with the formation of commissural

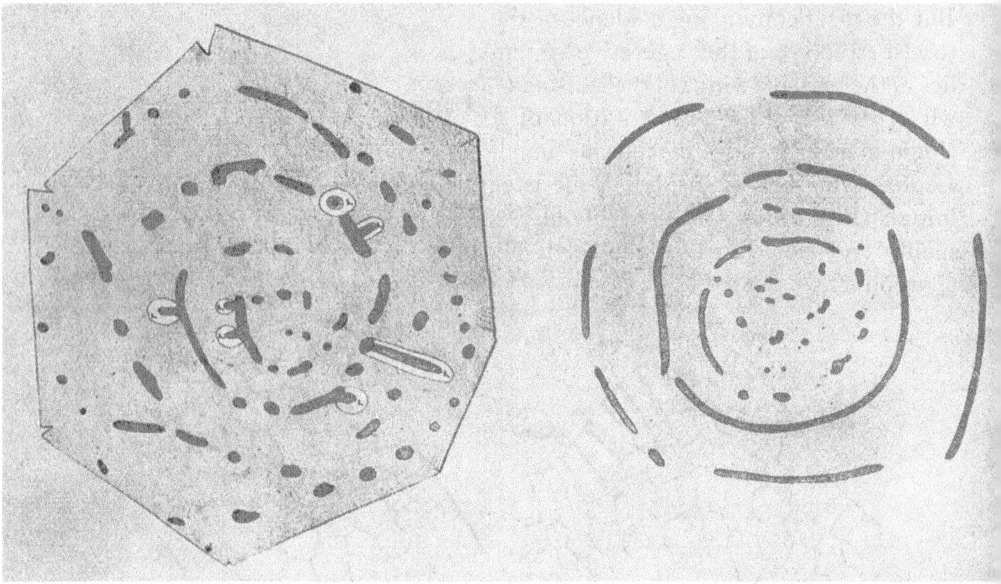

Fig. 397. *Angiopteris evecta*. Transverse sections of stem. *A* was taken from the region of its greatest diamete here the vascular system is polycyclic and highly segregated into separate strands. *B* was taken at a level abo *A*, from a region formed under less favourable conditions, with fewer leaves and less bulky axis. The vascu system is still polycyclic, but the strands are less highly segregated. (After Mettenius. Natural size.)

connections, that the complex structure of the adult is due. A striking feature of the Family is the absence of endodermis in the adult state. In the sporeling the stele is delimited as usual: but the endodermis is not continued upwards. The Marattiaceae share this detail with *Ophioglossum*, and it is noteworthy that in both the foliage is of a leathery texture. (See Vol. I, Chapter VII, p. 135.)

The monostele of the sporeling has been shown to be, as in other Ferns, protostelic at first in *Danaea alata* (West), and *Marattia alata* (Charles), and others. This is probably general. It expands upwards into a dictyostele with large leaf-gaps. The first leaf-traces are simple strands (Fig. 398, *l.t.* 1–6), but in the later leaves the trace consists of paired strands (*l.t.* 8). A root-trace appears with some regularity related to each of the early leaves. Presently a vascular commissure (Fig. 398, *c.s.*) traverses the pith linking together the opposite sides of the dictyostele, and this pursuing an upward course with fusions at the successive leaf-gaps initiates a medullary system that becomes more complex upwards. This state of the stelar structure, accompanied by further disintegration of the dictyostele and of the leaf-trace, and modified by the dorsiventrality of the rhizome in those genera, gives the relatively simple vascular system of *Archangiopteris* and of *Kaulfussia*. In *Marattia* and *Angiopteris* the initial steps are similar, though the final structure is more complex. As the stem passes to the adult state in *Angiopteris* three or four concentric-meshed zones of vascular tissue may be formed within the original cylinder, giving the most complicated structure recorded for the family (Fig. 397). From this brief description it will be apparent that there is no essential difference between the Marattiaceae and other Filicales as regards the origin and final distribution of the vascular strands in the stem, nor as regards the primitive vascular arrangement in the petiole. The most distinctive points which they show are the high degree of the disintegration, the absence of endodermis, and the origin of the root-traces from the internal strands of the stem. The ontogeny opens with a monostelic state and a solid xylem-core. It is in the later stages that polycyclic and disintegrating complications arise; it will be recognised that these vary in rough proportion to the size, and are most striking in large plants of *Angiopteris*.

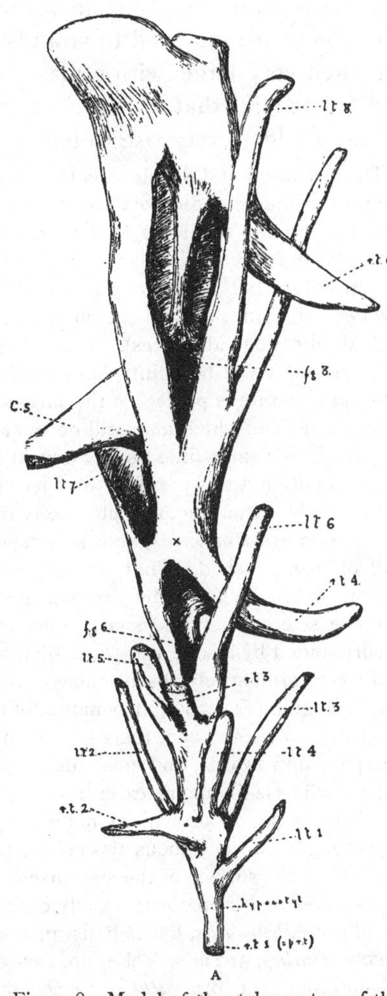

Fig. 398. Model of the stelar system of the rhizome of a young sporophyte of *Danaea alata* Sm. *l.t.* = leaf-traces; *c.s.* = commissure; *r.t.* = root-trace. (After West.)

The vascular supply to the adult leaves originates as many distinct strands from the dictyostele of the axis, and it is stated that it springs from the outermost zone only. The roots, on the other hand, derive their supply from the medullary system: in young plants it springs from the commissural strand, but in the adult mostly from various points on the internal system, though some roots are attached to strands of the outer zone. The roots themselves are often very large, with a massive cortex, and a well-marked stele: this is constructed like that of *Helminthostachys*, but in the Marattiaceae the number of protoxylems may rise to ten or even fifteen.

The structural and developmental details derived from living species are essential for the proper understanding of those fossil stems, often of large dimensions, which have been held as related to the living Marattiaceae. These are grouped under the title Psaronieae, from *Psaronius*, a fossil characteristic of the Lower Permian. Other stems, probably of similar nature but imperfectly preserved as casts or decorticated, have been designated *Caulopteris* (from the Coal Measures), and *Megaphyton* and *Ptychopteris* (from the Upper and Middle Coal Measures). All of these may very probably be Psaronieae. Naturally the specimens will be adult plants, and often they are of very large size. Where their internal structure is preserved the largest of them show extraordinary complexity of the vascular system, which may still be interpreted in terms of the living Marattiaceae, or in general on the same lines as in Ferns at large. But they show only the complex structure of the adult, and this can best find its elucidation through the simpler condition seen in the living Marattiaceae, and ultimately in their sporelings.

The stems of *Psaronius*, known as "starling-stones," are found with structure preserved, and often uncrushed. They are characterised by the size of the well-defined central region, which includes the vascular system proper to the stem, surrounded by a thin band of sclerenchyma. In large specimens this may be 4 to 8 inches in diameter. It is surrounded by a zone traversed by innumerable roots, from which, however, outgoing leaf-traces are entirely absent, though their origin is clearly marked in the structure of the central region (Fig. 399). The nature of the peripheral zone has been the subject of much discussion. It consists of roots similar in structure to those of the Marattiaceae, running obliquely downwards, and embedded in a dense filamentous tissue formerly interpreted as cortical. These filaments arose from the outer side of each root, and running radially outwards came in contact with an outer-lying root. The whole mass of roots was thus welded together into a continuous tissue. In the innermost region, however, similar filaments arose from the surface of the stem itself. Thus the whole zone is made up of adventitious roots embedded in filaments partly of cauline partly of radical origin. The structure is peculiar to *Psaronius*, but it finds an analogue in the hyphal plexus of Fungi and Algae (Scott, *Studies*, 3rd edn. Vol. i, pp. 273–5). A more general analogy is with the zone of adventitious roots of *Cyathea* and *Dicksonia*, which serve a like biological end. But there the roots are not welded together (compare Vol. I, Fig. 152). In either case the roots form an additional support for an upright stem destitute of secondary increase. A late origin of the zone in the ontogeny of *Psaronius* would explain the absence from it of leaf-traces, for it may be assumed that the leaf-stalks would have already fallen away before the outer zone was formed.

The central region of these fossils is the axis proper. It contains a vascular system the complexity of which is, as in other Ferns, roughly but not exactly in proportion to the size both being in extreme instances very great. The stem of *P. brasiliensis* (Fig. 399) will serve as a complicated but not extreme example. Embedded in parenchyma limited externally by a sclerotic band, which shows inward extensions and isolated tracts, is the vascular system composed of numerous meristeles disposed regularly; often these are of large and varying form. The leaf-arrangement in this plant was decussate, and leaf-traces are seen departing at *F* 1, *F* 2, though not exactly at the same level. *F* 2 shows the outgoing leaf-trace as a continuous curve open inwards. The long tracts *A*, *B* lying laterally are the leaf-traces of the next pair of leaves, *D*, *E* of the next higher pair. These have been called the "reparatory" meristeles to distinguish them from the four long and curved "peripheral" meristeles, *P* 1–*P* 4, that lie diagonally. These give rise to the root-supply, and after anastomosis with the reparatory tracts they form the leaf-traces. At *C* a tract

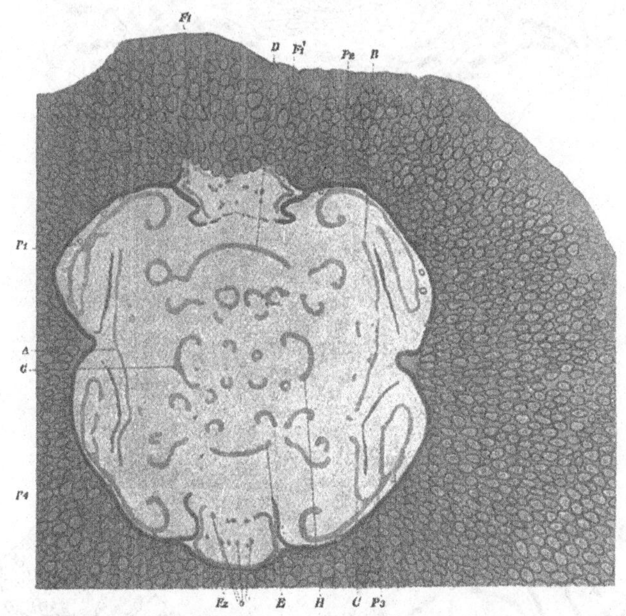

Fig. 399. *Psaronius brasiliensis*. Transverse section of stem. The whole of the true stem, containing the vascular bundles, is shown, together with a great part of the outer zone with innumerable adventitious roots. *F* 1, *F* 2, leaf-gaps. See Text. (Reduced, after Zeiller, from Scott.)

is seen separating from the peripheral meristele to pass to the reparatory tract, as a contribution to the leaf-trace. The meristeles lying internally form a cauline system, as in other large Fern-stems. It is thus seen that a large *Psaronius* corresponds in essentials to a very complicated polycyclic Fern. The type of structure is suggested, though on a very much smaller scale, by *Saccoloma* (Vol. I, Fig. 146, p. 153).

Stems which are evidently Psaronioid show a great variety in size, structure, and leaf-arrangement. On the latter point Zeiller distinguished three types: the *Polystichi*, with numerous vertical rows of leaves (Fig. 400, *A*): the *Tetrastichi*, with four series (Fig. 399): and the *Distichi* with only two (Fig. 400, *D*). Upon this feature the detail of the vascular system naturally depends. But it is also closely related to size. The actual dimensions are not always stated by authors, but in the following examples (p. 107) they are believed to be approximately correct.

In discussing such questions in fossils, the interest at once centralises on *P. Renaulti*, from the Lower Coal Measures (Fig. 401). Its dimensions are approximately those of any ordinary solenostelic rhizome among living Ferns. There can be no doubt of its Psaronial nature (compare Williamson, *Phil. Trans.* 1876, Pl. III). It contains a solenoxylic stele,

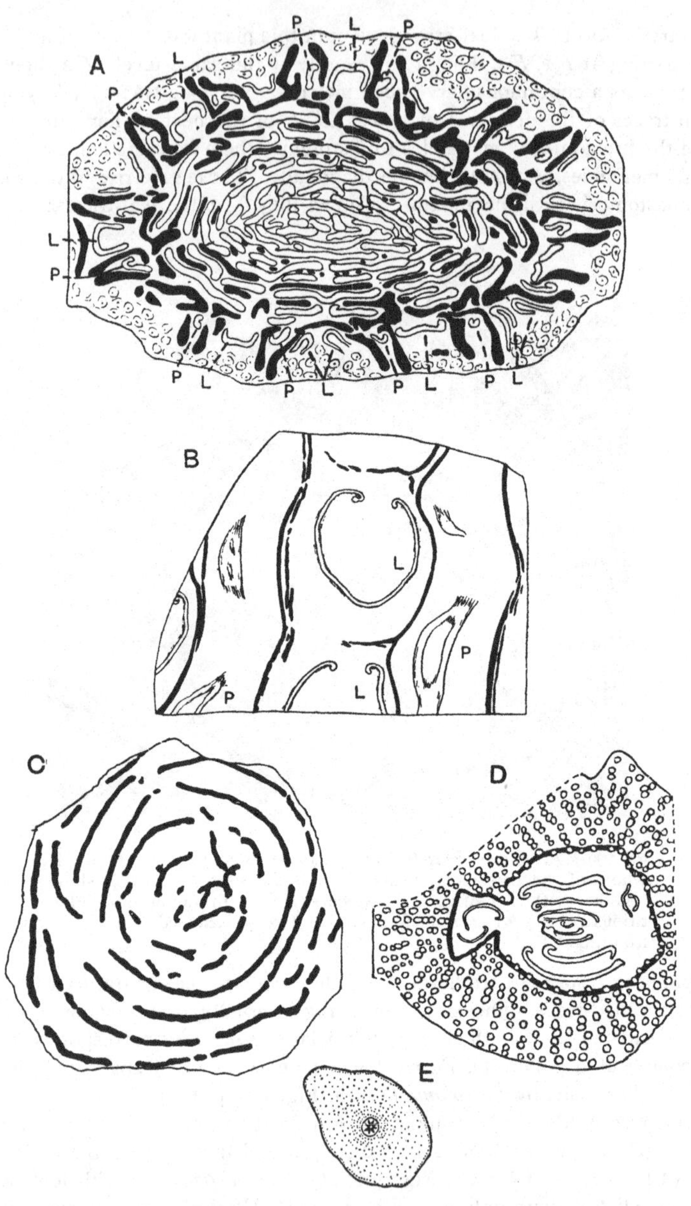

Fig. 400. Sections of stems of *Psaronius*. [*A–C* and *E*, after Zeiller: *D*, after Stenzel, from Seward.] *A*, *Psaronius infarctus*. *P*=peripheral steles; *L*=leaf-traces. *B*, *P. infarctus*, longitudinal tangential section through the peripheral region of the stem. *C*, *P. coalescens*. *D*, *P. musaeformis*. *E*, *P. asterolithus*, root[1].

[1] Solms-Laubach described in 1913 a large Brazilian fossil stem named *Tietea singularis*, of uncertain horizon, and suggested its relationship to the Psaronieae. It seems more probable that it is related rather to *Tempskya*. Notwithstanding its large size and complicated structure it would not be wise to bring it into the present comparisons, pending more exact information (*Zeit. f. Bot.* v. (1913), p. 673).

Name	Source	Diameter of stem	Remarks
P. musaeformis Corda	Stenzel, *Nova Acta*, xxiv, Fig. 2 (after Corda)	2·2 in.	As in Fig. 400, *D*
P. Demolei	Zeiller, Pl. XXIV, Fig. 3	1·25 in.	Three cycles
P. Demolei	Zeiller, Pl. XXIV, Fig. 1	2·5 in.	Four to five cycles
P. infarctus Unger	Potonié, *Lehrbuch*, p. 71. Fig. after Zeiller	about 4 in.	Slightly crushed, about eight cycles
P. brasiliensis	Zeiller, Pl. XXI	4 in.	Not crushed, structure as in Fig. 399
P. bibractensis	Zeiller, Pl. XVII	Crushed but diam. estimated as a cylinder at 5 in.	Five to six cycles

without endodermis apparently, and so it would be comparable to some Ophioglossaceae. It may be recognised as the earliest recorded solenostelic Fern. Structurally it appears to represent an ontogenetic stage which the larger Marattiaceae pass rapidly through. The next as regards complexity is *P. musaeformis* (Fig. 400, *D*), a species from the Lower Permian of Chemnitz. It is one of the *Distichi*, with leaves in two rows, the leaf-traces passing off

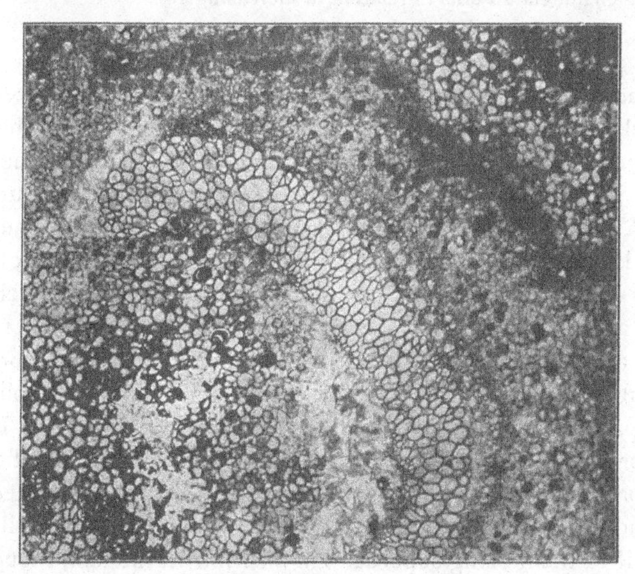

Fig. 401. *Psaronius Renaulti.* Part of transverse section of stem, showing a portion of the solenostele. On the left is the edge of a leaf-gap. In the stele, note the protoxylem-groups on the inner edge of the wood. The phloem, with large sieve-tubes, is well preserved on the outer side. The dark elements in the pericycle are secretory sacs. (× 13.) S. Coll. 2174. (From a photograph by Mr Tams, after Scott.)

right and left from the polycyclic stele, which is here seen as four pairs of continuous vascular tracts. The drawings of Zeiller of *P. Demolei* are specially interesting, since he shows two stems of different sizes: the smaller has only three vascular cycles, the larger has 4–5 cycles. In the next three species quoted in the table a highly polycyclic state accompanies their larger size: but they also show that the number of cycles is not directly proportional to the dimensions. The result of the whole comparison is to demonstrate in a general way that increasing vascular complexity tends to follow increase in size, but as in Leptosporangiate Ferns a direct numerical relation does not exist. These two series illustrate a homoplastic advance, and their similarity in principle though not in detail suggests that similar causal factors underlie them both.

The rather irregular cycles of vascular tissue in the Psaronieae no doubt correspond to the rings of isolated strands of the Marattiaceae: so also their frequently continuous leaf-trace corresponds to the series of separate strands of the Marattiaceous leaf-trace (Fig. 400, *B*). It appears that in these points the modern Marattiaceae are more advanced in vascular disintegration than the Psaronieae. But Mettenius (*Abhandl. K. Sachs. Ges.* Bd. vi, 1863, Taf. I, Fig. 2) has shown how in a plant of *Angiopteris* that died after a long struggle under cultivation the vascular rings are almost continuous in the upper region (Fig. 397, *B*). Presumably it had retrograded structurally and pathologically towards the earlier type. The general result of these comparisons is that the vascular construction of the Psaronieae and of the Marattiaceae is according to a common type, both being highly polycyclic: that the fossils show a less and the living Ferns a greater sub-division of the vascular tracts: and that both conform to the methods of construction already recognised in Leptosporangiate Ferns in relation to increasing size.

THE SPORE-PRODUCING MEMBERS

The sorus of the modern Marattiaceae is strictly circumscribed, and has no definitely organised indusium. It is true that hairs may be scattered round its periphery in *Angiopteris* (Fig. 402, *B, D*), and that the sporangia of *Danaea* may be partially enveloped by upgrowths of the surface-tissue (Fig. 402, *K*): but these hardly deserve the name of indusium. The sori are all constructed on the radiate uniseriate plan, where a single series of sporangia is disposed in a radiate fashion round a central receptacle. When the sorus is circular, as in *Christensenia*, the receptacle is a central point: when it is elongated, as in *Danaea*, the attachment is linear. All the sporangia of a sorus originate simultaneously, a character general for the Simplices. They may be separate or united synangially: they are massive with a broad base, and each produces a large spore-output. The dehiscence is in all of them by a slit or pore, which though varying in form lies always in the median plane.

The structure of the mature sorus in the five older genera is illustrated in Fig. 402, all of them being superficial in position on the leaf, though varying in their spread from the margin inwards. They vary also according as the sporangia are separate as in *Angiopteris* (Fig. 402, *A*), or fused synangially as in *Marattia*, *Danaea* and *Christensenia* (*E*, *G*, *J*). The sporangia are all of the same massive type, and are arranged so as to face centrally, each opening by a slit on the oblique inner face. Of those genera with separate sporangia, the sori of *Angiopteris* are near to the margin (Fig. 392, *A*), and each consists of about 20 sporangia. In *Macroglossum* also they are marginal, but they extend farther inwards, while in *Archangiopteris* (Figs. 392, *B*; 402, *C*) they are about half-way between margin and midrib, and extend far along the veins, each consisting of a much larger number of sporangia. A fossil type, *Danaeopsis* Heer, has lately been shown by Halle (*Archiv für Botanik*, xvii, No. 1. 1921) to have sori extending as in *Danaea* the whole distance from margin to midrib: but here as in *Angiopteris* all the sporangia are separate. The genera with synangial sori illustrate like differences. *Marattia* bears its synangia near to the margin (Fig. 392, *C*): in *Protomarattia* (Hayata) they are submarginal, and rather longer: while in *Danaea* (Fig. 392, *E*) they may extend along the whole length of the veins. Thus the

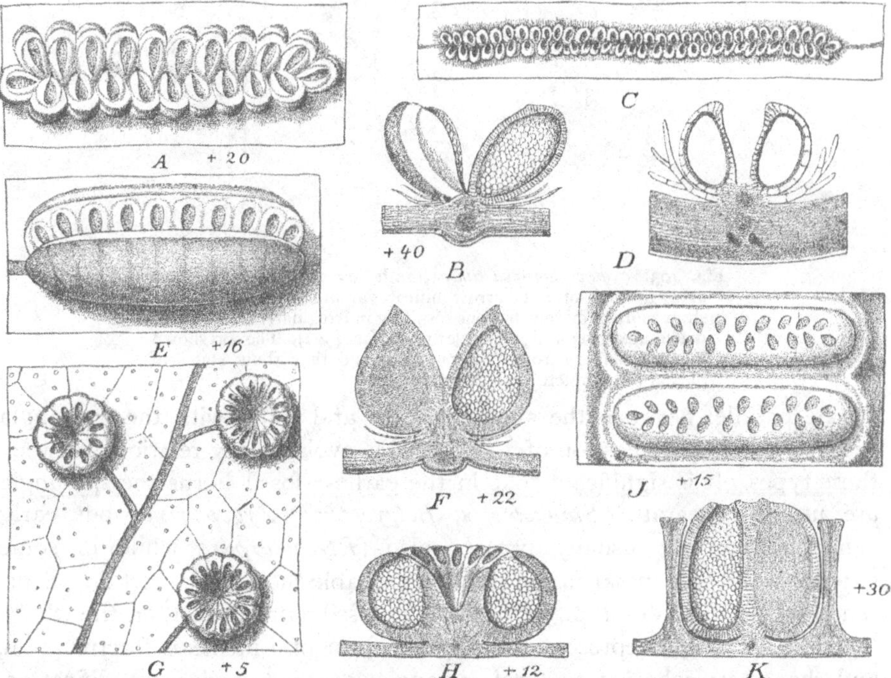

Fig. 402. Sori and sporangia of Marattiaceae. *A*, *B*, *Angiopteris crassipes* Wall. *A*, sorus: *B*, two sporangia, one in surface view from without, the other cut longitudinally. *C*, *D*, *Archangiopteris Henryi* Christ and Giesen. *C*, sorus; *D*, two sporangia in section. *E*, *F*, *Marattia fraxinea* Lin. *E*, synangium; *F*, same in section. *G*, *H*, *Christensenia* (*Kaulfussia*) *aesculifolia* (Bl.). *G*, part of the lamina seen from below, with three synangia; *H*, section through a synangium. *J*, *K*, *Danaea elliptica* Sm. *J*, two synangia; *K*, section through a synangium. (*A*, *B*, *E*, *G*, *J*, after Bitter. *C*, *D*, after Christ and Giesenhagen. *F*, *H*, *K*, after Hooker-Baker. From Engler and Prantl.)

degrees of extent of the sorus and its relation to the margin appear to be similar, whether the sori be synangial or with the sporangia all separate. Finally, in *Christensenia* (Fig. 402, *G*), which is a shade-loving, broad-leaved, reticulate type, the rosette-like sori are dotted over the broad under-surface of the pinna. Comparison with the leaves of *Danaea*, and especially with those which are only partially fertile, gives a clue to an explanation: for there the elongated sori are found to show occasional fissions, and the partial sori with circular outline like those of *Christensenia* appear isolated upon the enlarged surface (Fig. 403, *a*, *b*, *c*). It seems probable that the state seen in *Christensenia* was thus acquired during the descent of a gradually broadening leaf. The frequent occurrence of soral fissions on its leaves supports this view (Fig. 403, lower series). Thus this broadest-leaved living genus with its scattered sori may be brought into line with the rest, but as a derivative type.

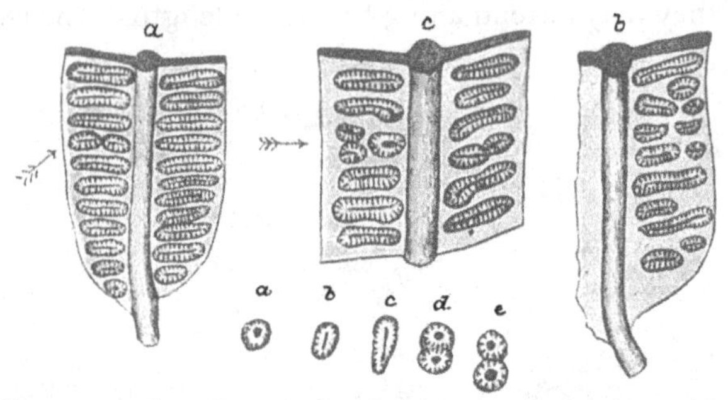

Fig. 403. *a*, *b*, *c*, *Danaea alata* Smith. *a* = a fertile pinna with many normal sori: the arrow indicates an abnormal fission. *b*, *c* show more numerous fissions resulting in irregularly formed sori, distributed over a slightly enlarged surface (× 2). The sori shown below (*a*–*e*) are from *Christensenia*, and they show states of partial or complete abstriction.

The parallel between the synangial sori and those with the sporangia separate raises the question of the probable evolutionary relations between these types. It is significant that in the earliest fossil Ferns the sporangia are usually separate (*Stauropteris*, *Botryopteris*, *Zygopteris*): but early synangial sori are usually present also (*Ptychocarpus*), while in some examples the sporangial fusion may be probable though less certain. Scott remarks (*Studies*, Vol. i, p. 366) that "the fossil data suggest, on the whole, that free sporangia represent the original form of Filicinean fructification, and that their cohesion to form synangia was a secondary modification, though one which, in certain groups, took place at a very early period." The evidence from living plants supports this view. The comparisons already drawn for the living Ophioglossaceae make it appear probable that in them evolution has led from the state of separate sporangia of *Botrychium* and *Helminthostachys* to the synangial state of *Ophioglossum*. Similarly, though with less cogent evidence, it may be held that the synangial state of *Marattia*, *Protomarattia*, and *Danaea*, a series parallel in soral form to that of *Angiopteris*, *Archangiopteris*, and *Danaeopsis*, has also resulted from a secondary fusion of their sporangia to form synangia.

As the development of the individual sporangium has been found to be essentially the same in the several genera, it will suffice to describe it for one only, and *Angiopteris* may be selected as being the most familiar. At an early stage the sporangia begin to project from the soral area as separate outgrowths, but a comparison of the four sporangia shown in Fig. 404, *A*, as seen from above, discloses no regular sequence of segmentations. Of those shown, that marked *b* is the most regular and usual type, and a single superficial cell (here already divided into two) gives rise to the essential part

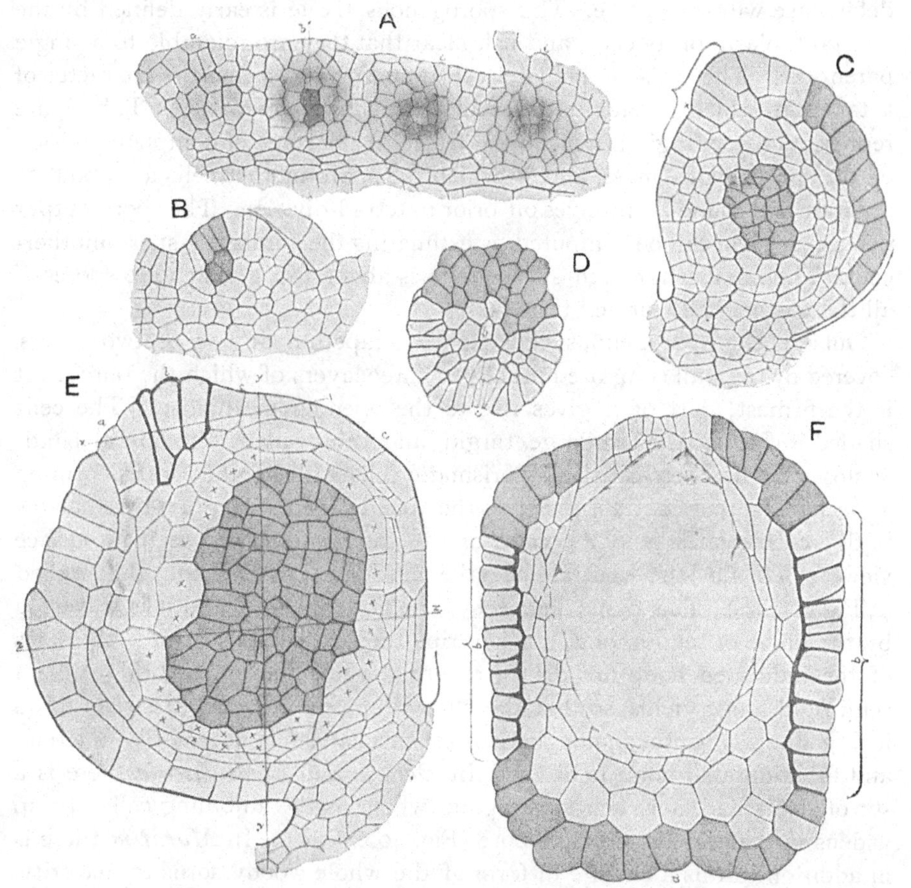

Fig. 404. *Angiopteris evecta* Hoffm. *A* = part of a young sorus seen in surface view from without. *B* = vertical (radial) section of a sporangium such as would be seen on cutting a sporangium (*b*) in Fig. *A*, along the line indicated. *C* = vertical section of an older sporangium showing the genetic grouping of cells. *D* = apex of an almost mature sporangium seen from above: such a section as along the line ×, ×, in Fig. *E*. *E* = vertical section of a sporangium with spore-mother-cells; the tapetum is marked ×. *F* = transverse section of an almost mature sporangium: *b*, *b* = annulus: *c* = region of dehiscence: *a* = thin-walled cells which contract. (× 200.)

of the sporangium. If a section were taken along a line ×, ×, through such a sporangium, after it had grown more convex and segmented further, it would appear as in Fig. 404, *B*, in which the cells shaded correspond to those shaded in *A*. A single cell has divided periclinally: the inner cell gives rise to the sporogenous tissue, and has already divided in *A*: the outer cell has also divided, and gives rise to part of the sporangial wall. As development proceeds the divisions are often sufficiently regular to allow of the genetic grouping of the tissues being clearly followed (*C*). Meanwhile certain cells at the apex enlarge to form the crest-like ridge above the annulus: from this downwards on the central side of the sporangium the dehiscence will take place. The sporogenous tissue is early defined by the rich protoplasm of its cells, and it is clear that they are referable to a single parent-cell. The cells immediately surrounding it assume the character of a tapetum, which is thus extra-archesporial in its origin (*E*). This figure represents a sporangium which has arrived at the stage of complete division of the sporogenous mass, and in which the spore-mother-cells are about to separate and round themselves off prior to tetrad-division. The spore-output per sporangium can be computed by estimating the number of spore-mother-cells. The number for each sporangium is about 360, giving 1440 spores if all the mother-cells formed tetrads.

Outside the sporogenous cells lies the tapetum of one or two layers, covered by the wall composed usually of three layers of which the outermost is the firmest; part of it gives rise to the opening mechanism. The cells shaded in Fig. 404, *F* are large, turgid, and thin-walled: the lateral bands enclosed by brackets consist of prismatic cells with lignified walls, forming a mechanical ring continuous across the apex of the sporangium by a narrow bridge corresponding to the crest of the sporangium, and seen in surface view in *D*. On the ventral face the cells are smaller and thin-walled (*c*, Fig. 404, *F*). This tissue defines the line of dehiscence, and it is traversed by the plane of section of *E*. When ripe the shaded cells shrink: the sides of the indurated hoop are drawn together, while the apical bridge like a semi-rigid hinge yields, so that the slit will gape widely, as it is seen to do in Fig. 402, *A*. In the synangial sori the mechanism is necessarily different, and the indurated hoop is absent. In *Danaea* and *Christensenia* there is a slit of dehiscence for each sporangium which as the adjoining cells dry up widens into an almost circular pore (Fig. 402, *G–K*). In *Marattia* there is in addition to this a change in form of the whole woody sorus at maturity. Its two sides, originally facing one another closely, move apart as in the opening of a book (Fig. 402, *E*).

The estimate of about 1440 spores as the product of a single sporangium of *Angiopteris* is relatively low for the Family. It may be compared with the somewhat smaller spore-counts actually made by Halle from sporangia of his recently discovered Rhaetic Fossil, *Danaeopsis fecunda*. His actual counts were 1100 to 1159. Such figures are below the estimates of 1750 for *Danaea*, 2500 for *Marattia*, and over 7000 for *Christensenia*, a number which is probably equalled also by the synangial fossil *Ptychocarpus*. It appears thus that the higher numbers fall consistently to the synangial types, and a like result comes from the Ophioglossaceae. At first sight this might suggest that the higher numbers are an indication that the synangial types are the more primitive: but for reasons already explained (p. 89) this will not hold for the Ophioglossaceae. It seems not improbable that a biological advantage follows on the synangial state, and this may explain what is apparently the fact, that it has originated along a plurality of evolutionary lines.

Passing now to the fossil Ferns having fructifications which may be ascribed to a Marattiaceous affinity, many of them have foliage of a Pecopterid type: the segments are relatively narrow, and in none of the earliest are there broad leaf-expanses. The sori are habitually disposed in two intra-marginal rows, one on either side of the midrib. The same arrangement is constant also for the living Marattiaceae, excepting *Christensenia*. This relation to the margin is significant for comparison with the Ophioglossaceae, and still more with the Botryopterideae. In both of these the sporangia or sori are actually marginal. *Corynepteris* is specially pertinent since its sori, so closely resembling those of certain Marattiaceae, nevertheless appear to be marginal on the very narrow segments (Fig. 405). Kidston

Fig. 405. *A* = *Corynepteris Essenghi* Andrae, from the Carboniferous (Westphalian). Fragment of a fertile pinna. (×6.) *B* = *C. coralloides* Gutbier, from the Westphalian. Fragment of a fertile pinna. (×4.) *B'* = sorus of the same species seen laterally. (×28.) (After Zeiller.)

remarks (Fossil Plants, *Mem. Geol. Survey*, Pt. 4, p. 300, 1923), "There can be no doubt that *C. coralloides* Gutbier sp. belongs to the Zygopterideae." It has been demonstrated in Vol. I, Chapter XII, how frequently in modern Ferns the sorus has slid phyletically from the margin to the surface of an enlarging leaf-area. The facts suggest very strongly that a similar transfer from the margin to the surface has taken place early in the evolution of the Marattiaceae from a narrower-leaved ancestry. All the facts relating to the sporophylls of modern Marattiaceae coincide with such an origin, subject to the sori spreading over an enlarging leaf-area, as explained in the foregoing descriptions of them. Such considerations point to a circular group of a small number of sporangia as constituting an early type of sorus: and this is what is actually prevalent in those early fossil Ferns which are referred to a Marattiaceous affinity.

The Ferns so recognised have been enumerated and described with full references to the literature by Seward (*Fossil Plants*, ii, p. 395). It will therefore suffice here to describe a few of the best-known examples with a view to their comparison with the fructifications of living Marattiaceae. The genus *Scolecopteris* includes plants with sessile or shortly pedunculate sori of three to six sporangia, which are united below but separate above, and extended into more or less elongated beaks (Fig. 406, *D*). The Marattiaceous character of this fructification is unmistakeable, and it has been pointed out by Strasburger (*Jenaische Zeitschrift*, 1874, p. 87) that *S. elegans* Zenk. shows features connecting it with *Marattia* as regards the form of the sporangia, and with *Christensenia* in their circular disposition in the sorus, while the outline of their upper free portion would point to *Angiopteris*: in dehiscence it compares with all three, but more especially with *Marattia*. It is in fact a type which unites in itself the characters of various living genera, though the number of sporangia is smaller than in any of them. *S. elegans* Zenk. is from the Lower Permian of Saxony, *S. polymorpha* Brongn. is from the Upper Coal Measures.

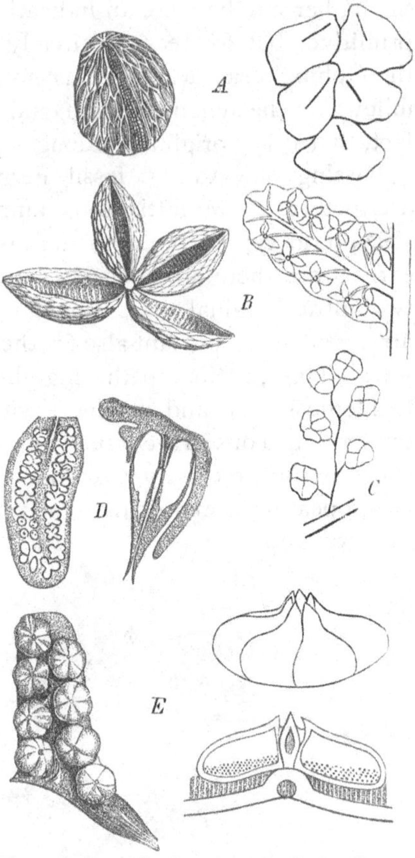

Fig. 406. Fructifications from the Carboniferous formation. *A*, *Senftenbergia ophiodermatica*: to the right the position of the sporangia on both sides of the median nerve: to the left a single sporangium, seen from above. *B*, *Hawlea miltoni*: to the right a pinna with sori on the extremities of the lateral nerves: to the left a single sorus more highly magnified. *C*, *Oligocarpia lindsayoides*, showing position of the sori. *D*, *Scolecopteris polymorpha* Brongn.: to the left a pinnule showing the position of a sorus in transverse section: to the right a longitudinal section of a sorus in which the sporangia are united below with a columnar receptacle. *E*, *Asterotheca Sternbergii*: to the left the pinnule with sori: to the right a side view of a sorus, and a sorus in radial section. *D* and *E*, diagrammatic. (All Figs. after Stur. from Solms-Laubach.) It is possible that some of these may really be Pteridosperms.

In *Asterotheca*, a fossil from the Coal Measures which is believed to be identical with *Hawlea* (Fig. 406, *B*, *E*), each sorus has 3–8 sporangia in close apposition whilst young, but separating and diverging widely when mature. The peripheral wall is strongly convex and shows no annulus. The mode of dehiscence was by radial slits.

One of the most striking of these fossils is *Ptychocarpus* (*Pecopteris*) *unitus* Brongn., from the Upper Coal Measures of France. Here on the lower surface of the pinnules of a Pecopterid leaf the sori are disposed in a row on either side of the midrib. Each is a synangium composed of about seven sporangia united upon a common receptacle. The form is that of a truncated cone with a slight terminal dimple. Centrally there is a vascular strand connected with the system of the leaf. The sporangial wall consisted of about four layers of cells, and there was no annulus. Dehiscence appears to have been by terminal pores, and the number of spores in each sporangium was very large, probably equal to that of the modern *Christensenia* (Fig. 407). The correspondence with this genus is very close indeed, the differences being of degree only. But the leaf-blades that bore them were very different.

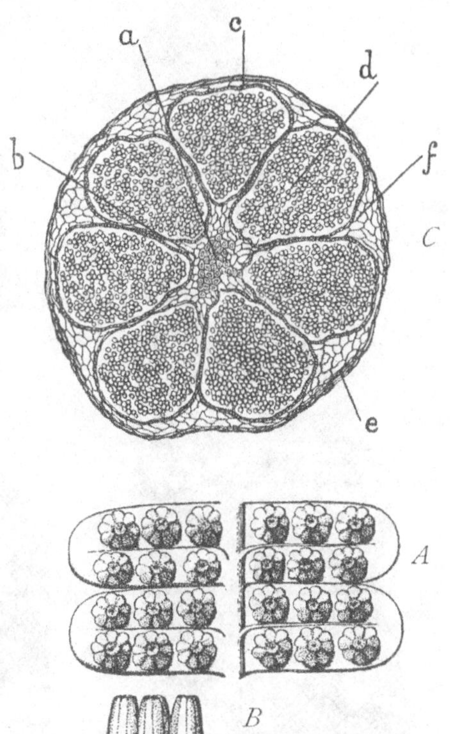

Fig. 407. *Ptychocarpus unitus*. Fructification. *A*, part of a fertile pinnule (lower surface), showing numerous synangia. *B*, synangia in side-view. (× about 6.) (After Grand 'Eury.) *C*, a synangium in section parallel to the surface of the leaf, showing seven confluent sporangia. *a*, bundle of the receptacle: *b*, its parenchyma: *c*, tapetum: *d*, spores: *e*, *f*, common envelope of the synangium. (× about 60.) (After Renault, from Scott's Studies.)

Certain Pecopterid forms from the Coal Measures have been included under the name *Danaeites*, and their sori are described as synangial and linear, consisting as in *Danaea* of two rows of fused sporangia each dehiscing by an apical pore. But Seward (*l.c.* p. 398) discredits this conclusion on the ground of lack of detail. On the other hand, Halle's description of the Rhaetic Fern *Danaeopsis* (*Arkiv for Botanik*, Band 17, No. 1) shows that notwithstanding its name the sori of this Fern are composed of separate sporangia. The venation of the lanceolate pinna is forked, but the branches anastomose. The sporangia are arranged in double rows facing one another, dehiscence being by slits as in *Angiopteris*. But it appears that the relation between the sori and the veins is not so regular as in the recent Marattiaceae. The spore-counts approach the estimate for *Angiopteris*. But Halle sees in *Danaeopsis* a resemblance more nearly to *Archangiopteris* (Halle, Figs. 2, 7, 12) (Fig. 408).

It thus appears evident that the Marattiaceous type of sorus was already existent in Palaeozoic times, and that it was then represented both by types having separate sporangia as in *Angiopteris*, and also by synangial types as in *Marattia* or *Christensenia*. Both the ancient and the modern representatives of the sorus are strictly circumscribed. But those of the Palaeozoic Age appear

Fig. 408. *Danaeopsis fecunda* Halle, from the Rhaetic coal mines of Billesholm, after Halle. *A*, a fertile pinna. (Nat. size.) *B*, sporangia partly opened. (× 12.) *C*, ditto. (× 25.)

to have been for the most part circular, with their few sporangia forming a rosette: and this goes with the narrow Pecopterid-type of pinnule. The sori borne on these compare in the number and disposition of their sporangia with those of the Coenopterids, and especially of *Etapteris* (Fig. 333), and *Corynepteris* (Fig. 405).

The facts derived from the fossils and their comparison with the living Ophioglossaceae and Marattiaceae suggest in the first instance a marginal origin of the sporangia upon the leaf or segment. A slide of the sorus from the margin to the surface of a broadening leaf-area probably took place: also

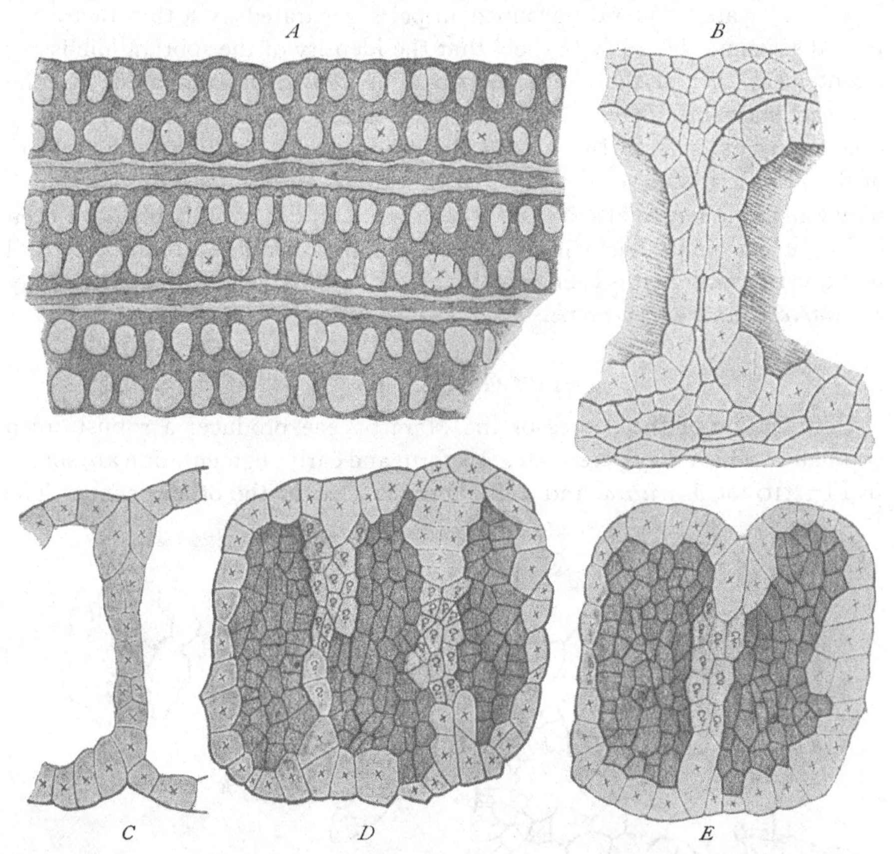

Fig. 409. *Danaea elliptica* Smith. Drawings illustrating the partial septations of the sporangia. *A*, tangential section through three sori, showing loculi in ground plan: the septa are often thin, so that pairs of loculi are in close relation: the loculi marked (×) are large and show one or more partial septa. (× 20.) *B*, *C*, *D*, *E*, show individual loculi with partial septa in greater detail. In *D* and *E* it is difficult to decide whether the cells marked (?) will develop as tapetum or as spore-mother-cells. (× 150.)

an extension of the sorus from a circular to an elongated form, so as finally to occupy the whole space between the margin and the midrib: and sometimes fissions of the elongated sori into rosette-like secondary sori. The synangial fusion of the closely related sporangia appears to have taken place as a step not dependent upon any one of these changes, and probably it happened repeatedly, appearing as a secondary and biologically intelligible amendment on the separate state of closely ranged sporangia. It is interesting to note that in *Danaea* and *Christensenia* the synangial state is accompanied by flattened scales on the rhizome, which may also be held as a feature of advance. A line of evidence bearing on the further longitudinal extension of the synangial sorus is seen in *Danaea.* Sections superficially through the elongated sorus show great variations in size of the sporangia (Fig. 409). The smaller are frequently grouped in pairs, separated by a thin or even a partial septum. These facts show that the identity of the sporangium is not maintained. When the elongated sorus of *Danaea* is compared with the compact sorus of *Marattia* these irregularities suggest that septation may accompany extension. In any case they introduce into the whole problem of the Marattiaceous sorus and its modifications an elasticity of view which may well aid in any final solution. The original type may probably have been a simple, marginal sorus with separate sporangia arranged round a central point, very much as it is seen in so many of the early fossils, and particularly in *Etapteris* and *Corynepteris.*

The Gametophyte

Germination of the spores of the Marattiaceae produces a robust green prothallus, which may perennate. Its form and early segmentation are shown in Fig. 410 for *Marattia*, and a like form is seen in the other genera. The

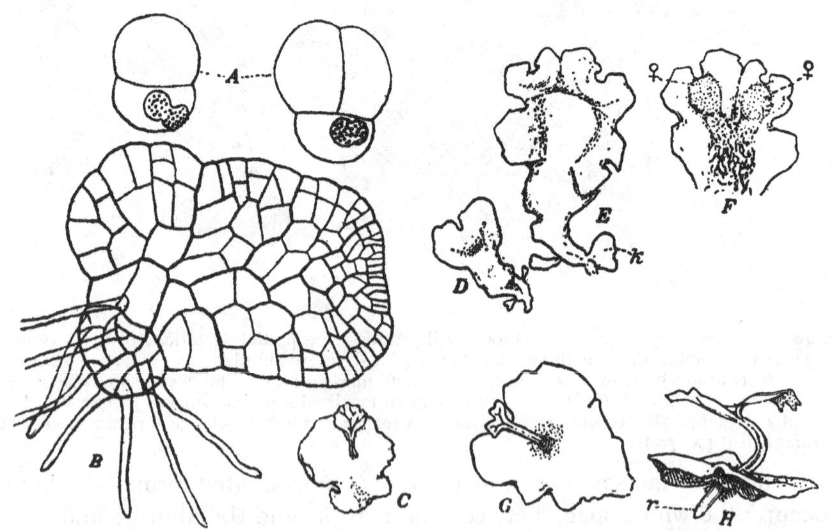

Fig. 410. *A*, two germinating spores of *Marattia fraxinea* Smith (× 200): the remains of the spore-membrane can be seen. *B*, young gametophyte of same (× 75). (*A*, *B*, after Jonkman.) *C*, *M. sambucina* Blume (× 15). *D–H*, *M. Douglasii* Baker. *D–F* (× 1·5): *G*, *H* (× 3) show the young sporophyte. *D*, *E* represent the same prothallium after an interval of about a year. *F* is the ventral view of *E*; *r*=root, *k*=adventitious bud. (After Campbell.)

prothallus does not start with a filamentous stage, but at once proceeds to a flattened expansion, soon forming the massive cushion. In old prothalli apical dichotomy may appear, as in many Liverworts (*E*, *F*). Campbell recognises a near resemblance to the prothalli of *Osmunda*, and especially in the tendency of both for the prothallus to be more than one cell thick from the first. In the Marattiaceae this becomes very conspicuous with age, extending almost to the margin of the thallus. Further, adventitious buds originate from the margin, often forming independent secondary prothalli. The rhizoids have been described as being septate: but Campbell finds in *Christensenia* that this does not result from a true cell-division, since only one nucleus is found in the whole rhizoid.

In all the genera Campbell found an endophytic fungus occupying the central regions of the prothalli, as in the Ophioglossaceae. The mycelium resembles that of the latter, but in the Marattiaceae there is no evidence of digestive cells containing the characteristic varicose mycelium. It appears that here there is no destruction of the fungus by the host: it seems to live more nearly as a true parasite than in the saprophytic Ophioglossaceae.

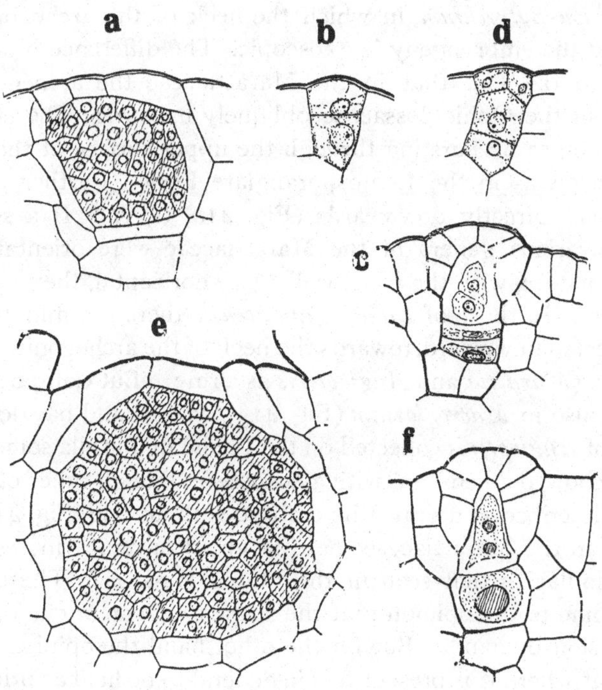

Fig. 411. *a—d*=*Marattia Douglasii*, after Campbell. *e—f*=*Ophioglossum pendulum*, after Lang. *a*=antheridium, with divisions of spermatocytes (32) in section, perhaps not complete; *b*=young antheridium; *c*=archegonium; *d*=young archegonium; *e*=antheridium with 88 spermatocytes in section; *f*=archegonium.

The gametangia are borne as a rule upon the lower surface, though occasionally observed upon the upper also (Jonkman). They are of a sunken type as usual in Eusporangiate Ferns. The antheridia agree very closely with those of *Ophioglossum*, but with fewer spermatocytes (Fig 411), and they open by an opercular cell, which is usually thrown off when the antheridium ruptures. The spermatozoids of *Angiopteris* and *Marattia* closely resemble those of Leptosporangiate Ferns, but those of *Christensenia* are more like those of *Ophioglossum*. The archegonia of the family are also like those of *Ophioglossum*, having a very slightly developed neck. The central series is as usual in Ferns; but the ventral-canal-cell, difficult to see in *Ophioglossum*, is large and conspicuous, except in *Danaea* where the same difficulty arises.

EMBRYOLOGY

A marked feature of the young sporophyte of the Marattiaceae is that the basal wall is always in a plane at right angles to the axis of the archegonium. The polarity of the embryo is defined by its very first segmentation, and the embryogeny is endoscopic. This is the exact reverse of that in *Ophioglossum* and *Eu-Botrychium*, in which the neck of the archegonium points upwards, and the embryogeny is exoscopic. The difference probably arises in relation to the fact that in the Marattiaceae the archegonium faces downwards, in the Ophioglossaceae obliquely upwards. The effect is that the shoot emerges by bursting through the upper surface of the prothallus, instead of below as in the Leptosporangiate Ferns. Further, the root will naturally point directly downwards (Fig. 410, *G*, *H*). If a series of the embryos of various genera of the Marattiaceae were orientated as they would be in nature, with the basal wall (*b*, *b*) horizontal, they would appear as in Fig. 412. In those of *Danaea jamaicensis* there is a minute suspensor, which is directed downward, towards the neck of the archegonium. In others there is none (*Marattia* and *Angiopteris* as a rule). But Campbell has found a suspensor also in *Macroglossum* (Fig. 413), while Land has described how in material of *Angiopteris* collected on the Island of Tutuila some specimens, but not all, showed an embryo with a suspensor, though other observers had repeatedly described and drawn its embryo without one (Fig. 414). Thus it is clear that in the Marattiaceae there is an inconstancy in respect of the suspensor similar to that seen in the Ophioglossaceae. These facts have prompted some to the opinion that the suspensor is an organ formed as biological occasion demands. But on the other hand the opinion seems more probable that where it is present it is in general to be held as primitive; and in particular it is to be remembered that it is absent from all Leptosporangiate Ferns. A still wider comparison has given ground for the opinion developed in Vol. I, Chapter XV, that the suspensor is a vestigial organ, and that plants which possess it are in this respect archaic. Many instances might be quoted where a vestigial part is inconstant in its occurrence, in the genus, species, or even the individual. Accordingly those embryos which possess a suspensor

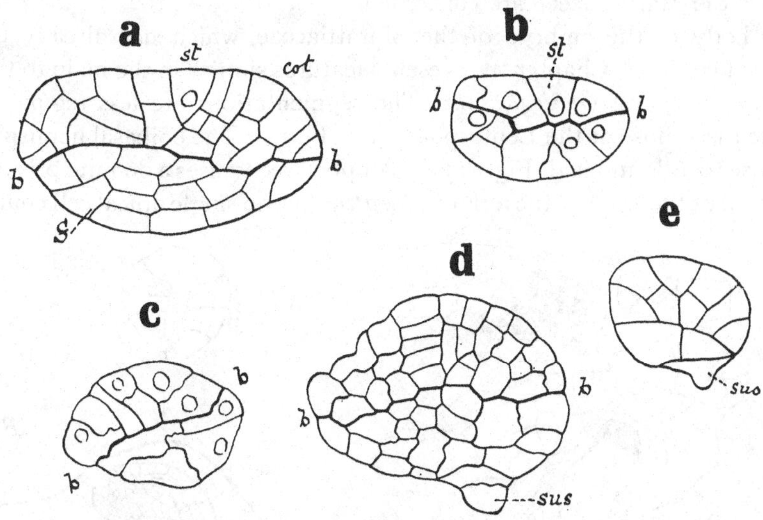

Fig. 412. Embryos of Marattiaceae, all orientated with the archegonial neck downwards, as in nature. a = *Angiopteris*: b = *Christensenia*: c = *Marattia*: d, e = *Danaea jamaicensis*: *sus* = suspensor: *st* = stem: *cot* = cotyledon: f = foot: b, b = basal wall. (All after Campbell.)

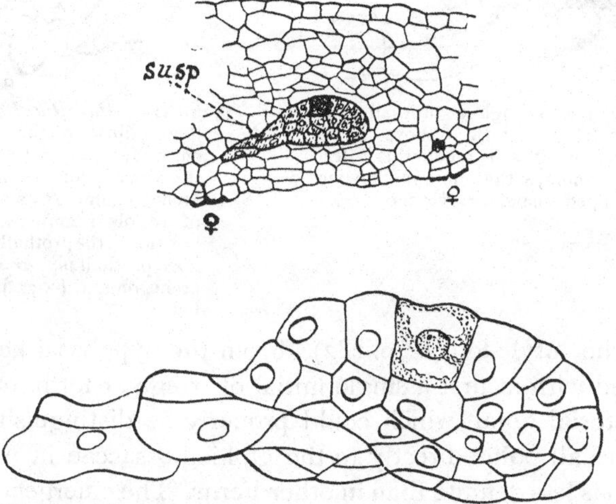

Fig. 413. Embryos of *Macroglossum* after Campbell, showing the natural orientation, and the relation of the suspensor to the archegonium.

are here held to be relatively primitive in respect of that feature, while those without it are held as relatively advanced. The Ophioglossaceae and Marattiaceae are the only two families of Ferns in which this organ has been observed. Though the feature is variable in both of them, it may be held as confirming their primitive position, which is itself based upon many other lines of comparison, as well as upon the positive data of Palaeontology as far as the Marattiaceae are concerned.

The body of the embryo of the Marattiaceae, which has already been discussed in Vol. I, Chapter XV, presents features similar in the main to those of *Botrychium* and *Ophioglossum*. The segmentations are less regular and definite than those of the Leptosporangiate Ferns. The epibasal hemisphere gives rise to axis and leaf (Fig. 415). Campbell recognises a definite prismatic initial cell at the apex of the axis of *Marattia*, but no single apical cell could be

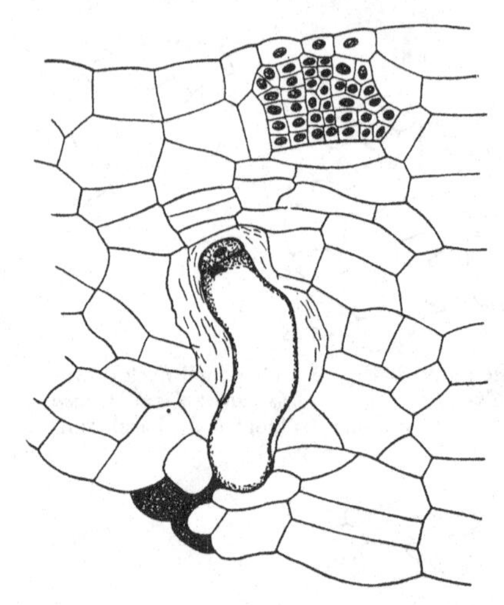

Fig. 414. Section through a prothallus of *Angiopteris* (after Land). Above is an antheridium, deeply sunk: below is a fertilised archegonium, with an embryo, showing a well-developed suspensor. (× 200.)

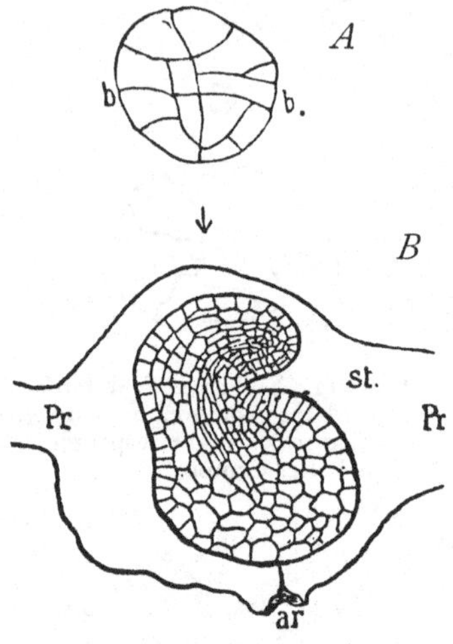

Fig. 415. *Marattia Douglasii*. *A* = longitudinal section of a young embryo (× 225): *b*, *b* = the basal wall: the arrow points to the neck of the archegonium. *B* = a similar section of an older embryo, showing its position in the prothallus: *st* = stem: *Pr* = prothallus: *ar* = neck of the archegonium. (× 72.) (After Campbell.)

made out in the cotyledon (*l.c.* p. 282). From the hypobasal hemisphere the root is formed with at first a single initial of irregular form, but there is no distinct haustorial organ which could properly be distinguished as a foot. These features all point directly to the Ophioglossaceae in which also the segmentation is less definite than in other Ferns. The emergence of the shoot through the upper surface of the prothallus is constant for the family so far as described, and it is in strong antithesis to the prone embryology of all Leptosporangiate Ferns. Even in *Macroglossum* it is the same: but in this genus the oblique position of the embryo suggests a possible step from the erect embryology of the Eusporangiatae, towards that prone embryology so characteristic of all the later and derivative Leptosporangiatae, where the embryo bursts through the lower surface of the prothallus (Fig. 413).

Comparative Treatment

The relatively primitive character of the Marattiaceae follows naturally from the features of form and structure above detailed, as well as from the special comparisons with the Ophioglossaceae, and ultimately with the Coenopteridaceae. The fossil record proves the great antiquity of the type, and the wonder is that the features of the sorus have been preserved through the ages with such precision. Nothing in Palaeobotany is more striking than a detailed comparison of the Palaeozoic sori of *Ptychocarpus* or of *Scolecopteris* with those of the living *Christensenia* and *Marattia*.

Nevertheless, it is difficult to place the Marattiaceae with certainty in relation to other Ferns. There appears to be no other family of living Ferns that could be directly referred to them as its ancestral source. They appear as survivals to the present day of a Palaeozoic and Mesozoic stock that was not readily adaptable except in minor features. But as regards the structure of the sorus certain analogies do exist with the Gleicheniaceae and Matoniaceae: these receive no support, however, from the vegetative system of those Ferns. On the other hand, in *Corynepteris*, which Kidston regarded as belonging undoubtedy to the Zygopterideae, we already see a sorus of a simple Marattiaceous type: and it has been shown how the various forms of sorus of the living Marattiaceae might readily have been derived from such a type, first by a slide of the sorus from the margin of the narrow pinnule to its broadening surface; and then as the surface became further expanded (probably with progressive webbing as the venation clearly suggests) the sorus may be held to have extended along the veins. Further, in certain types the sporangia, originally distinct from one another, became fused to form synangia, with as a natural consequence the loss of the annulus, which could no longer be mechanically effective. All this appears to be biologically probable, and morphologically reasonable. It may be the actual truth, but there is no clear evidence that it actually is so; though the presence of scales in place of hairs in *Danaea* and *Christensenia*, *Protomarattia* and *Archangiopteris*, is a significant parallel, suggestive of advance.

The problem of the phyletic position and arrangement of the living members of the Family among themselves presents hardly less formidable difficulties than appear in placing the Family as a whole. They arise partly from the absence of exact parallelism of the various characters, partly from differences of opinion as to their relative value. Priority should be given to the characters of the sorus, its position, the number of its sporangia, and the degree of its extension along the veins: also to the relation of the sporangia to one another. A prototype may probably be held to have possessed a marginal sorus, of few sporangia seated round a central point, as in *Etapteris* or *Corynepteris*. The nearest to this of living genera is *Angiopteris*, while stages of secondary spread of the sorus along the veins would lead to the state seen in *Macroglossum* and *Archangiopteris*. The Rhaetic fossil *Danaeopsis* shows that the spread from the margin all the way to the midrib was arrived at early when the sporangia were still separate. A like spread from the margin carried out in sori with the sporangia fused into synangia may have given a parallel sequence, as seen in *Marattia*, *Protomarattia*, and *Danaea*. The sole living representative of a third line is seen in *Christensenia*, which is clearly specialised for growth under forest shade, its broad pinnae showing advanced reticulation. The mode of origin of the scattered rosette-like sori is suggested by half-fertile pinnae of *Danaea*. Thus, however nearly they may resemble the sori of *Ptychocarpus*, it seems probable that the individual synangia of *Christensenia* are really only fragments of an originally intra-marginal sorus.

The characters of the conducting system do not follow parallel with those of the sorus. But in this Size has doubtless been one determining factor. *Angiopteris* with a relatively primitive sorus grows into a large plant, and has the most elaborate vascular system, while *Danaea* with an advanced sorus is relatively small, and has a less complex stock. Nor is even the presence of a suspensor a secure guide. *Danaea*, *Macroglossum*, and *Angiopteris*, in which a suspensor has been seen, occupy divergent positions according to soral characters. All the living genera except *Christensenia* have dichotomous and open venation. But the fossil *Danaeopsis* has vein-fusions. Among living forms the reticulation in *Christensenia* clearly supports the derivative character of its sorus. The simple dermal appendages of *Angiopteris* and *Marattia* mark them off as primitive in this feature, as against the scales of *Protomarattia*, *Archangiopteris*, *Danaea*, and *Christensenia* which, being held as derivative, support the conclusion that the sori of these genera are derivative also. From such comparative considerations a tentative phyletic grouping of the living Marattiaceae may be indicated as follows:

MARATTIACEAE[1]

I. SPORANGIA SEPARATE.

(1) *Angiopteris* (Hoffmann, 1796) ... ... 62 species.
(2) *Macroglossum* (Campbell, 1914) ... ... 1 species.
(3) *Archangiopteris* (Christ, 1899) ... ... 1 species.
(4) *Danaeopsis*—a fossil (Halle, 1921).

[1] A protest should be entered against names that assume too much, such as *Archangiopteris* and *Protomarattia*. The assumption has been that these Ferns are primitive as compared respectively with *Angiopteris* and *Marattia*. But both bear sori more extended than those of the older genera, and both bear scales. If, as argued in the text above, the original sorus was rosette-like and marginal, then their names are misleading. That their sori are highly derivative is the conclusion clearly indicated by comparison with the early fossils, such as the Coenopteridaceae.

II. SPORANGIA UNITED INTO SYNANGIA.
- (5) *Ptychocarpus*—a fossil.
- (6) *Marattia* (Swartz, 1788) ... ... 28 species.
- (7) *Protomarattia* (Hayata, 1919) ... ... 1 species.
- (8) *Danaea* (Smith, 1793) ... ... 26 species.

III. SORI SUB-DIVIDED.
- (9) *Christensenia* (Maxon, 1905) [= *Kaulfussia* (Blume, 1828)] ... ... 26 species.

BIBLIOGRAPHY FOR CHAPTER XX

380. ENGLER & PRANTL. Natürl. Pflanzenfam. i, 4, p. 422, etc. Here the literature is fully cited.
381. ZEILLER. Bassin Houiller d'Autun et d'Épinac. Fasc. ii, Part i.
382. STRASBURGER. *Scolecopteris elegans*. Zenk. Jenaisch. Zeitsch. 1874, p. 87.
383. BOWER. Studies III. Phil. Trans. Vol. clxxxix, p. 35. 1897.
384. POTONIÉ. Lehrbuch d. Pflanzenpalaeontologie. 1899.
385. CHRIST & GIESENHAGEN. *Archangiopteris*. Flora. Vol. lxxxvi, p. 72. 1899.
386. RUDOLPH. Psaronien und Marattiaceen. Wien. 1905.
387. GWYNNE-VAUGHAN. *Archangiopteris Henryi*. Ann. of Bot. xix, p. 259. 1905.
388. BOWER. Origin of a Land Flora. Chapter xxxiii. 1908.
389. SEWARD. Fossil Plants. Vol. ii, Chapter xxii. 1910. Here the literature on the fossils is fully cited.
390. CAMPBELL. Eusporangiate Ferns. Washington. 1911.
391. CHARLES. Sporeling of *Marattia alata*. Bot. Gaz. Vol. li, p. 81. 1911.
392. P. BERTRAND. Progressus Rei Bot. Band iv, p. 281. 1912.
393. VAN LEEWIN. Veg. Vermehrung v. *Angiopteris*. Buit. Ann. 2, Ser. x, p. 202. 1912.
394. CAMPBELL. *Macroglossum alidae*. Ann. of Bot. xxviii, p. 651. 1914.
395. WEST. Secretory tissue in the Marattiaceae. Ann. of Bot. p. 409. 1915.
396. WEST. Study of the Marattiaceae. Ann. of Bot. xxxi, p. 361. 1917.
397. WEST. *Stigeosporium marattiacearum*. Ann. of Bot. xxxi, p. 71.
398. CAMPBELL. Mosses and Ferns. 3rd edn., p. 273. 1918. Here the literature is fully cited.
399. VON GOEBEL. Organographie II, ii. 1918.
400. HAYATA. *Protomarattia*. Bot. Gaz. lxii, p. 84. 1919.
401. SCOTT. Studies in Fossil Botany. 3rd edn., Vol. i, Chap. viii. 1920.
402. HALLE. Sporangia of Mesozoic Ferns. Arkiv. for Botanik. Bd. xvii, p. 1. 1921.
403. BLOMQUIST. Vascular Anat. of *Angiopteris*. Bot. Gaz., p. 181. March, 1922.
404. SEWARD. Hooker Lecture. Journ. Linn. Soc. xlvi, p. 218. 1922.
405. LAND. Suspensor in *Angiopteris*. Bot. Gaz., p. 421. 1923.

CHAPTER XXI

OSMUNDACEAE

THE family of the Osmundaceae comprises three genera and 17 species of living Ferns, according to Christensen's *Index*: viz. *Todea* (1 species), *Leptopteris* (7 species), and *Osmunda* (9 species): but some writers have regarded *Leptopteris* as a section of the genus *Todea*, thus recognising only two genera. The Family is of wide geographical distribution, and it extends to both hemispheres. *Osmunda* is cosmopolitan, but *Todea* and *Leptopteris* are native in South Africa and Australasia. In point of time the Family extends back to the Palaeozoic Period. The evidence for this is derived from beautifully preserved stems such as those of *Zalesskya* and *Thamnopteris* from the Upper Permian of Russia, first described by Eichwald (*Lethaea Rossica*, i, Stuttgart 1860), and thoroughly examined by Kidston and Gwynne-Vaughan (Fossil Osmundaceae, *Trans. R. S. Edin.* 1907 to 1914). These show undoubted Osmundaceous characters, but their fructifications are unknown. On the other hand many sporangia of structure resembling those of living Osmundaceae have been described from the Upper Carboniferous strata; but as Seward remarks (*Fossil Plants*, ii, p. 325) it is not safe to assume that these were borne on plants possessing the anatomical characters of the Osmundaceae rather than of the Botryopterideae. However, fructifications of undoubted Osmundaceous character have been traced to the Jurassic Period, and even referred by Raciborski to the genera *Osmunda* and *Todea* (Engler's *Jahrb.* xiii, p. 7). He adds that they are so highly differentiated that their origin probably dates back still earlier. Such comparisons both of anatomical structure and of the fructifications suggest an early origin of the Osmundaceous stock. They stimulate comparison with the Botryopterideae, and point to a probability that a family of such antiquity as the Osmundaceae should prove a valuable basis for comparison also with more modern Ferns. It will be found that it does so, and indeed the living Osmundaceae have long been recognised as holding a unique position intermediate between Eusporangiate and Leptosporangiate Ferns. The description of them in detail will fully justify this preliminary forecast.

EXTERNAL CHARACTERS

The living Osmundaceae are all perennials, with an upright, usually short, radially constructed stock, which bifurcates occasionally (Frontispiece). The axis is covered by the persistent bases of leaves, which are disposed upon it in a dense spiral. It is attached to the soil by numerous dark-coloured roots, which originate in close relation to the bases of the crowded leaves, and form with them a massive investment which hides the upright and branching stock. This may sometimes extend to a height of several feet in *Todea barbara*: the habit is then that of a stunted tree-fern. Compton states (*Linn. Journ.* xlv, p. 456) that *Leptopteris Wilkesiana* Christ may reach a height of 10 feet in New Caledonia. The leaves expanded in the current form together a basket-shaped group, the outermost of which are often sterile and the inner fertile. Some of the leaves at the outside of the winter bud have an abortive lamina, while their broad persistent basal region acts as a protection. The leaves have the usual circinate vernation, and are covered while young with a dense mat of hairs, which fall off at maturity.

The leaves are pinnate with varying degrees of branching: but the juvenile leaves show clearly how that state is referable to sympodial development of dichotomy (Fig. 416). This same method continued in

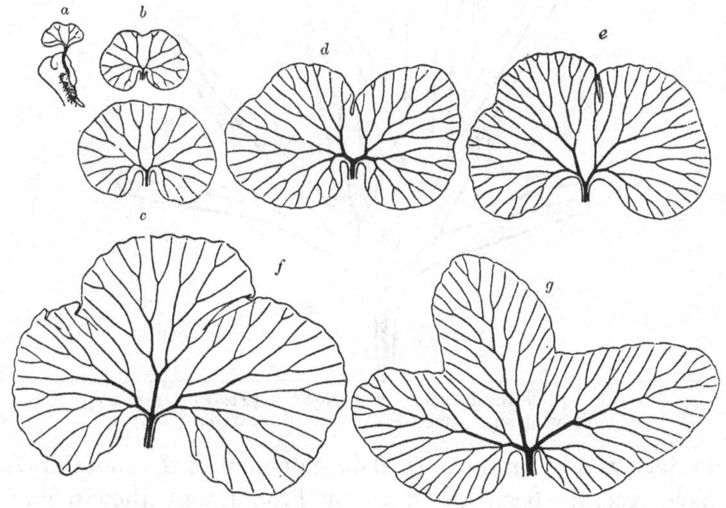

Fig. 416. Successive juvenile leaves of *Osmunda regalis*, showing steps (*a*—*g*) of progression from equal dichotomous venation to sympodial branching, and the establishment of a terminal lobe. (× 2½.)

successive branchings will explain the structure of the most complicated leaves. The broad protective leaf-base is a feature of importance for comparison. The wings of the petiole are traceable downwards into broad flaps, or "stipular leaf-bases," forming together with the mucilaginous hairs a very perfect protection to the young parts. In *Todea* their development extends also to a transverse commissure across the face of the leaf-stalk. Commissures are unknown in Leptosporangiate Ferns: but they occur in the Ophioglossaceae and Marattiaceae, while like structures appear also among the Cycads. Such stipular leaf-bases are archaic features. Kidston and Gwynne-Vaughan have shown how similar leaf-bases are characteristic also of the fossils referred by them to an Osmundaceous affinity (*Fossil Osmundaceae*, Part I, Plate VI). The texture of the leaf-blade may be leathery or filmy.

The former is characteristic of *Osmunda* and *Todea barbara*, and it is probably a primitive feature (Chapter II, p. 40). The latter is seen in *Leptopteris*, and is the basis of its distinction as a genus. Here there is a thin pellucid structure of the pinnules. It is, however, a question whether the difference deserves generic value, since it is probably a relatively late adaptation to life under conditions of excessive moisture. It is found in species from Australasia, and the South Sea Islands. The pinnules, which are cut into fine segments, each with a single vein, approach in their filmy structure that of the Hymenophyllaceae (Fig. 417).

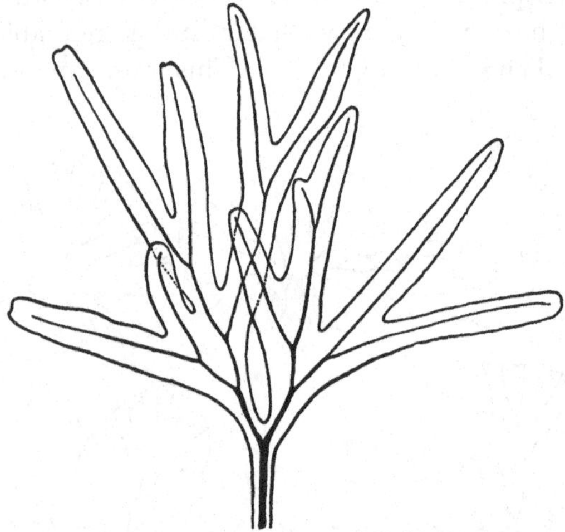

Fig. 417. Juvenile leaf of *Leptopteris* (*Todea*) *superba*, showing single-nerved segments all separate. (× 3.)

Certain deviations from the normal development of the shoot in *Osmunda regalis* have recently been described by Prof. Lang; they seem to bear a definite comparative interest. They followed on conditions of disturbed development acting on young plants, and they resulted in transitions of what can only be regarded as leaf-rudiments into shoots. Two examples are quoted in Fig. 418, *A*, *B*. In one of them (*A*) a bud appears on the ventral face of a leaf (*b*), both being supported on a common stalk: in the other (*B*) a leaf is represented by a cylindrical structure, of a deep green colour, which ends without any lamina in a bud (*b*). Such indeterminate members, neither typically leaf nor bud, have an interesting relation to the morphological question as to the real nature of these parts. This interest is considerably increased by the fact that similar indeterminate states have been found in plants of *Plagiogyria* grown under natural conditions in the open (see Chapter XXXI).

In *Todea* there is no marked difference between sterile leaves and sporophylls, and the sporangia appear on the under surface of leaves not

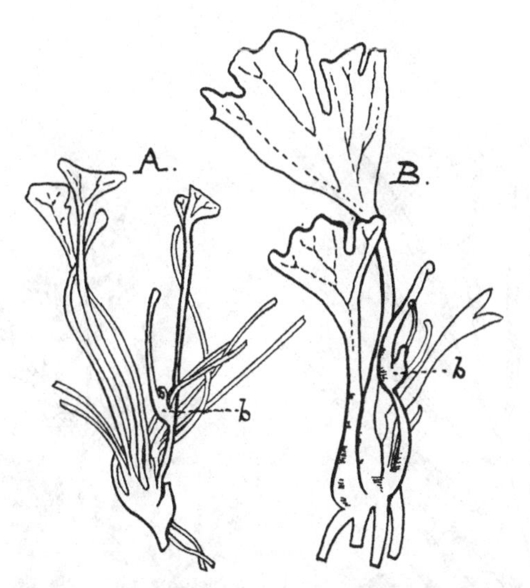

Fig. 418. Abnormal sporelings of *Osmunda regalis* showing indeterminate structures between leaf and shoot. In *A*, a leaf-blade and a bud (*b*) are borne on a common stalk; in *B*, there is no leaf-blade, and the bud (*b*) is as before raised upon an indeterminate stalk. ($A \times 3$; $B \times 4$.) (After Lang.)

specialised. But in *Osmunda* the sporangia are localised on various fertile parts of the leaf which then show a smaller expanse of surface, while the sporangia are seated at or near to the margin. This fertile region may be distal as in *O. regalis*, or intermediate as in *O. javanica* and others (Fig. 419). The sporangia, which are numerous and large, are of pear-like form, and those in near proximity to one another originate simultaneously: there is no interpolation or marked sequence in origin. In *Osmunda* they are borne in approximately marginal tassels without any regularity of orientation. Sometimes intermediate states between sterile and fertile pinnules occur, with larger leaf-surface, and with the sporangia attached superficially (Fig. 420). This leads to the state seen in *Todea*, where the sporangia are borne only on the under surface: their arrangement has some relation to the veins, but again there is no common rule of orientation. These Ferns are then non-soral. The conditions seen would be consistent with the sporophyll having been originally narrow, and the sporangia borne at or near to the margins; but that with increasing leaf-area they passed to a superficial position. Comparison with the Coenopterids would support this hypothetical interpretation. The sporangia, which are large and thick-stalked, consist at maturity of a single layer of cells forming the wall, but with a few tabular cells remaining within. A group of polygonal thick-walled cells in a lateral position near the distal end is recognised as an annulus. It is related to the slit of dehiscence so that the latter passes from the centre of the annulus

Fig. 419. *A*, a leaf of *Osmunda Presliana* J.Sm. showing a middle zone of fertile pinnae. *B*, a young plant of *Osmunda regalis* L. *C*, a leaf of an adult plant of *Osmunda regalis*, in which the fertile region is distal. (From Engler and Prantl.)

over the distal end, and leads downwards to the stalk on the opposite side of the sporangium. The line of dehiscence, defined structurally by narrow thin-walled cells, gapes widely at ripeness (Fig. 421). This requires elbow-

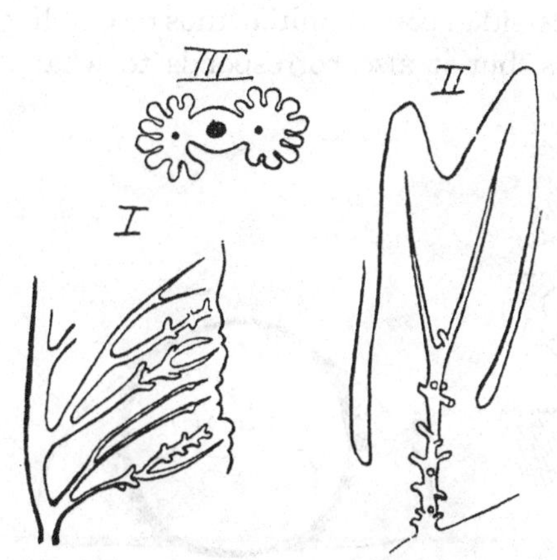

Fig. 420. *I*, *Osmunda regalis*, part of a metamorphosed sporophyll with aborted sporangia attached superficially. *II*, *Todea pellucida*, part of a pinnule with sporangial stalks. *III*, transverse section of a fertile pinnule of *Osmunda*, showing normal insertion of sporangia. (All after von Goebel.)

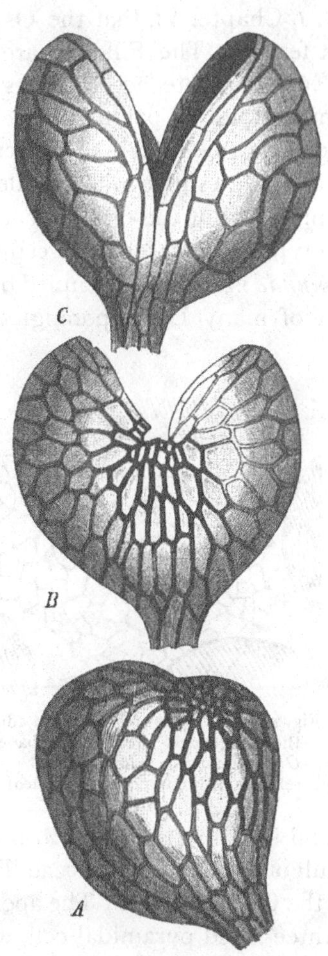

Fig. 421. *Todea barbara* Moore, sporangium. *A*, in side view, closed; *B*, seen from behind; *C*, from in front, in both cases after dehiscence: the annulus is darkly shaded. (×80.) (After Luerssen.)

room, which the lax arrangement of the sporangia readily allows. The spore-output is large, ranging in the family from typical numbers of 128 to 512.

CELLULAR CONSTRUCTION

It is not often that a comparative argument can be effectively applied from the facts of cellular construction: but it has already been shown in Vol. I, Chapter VI, that the Osmundaceae hold an exceptional position in that feature. The Filicales are broadly distinguished as Leptosporangiate or Eusporangiate, according as the sporangium arises from a single cell or from a group of parent cells. It has been shown that a similar distinction holds as a rule for all the apical meristems, the Leptosporangiates having in each case a single initial, while the Eusporangiates have as a rule a plurality of initials. But the Osmundaceae take an intermediate position between the two types as regards their cellular construction. The apex of the stem of *Osmunda* is, it is true, occupied by a three-sided conical initial, thus resembling that of many Leptosporangiate Ferns, but it also corresponds to what is

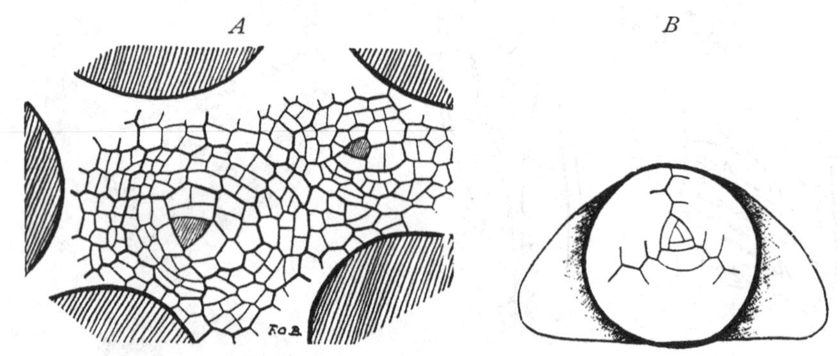

Fig. 422. *A*, apical region of an adult plant of *Osmunda regalis*, seen from above, showing the three-sided initial cells (shaded) of stem and leaf. (× 33.) *B*, apex of young leaf of *O. cinnamomea*, before the first pinnae appear, showing the three-sided initial cell, and the relation of the lateral stipular leaf-bases to its segmentation. (× 35.)

found in young stems of *Marattia* (Charles, *Bot. Gaz.* li, p. 94), and in the adult of the Ophioglossaceae (Fig. 422, *A*). As far as recorded, this is constant in the Osmundaceae. The apex of the leaf, both in *Osmunda* and *Todea*, has a three-sided pyramidal cell, which is a more complex organisation than the two-sided type of all observed Leptosporangiate Ferns: this accords with the relatively robust structure of the Osmundaceous leaf-stalk (Fig. 422, *B*). A very distinctive feature is the wing of the leaf or pinna, which in Leptosporangiate Ferns shows definite segmentation from a single row of marginal cells (see Vol. I, Figs. 105, 107). But in *Osmunda* and *Todea barbara*, as in the Marattiaceae, the massive wing has no single row of marginal initials (Fig. 396, *B*, p. 101). Similarly with the roots, though small roots may have the usual tetrahedral initial cell, large roots both of *Osmunda* and *Todea* may exhibit various irregularities of segmentation; or they may settle down to a structure with four truncated prismatic initials, as is seen in the Marattiaceae (Fig. 423, *A*, *B*). Lastly, the sporangium is very

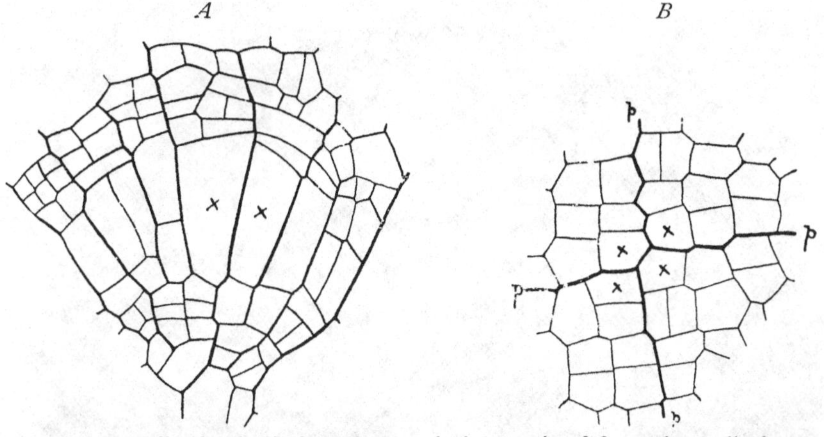

Fig. 423. *A*, median longitudinal section through the root-tip of *Osmunda regalis* showing two truncated initials (×), segments being cut off from both ends as well as from the sides. (×200.) *B*, transverse section through the root-tip of *Todea barbara*, showing four initials (×), and the principal walls *p.p.* (×200.) Both were taken from large roots of adult plants.

inconstant in its cleavages, and is not always referable to the segmentation of a single cell. Moreover, the sporogenous cell may sometimes be cubical, as in the Eusporangiatae, but more frequently it is tetrahedral, as in the Leptosporangiatae (Fig. 424). All these facts show that the Osmundaceae are unstable in their segmentation: that they conform decisively neither to the Eusporangiate nor to the Leptosporangiate types, but hover between the two. This conclusion acquires special significance when the very early fossil history of the Family is remembered.

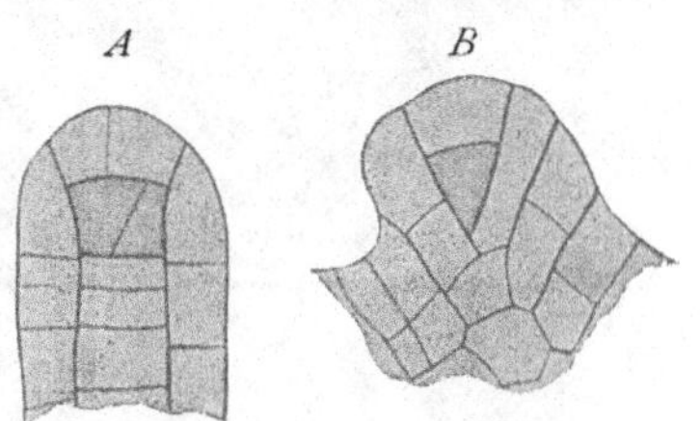

Fig. 424. Young sporangia of *Todea barbara* in longitudinal section, showing different modes of segmentation. (×363.)

ANATOMY

The massive upright stock of the Osmundaceae, so closely invested by leaf-bases each with its sclerotic sheath and containing a densely sclerosed cortex, is very difficult to cut, and its appearance in section as a whole is better known from sections of the related fossils than from published figures of living stems. A large stock of *Todea* (*Leptopteris*) *Fraseri* from the Blue Mountains, Australia, was cut transversely, and its surface polished as an opaque object. Its appearance is seen in Fig. 425. At the periphery are the brown remains of the outer leaf-bases together with matted roots. Next follows a broad band made up of leaf-bases each with its C-shaped meristele embedded in soft parenchyma, which is delimited by a dark sclerotic sheath.

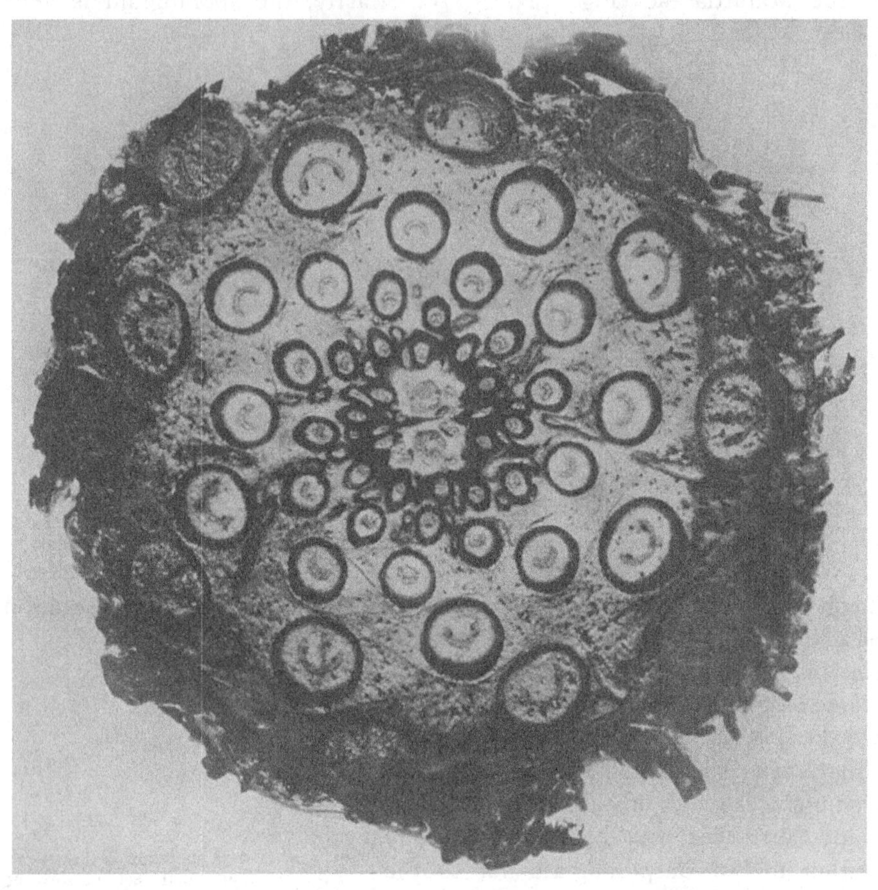

Fig. 425. Transverse section of stem of *Todea Fraseri*, showing the stele (recently bifurcated), leaf-traces each surrounded by a ring of sclerenchyma, and the rhomboidal bases of the leaves faintly outlined. (×2.) (From a photograph by the late Dr R. Kidston.)

This is not the outer surface of the leaf-base: external to it is a broad band of pale soft parenchyma, which is a regularly recurring feature of Osmundaceous leaf-bases. But these fit so closely together that in transverse sections the appearance is given of a continuous soft matrix in which the leaf-bases are embedded: and it is only when these split apart, as they readily do, that the rhomboidal form of the leaf-base is recognised. Passing inwards, a dense brown sclerotic zone of the true outer cortex is reached, which again passes into a softer and paler inner cortex, through both of which the narrowing meristeles of the numerous leaf-traces pass to the centrally-lying, relatively small and pale-coloured stele. Roots originating laterally from the leaf-bases may be occasionally seen taking a horizontal course outwards. It is essential to be clear as to the structure thus described, which is general for living Osmundaceae; for it corresponds with singular fidelity of detail to that seen in the Permian fossil *Thamnopteris Schlechtendalii* (Eichwald), and in other fossil Osmundaceae. It is thus a very ancient type of shoot-construction (Kidston and Gwynne-Vaughan, *l.c.* iii, p. 656).

Turning to the vascular tissue of living Osmundaceae, and especially to that of the central stele, it is found to be constantly and completely circumscribed by endodermis: even when a leaf-trace passes off, the endodermal sheath does not open. The only exception to its completeness is where a "ramular gap" is formed, and even then an opening is only of occasional occurrence, as in *O. cinnamomea* (Vol. I, p. 134). Otherwise the endodermal investment of the vascular system is complete in the living Osmundaceae.

The description of the vascular system of the shoot may start from the leaf-blade. The numerous peripheral forked veins fuse on passing downwards from pinna and pinnule to the leaf-stalk, forming a single half-cylindrical strand, its concave channel being directed inwards. This strand consists of

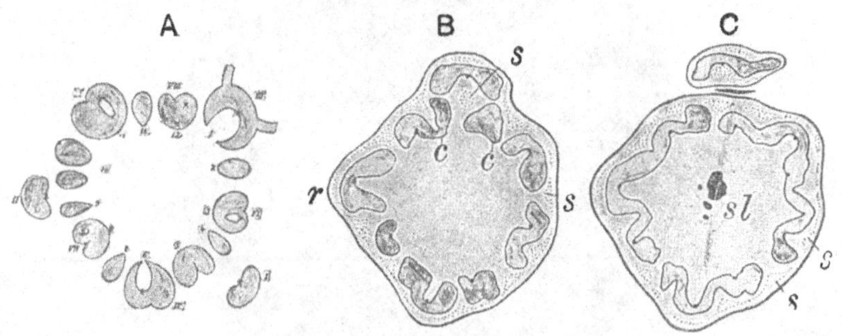

Fig. 426. *A*, diagram showing the arrangement of the xylem-strands in the axis of *Osmunda regalis*. (After Zenetti.) The phloem and endodermis are omitted. *B*, *C*, transverse sections of the stele of *Todea barbara*, with leaf-trace (after Seward and Ford), showing a greater continuity of the xylem (shaded) than in *Osmunda*; *s*=phloem, *sl*=sclerotic tissue.

a continuous band of metaxylem, with numerous protoxylem-groups at its concave limit, which alternate in position with groups of mucilage-sacs. Peripherally is a mantle of phloem, and surrounding the whole is a continuous endodermis. As it passes to the base of the petiole the strand contracts, and the protoxylems unite into a single one in a median position, while in transverse section the whole strand assumes a deep U-shaped outline (Fig. 426, *A*, *B*). It is here that the vascular supply to the roots is given off laterally from the leaf-trace (Fig. 426, *A*, III). The strand thus contracted then enters the stele, and takes its place in a ring of similar traces surrounding the central pith. As it does so its endodermis becomes continuous with that which completely surrounds the stelar system (Fig. 426, *B*, *C*). The stele, as seen in transverse section, is composed of several layers of parenchyma at its periphery: then follows a band of phloem which is continuous, but may be uneven in width, extending inwards at the xylic gaps. Within this are the xylem-strands, which vary greatly in number. *O. claytoniana* may have as many as 40, *O. regalis* about 15 (in Fig. 426, *A*, there are 14), *Todea barbara* 8 or less (in Fig. 426, *B*, there are 8, in *C* there are only 3), while in *T. superba* the xylem may sometimes form an unbroken cylinder. The position of the protoxylem also varies: in *Osmunda* it is nearly on the outer edge of the metaxylem, but in *Todea* the xylem is mesarch, or in *T. hymenophylloides* the strands are almost exarch. Centrally lies the pith: sometimes an internal endodermis is present (*O. cinnamomea*, *T. hymenophylloides*), while in the former species some internal phloem has been found locally (see Vol. I, p. 133; also Jeffrey, *Phil. Trans.* cxcv, p. 119, etc.; Faull, *Bot. Gaz.* 1901, p. 381 (Fig. 427).

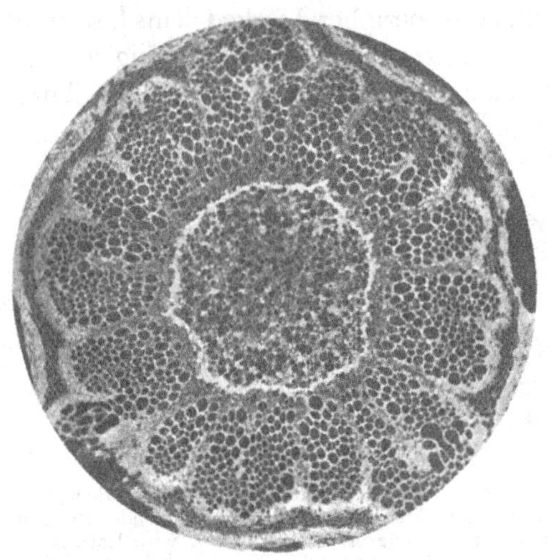

Fig. 427. Stele of a full-grown stem of *Osmunda cinnamomea*, from a photograph by Gwynne-Vaughan. For details see Vol. I, p. 133. (×25.)

If the course of the several xylem-strands be followed further, they are found to fuse downwards according to a regular scheme, so that they form a cylindrical network of which the meshes are very long and narrow. The number and proportions of these vary, but in all the intercommunication of the whole dictyoxylic cylinder is close and effective. The scheme is represented for *Osmunda* in Fig. 428, *A*, as flattened into a single plane; and for *Todea* where the number of the strands is less in Fig. 428, *B*. In the latter, the xylic gaps are shorter than they are in *Osmunda*. The structure thus described is not far removed from the protostelic state: moreover, the leaf-trace, being undivided and containing at its insertion on the stele only a single group of protoxylem, indicates a relatively simple vascular structure. There are two possible views of such facts: either that the state we see is the result of reduction from a more complex vascular condition; or that the system itself is in the up-grade of a relatively primitive evolution, in fact that the living examples illustrate approximately the limit of such development as their direct ancestors ever attained. The former opinion has been elaborated by Jeffrey (*Phil. Trans.* cxcv, p. 119, etc.) and by Faull (*Bot. Gaz.* 1901, p. 381). They hold the Osmundaceous stele to be a reduced form of "amphiphloic siphonostele," and in support of their opinion they adduce the presence of an internal endodermis in *O. cinnamomea* and *T. hymenophylloides*, and the occasional presence of internal phloem also locally in the neighbour-

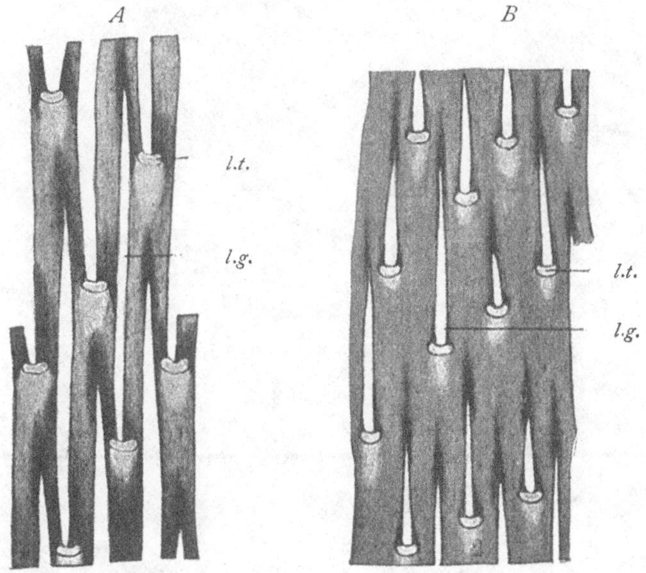

Fig. 428. *A*, scheme showing part of the xylem-ring of *Osmunda regalis* as it would be seen from without; *l.t.* = cut end of a departing leaf-trace; *l.g.* = xylic-gap. (After Lachmann, from Kidston and Gwynne-Vaughan.) *B*, a similar scheme for *Todea barbara*, seen from without. Lettering as above. (After Seward and Ford, from Kidston and Gwynne-Vaughan.)

hood of the branchings of the axis. There are good grounds for doubting whether the local and inconstant occurrence of endodermis and internal phloem will bear the weight of a far-reaching theory of reduction. It would appear less likely that a robust and large-leaved phylum of Ferns should show a reduced vascular system in its stock than that the stock should retain a primitive though perhaps imperfectly efficient system.

The contrary opinion, viz., that the stele of *Osmunda* is in the up-grade of development, though showing a relatively primitive structure, was entertained by Zenetti, who made a careful study of its anatomy (*Bot. Zeit.* 1895, pp. 72–76). A good basis for an opinion opposed to a theory of reduction is found in the ontogeny of the living species: for in their sporelings there is at first a prostele without internal complications, which expands later and becomes medullated. The first seven or eight leaves have a protostelic departure: after the medullation of the xylem the ring is breached by a xylic gap at the departure of each trace, but there are never true foliar gaps with continuity of pith and cortex. The changes are all intra-stelar, and remain so to the adult state. Thus the ontogeny suggests a progressive

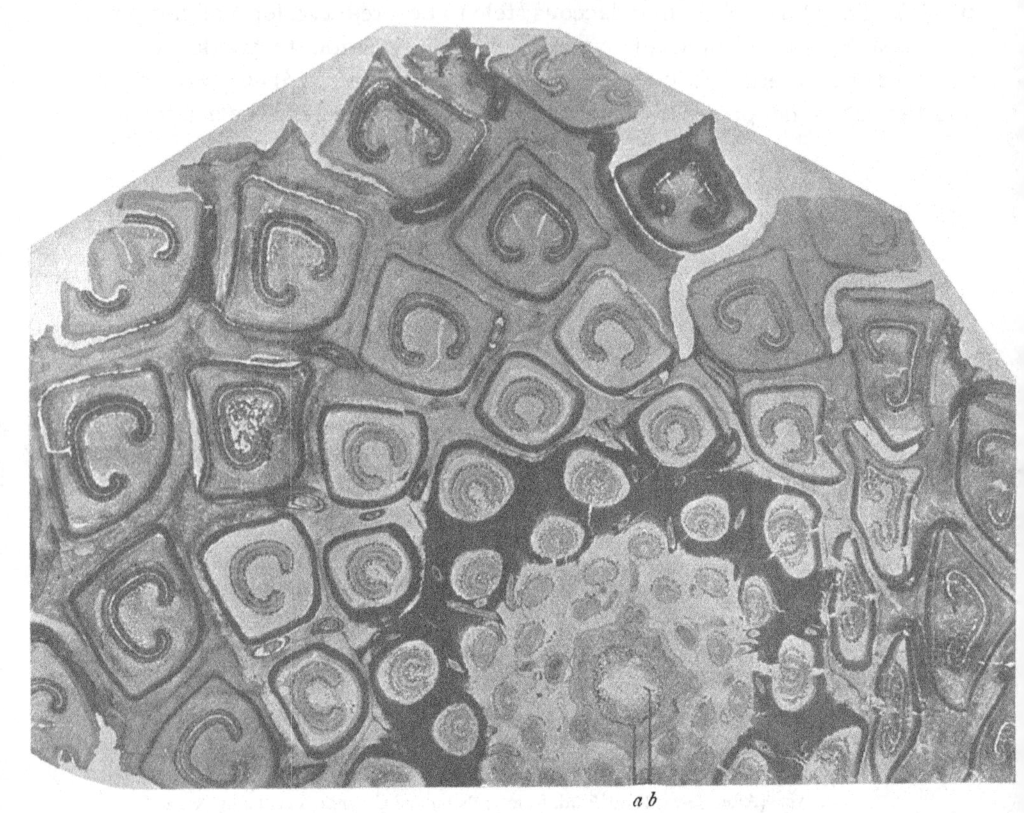

Fig. 429. Part of a transverse section of a Permian Osmundaceous Fern stem, *Thamnopteris Schlechtena* (Eichwald); *a*, outer xylem; *b*, inner xylem. (After Kidston and Gwynne-Vaughan, very slightly redu From Seward.)

evolution from the protostele which stops short at the state attained by the adult (Faull, *Trans. Can. Inst.* viii, p. 515, 1909; Gwynne-Vaughan, *Ann. of Bot.*, xxv, p. 525, 1911).

More cogent evidence than this is derived from the study of the structure seen in related fossils. This work has been carried out by Kidston and Gwynne-Vaughan, and the demonstration is a very convincing one. (*Trans. Roy. Soc. Edin.* No. I, xlv, p. 759, 1907. No. II, xlvi, p. 213, 1908. No. III, xlvi, p. 651, 1909. No. IV, xlvii, p. 455, 1910. No. V, l, p. 469, 1914.) If the stelar structure of the living Osmundaceae be of a reduced type, the fossil correlatives should show a progressively more complex structure of the stele as they are followed to earlier horizons. But it will be seen that, in a general way though not always in consecutive detail, the trend is the reverse of this. Six salient stages of stelar complexity are involved in the series as recognised by Kidston and Gwynne-Vaughan. They may be seriated from the simplest to the most complex, thus:

(1) *A solid homogeneous xylem (protostele).*

This is seen in the basal region of all young sporelings, but it is not described for any living species in the adult stage. Certain Botryopterideae, however, and especially *Grammatopteris* from the Permian of Autun, were regarded by Kidston and Gwynne-Vaughan as possessing structure primitively Osmundaceous (*l.c.* No. 1, p. 778). It was thought by Dr Stopes that a certain cretaceous fossil from Queensland, named *Osmundites Kidstoni*, showed this structure in the adult together with secondary thickening, as in *Botrychioxylon* (*Ann. of Bot.* xxxv, p. 55, 1921). But re-examination of the specimen has yielded a quite different interpretation (Posthumus, *Ann. of Bot.* xxxviii, p. 215, 1924).

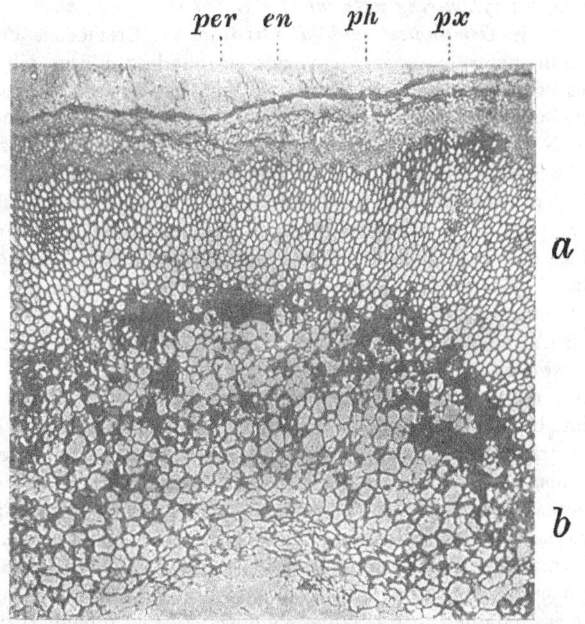

Fig. 430. *Thamnopteris Schlechtendalii* Eich. Part of a stele. *per*=pericycle; *en*=endodermis; *ph*=phloem; *px*=protoxylem. *a*=outer xylem; *b*=inner xylem. Permian Age. (After Kidston and Gwynne-Vaughan. From Seward.) (× 13.)

The incident is mentioned here so as to correct the statements made in Vol. I, pp. 129, 137, which were based on Dr Stopes' paper.

(2) *A solid pithless xylem, but heterogeneous in structure.*

This appears in two species of *Zalesskya* from the Upper Permian of Russia: but the characteristic structure is particularly well preserved in *Thamnopteris Schlechtendalii* from the same source. The general construction of the shoot corresponds remarkably with that of modern Osmundaceae: but the stele is simpler (Fig. 429). There are two zones of xylem which are seen to be quite distinct structurally (Fig. 430). The protoxylems are sunk in the continuous zone of outer xylem, while the leaf-traces separate from its outer surface in the protostelic fashion, expanding as already described into the characteristic Osmundaceous curve (Vol. I, Fig. 155).

(3) *A continuous xylem-ring surrounds a central pith.*

Osmundites Dunlopi from the Jurassic of New Zealand possesses a continuous xylem-ring surrounding a pith badly preserved, but probably parenchymatous. "Most of the leaf-traces, if not all, depart without in any way interrupting the continuity of the xylem-ring" (*l.c.* No. I, p. 760).

(4) *An interrupted xylem-ring surrounds pith only.*

In *O. Dowkeri*, from the same beds as *O. Dunlopi*, the pith is well preserved, and is surrounded by a xylem-ring which is broken up by xylic gaps into about 30 distinct strands. *O. Gibbeana* is similar, but the strands number about 20. *O. Kolbei* from the Neocomian (Wealden) of Cape Colony supplies the transition from the inner xylem to a true parenchymatous pith by possessing a "mixed pith" with wide tracheides. It thus appears that the establishment of a true pith and the interruption of the xylic-ring date from early Mesozoic times.

(5) *An interrupted xylem-ring with internal endodermis and phloem.*

This is first seen in *Osmundites skidegatensis* from the Cretaceous Period. The large stele, 24 mm. in diameter, has some 50 strands surrounding a large pith partly sclerotic, which becomes continuous with the cortex at each leaf-gap. No layer resembling an endodermis has been distinguished, and it is practically impossible to set a definite limit to the stele (*l.c.* No. I, p. 771). But the most remarkable feature of this fossil is the internal phloem which lines the inner surface of the ring, and is connected with the outer through each leaf-gap. The departure of the leaf-trace interrupts the continuity of the whole vascular ring (*l.c.* p. 770). The interest of this is increased by the fact that a similar internal phloem exists in *O. cinnamomea*, but with a well-defined internal endodermis (Fig. 427, p. 136). In this plant, however, as also in *Todea hymenophylloides*, the internal phloem is local rather than general throughout the stem (Vol. I, p. 134).

(6) *A dictyostelic ring, with cortex and pith continuous through each leaf-gap: the meristeles surrounded by endodermis.*

This is seen in the largest of all the known Osmundaceae, *Osmundites Carnieri*, from Paraguay, of late but uncertain horizon (*l.c.* No. V. p. 475). The stele is 35 mm. in diameter, with 35 distinct strands of xylem of the Osmundaceous type. It is delimited by a layer held to be endodermis, while at the departure of each leaf-trace there is a wide gap, giving continuity between cortex and pith. The phloem is badly preserved, but it seems probable that it followed the endodermis round the inner surface of each meristele. This appears to be an advance in a very large Osmundaceous type along lines of its own to dictyostely.

The leading facts relating to the fossil Osmundaceae may be tabulated as on p. 141.

From the facts contained in this table it follows: (i) that though there is no constant numerical proportion between size and the complexity of vascular construction still, speaking generally, the small fossils are simple, and the largest the most complex; (ii) that the progression in time runs substantially parallel with the progression in stelar complexity up to the Mesozoic age, at which time its highest point had been attained for the Family; and (iii) that the ordinary type of modern Osmundaceous structure was acquired in Jurassic time, and is there represented by examples which, though small as compared with other related fossils, correspond very nearly to the dimensions of the living species. But the largest fossils referred to the family became actually dictyostelic, a state not represented in living Osmundaceae. The phyletic application of these facts and conclusions is deferred till all the data material for that discussion have been considered.

The origin of the leaf-trace and its ontogenetic expansion into the characteristic C-shaped meristele has already been described for *Thamnopteris*

Name	Horizon	Diameter of xylem	Medullation	Character of xylem	Remarks
Zalesskya diploxylon	Upper Permian	7 mm.	Absent	Solid: central elements short and thin-walled	Xylem differentiated but continuous, *i.e.* not dictyoxylic
Zalesskya gracilis	Upper Permian	13 mm.	Absent	ditto	ditto
Thamnopteris Schlechtendalii	Upper Permian	11 mm.	Absent	ditto	ditto
Osmundites Gibbeana	Jurassic	2·5 to 4·5 mm.	?	Continuous ring of normal tracheae	Not dictyoxylic
Osmundites Dunlopi	Jurassic	5 mm.	?	Separate xylem-strands	Dictyoxylic as in modern Osmundaceae
Osmundites Kolbei	Wealden of Uitenhage (Neocomian)	?	Mixed pith	Over 50 separate strands	Dictyoxylic, but with "mixed pith" (compare *Osmunda*, I, Fig. 120)
Osmundites skidegatensis	Lower Cretaceous	24 mm.	Parenchymatous	Separate strands with internal phloem	Incipient dictyostely
Osmundites Carnieri	Uncertain, "between Jurassic and Tertiary"	33 mm.	Parenchymatous	Separate strands with internal phloem and endodermis	Complete dictyostely, but with Osmundaceous structure

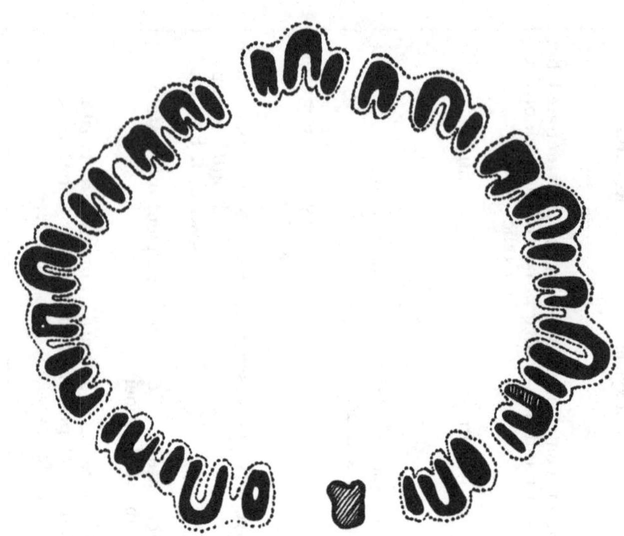

Fig. 431. *Osmundites Carnieri* Schuster. Arrangement of meristeles. The endodermis is shown by dotted lines. (After Kidston and Gwynne-Vaughan.)

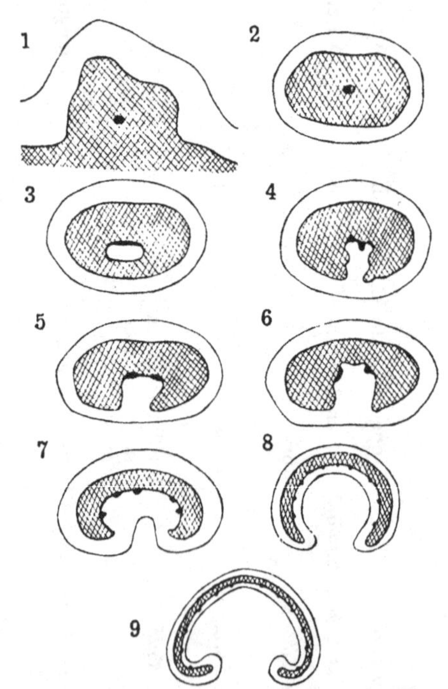

Fig. 432. Diagrams illustrating the departure of the leaf-trace in *Thamnopteris Schlechtendalii* Eich. (After Gwynne-Vaughan.) (Compare text.)

(Vol. I, p. 162). In that Mesozoic fossil the trace is at first mesoxylic (2, Fig. 432): it is only after forming an internal island of parenchyma (3) that it opens adaxially, so that the single protoxylem-group lies at the base of an involution (4). This is the condition shown by the leaf-trace of the living Osmundaceae at its first origin (compare Fig. 427, p. 136). In fact the most rudimentary stages seen in the fossil have been eliminated in the fossils of the later Mesozoic, and in the living representatives of the family. But the subsequent subdivision of the protoxylem in the departing leaf-trace, and the spreading of the whole trace into a C-shaped curve with the numerous protoxylems ranged along its concave face is seen in the same way in the living Osmundaceae as in *Thamnopteris*. The extra-marginal origin of the pinna-traces is probably a consequence of the strong curvature of the parent meristele. Finally the open venation of the blade accords with the proved antiquity of the Family.

The root-structure conforms to the ordinary Leptosporangiate type. As a rule the stele in the root is diarch, but in *Osmunda* triarch roots are occasionally found (Faull), and this may be held as a reminiscence of the more complicated structure common in Eusporangiate Ferns.

There are two types of hairs in the Osmundaceae, but both are composed only of a simple row of cells. Woolly hairs are found on the leaf-blade, but glandular hairs cover the widened leaf-bases. These have been studied by Gardiner and Ito (*Ann. of Bot.* i, p. 41). The secretion originates from the peripheral cytoplasm in a number of the cells of each hair: on access of water the accumulated mucilage swells, the cells burst, and the secretion forms a general protection for the leaf-base. We have learned by experience of many early Ferns that such simple uniseriate hairs are a primitive type, and these may also be held as such notwithstanding their specialised secretory function.

Spore-Producing Members

The variability in position of the sporangia, essentially marginal in *Osmunda* and superficial in *Todea*, has already been noted (Fig. 420, p. 131). In their development they are substantially alike whatever their position. The development differs from that of most Ferns in the variety of its details in different individual sporangia, even when they may be in close juxtaposition on the same pinnule. The sporangia fluctuate between two types, as shown by the details both in *Osmunda* and *Todea*: these are illustrated in Fig. 424, p. 133, which represents drawings from two actual sporangia of *Todea barbara*. In the one type the segmentation results in a square-based archesporial cell as seen in the Eusporangiate Ferns, the other shows the conical type characteristic of the Leptosporangiates. The latter is commoner in the Osmundaceae. The differences of individual detail start from the very first, as appears from Fig. 433, *A*, in which two sporangia are shown already differing in segmentation. The cells marked (×) do not compose the whole outgrowth: adjoining cells contribute to it, so that strictly speaking the whole sporangium is not referable in origin to a single cell: it is not truly Leptosporangiate. How various the subsequent cleavages may be will be seen from Fig. 433, *A* to *E*. A comparison of *C* and *D* also brings out clearly the very great differences in bulk occasionally seen in sporangia of the same age. The large cell in the centre divides usually by three anticlinal walls corresponding to those seen in the usual Leptosporangiate type, though

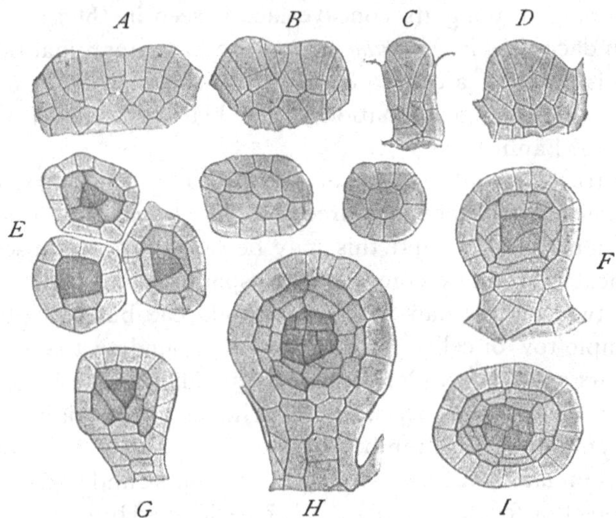

Fig. 433. *Todea barbara* Moore. *A*, small part of section of a pinnule showing two young sporangia (×, ×); *B*, *C*, *D*, examples of variety of sporangial segmentation, as seen in vertical sections; *E*, older sporangia cut transversely, showing difference in sporangia side by side; *F*, vertical section of a sporangium of like age, with square-based sporogenous cell; *G*, a similar sporangium with tetrahedral sporogenous cell; *H*, *I*, vertical and transverse sections of older sporangia. The central figures show two unequal sporangial stalks in transverse section. (All × 200.)

the cell which remains in the middle may still be either truncate or pointed at the base. But sometimes it appears that four lateral cells may be cut off by anticlinal walls, as in the largest sporangium in *E*, thus occasionally conforming to the Eusporangiate type. Then follows the periclinal division to separate the cap-cell. The archesporial cell thus surrounded by tissue which will form the sporangial wall undergoes further divisions to form the tapetum (*E*, *F*). Where the archesporium is truncate at the base the cell or cells below it take part in completing the tapetum (*F*). The division of the tapetum follows into two, or partially into three layers (*F* to *I*) together with the subdivision of the sporogenous cell to form the spore-mother-cells.

The drawings not lettered in Fig. 433 are added to show the differences of thickness and of segmentation seen in the sporangial stalks.

The origin and structure of the young sporangium in the Osmundaceae is thus seen to be more robust than in ordinary Leptosporangiate Ferns. This may be connected with the greater number of the spore-mother-cells produced, and consequently the greater potential output of the spores. In *Osmunda* Russow long ago estimated the number of spores in a single sporangium as over 500, and assumed therefore the number of spore-mother-cells to be 128 (*Vergleichende Untersuchungen*, p. 87, 1872). An estimate of the number of spores may be based on the number of spore-mother-cells seen in a median section (Fig. 434), which is 30 to 32. As the sporogenous mass is approximately spherical and the diameter of each cell about one-sixth that of the whole sphere, the total number of spore-mother-cells would approach 128. Actual countings of spores showed for *Osmunda* figures midway between 256 and 512, distinctly below Russow's estimate. In *Todea barbara* the numbers are often nearer to 256: but in *T. superba* and *hymenophylloides*, which are "filmy" in habit, the output is still lower, approximating in the last-named species to 128. The bearing of these facts will be discussed later: the main results readily accord with the relatively robust structure and the variability of the sporangial development.

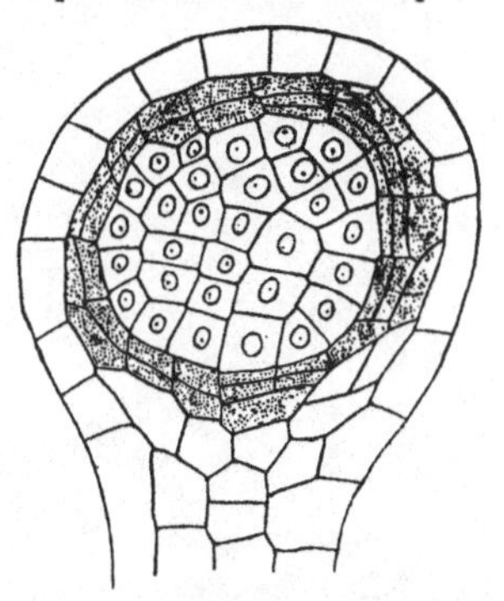

Fig. 434. Sporangium of *Osmunda regalis* containing a large sporogenous tissue surrounded by a tapetum consisting in parts of three layers of cells. (After von Goebel.)

THE GAMETOPHYTE[1]

The spores of the Osmundaceae contain chlorophyll, and soon lose their power of germination, the first stages of which are variable. An apical cell is soon established, as in most Leptosporangiate Ferns, but it is replaced as a rule later by a series of marginal cells. A downward-projecting midrib becomes conspicuous in the maturing thallus. The growth thus established may be continued for years, and the thallus may attain a length of four centimetres (Goebel). Forking occurs sometimes, and adventitious buds are often found, as in the Marattiaceae. The prothalli, though of the cordate type, are large, fleshy, and dark green in colour, resembling certain Liverworts. Altogether they are more massive than those of ordinary Leptosporangiate Ferns. The gametangia are essentially of the same type as these, but the antheridia produce a larger number of spermatocytes in rough parallelism with the large number of spores in the sporangium (see Vol. I, p. 292). The archegonia are borne on the sides of the projecting midrib, and Campbell

[1] See Campbell, *Mosses and Ferns*, 3rd edn., pp. 346–356, 1918.

describes their position as "horizontal" (*l.c.* p. 355). Their structure differs but little from that of Leptosporangiate Ferns: the central series of cells is constituted as in them, but each of the four rows of the neck consists of six cells in place of the usual four (Fig. 435).

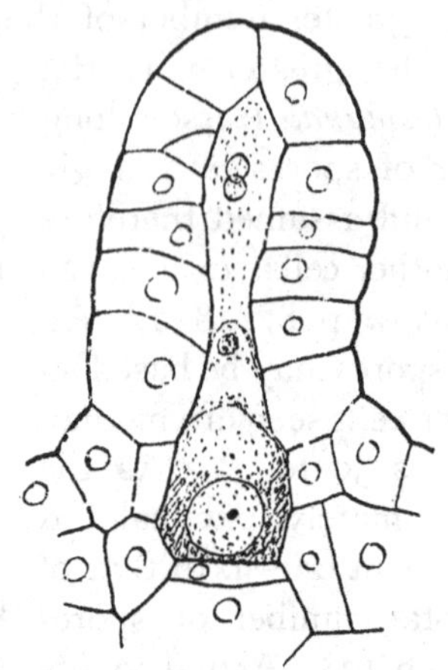

Fig. 435. Archegonium, nearly ripe, of *Osmunda cinnamomea*. (After Campbell.) (× 525.) This figure was wrongly named *Matteuccia struthiopteris* (× 250) in Vol. I, p. 288, Fig. 279.

The Embryo

Campbell (*l.c.* p. 356) gives the best account of the embryo of the Osmundaceae, which is developed from the horizontally directed archegonium. The first division of the zygote is by a basal wall parallel to its axis: but the quadrant walls are also parallel with it instead of being transverse, although their position with reference to the axis of the prothallium is the same: so that the embryo-quadrants and the organs derived from them are situated like those of the Polypodiaceous embryo with reference to the prothallium, though not to the archegonium. There is somewhat less regularity in the later divisions, and in this respect *Osmunda* appears to be intermediate between the Polypodiaceae and the Eusporangiate Ferns. Moreover, the embryo retains for a longer time than in the former its original nearly globular form, and the cotyledon does not break through the venter of the archegonium until later than in the Leptosporangiate Ferns, showing in this respect again a primitive character (see Campbell, *l.c.* Figs. 198–202). The foot is very large, and penetrates deeply into the prothallus, sometimes extending its superficial cells as haustoria, as in *Anthoceros*, or *Tmesipteris*. Frequently more than one embryo is initiated on one prothallus, but only one has been observed to come to maturity. Thus the embryo in the Osmundaceae, though it resembles that of the Leptosporangiate Ferns in its essential features, differs in many details from them, with a cumulative suggestion of a primitive character of the Family. Nevertheless, the embryo has the prone position, as against the erect habit of the Marattiaceous embryo.

COMPARISON

Ever since the distinction between Eusporangiate and Leptosporangiate Ferns was drawn by Goebel it has been growing more and more evident that the Osmundaceae hold an intermediate position between those broadly distinct types. It was shown relatively early that the general constitution of the sporophyte, as revealed by the apical segmentation, placed the Family in a middle position (Bower, 1884, 1885, 1891), and that the spore-output of the relatively large sporangium was high (Russow, 1872). The vascular structure of *Osmunda* was held to be relatively primitive by Zenetti (1895). But point was added to all such comparisons when Campbell definitely stated that the Eusporangiate type was relatively primitive for Ferns, and the Leptosporangiate derivative (1890), and still more when it was shown by Palaeontological evidence that the former antedated the latter in geological time (Bower, 1891). How important then becomes the position of the Osmundaceae in the whole system of Ferns when viewed phyletically, and how significant the fact that the Family dates back to the Palaeozoic Period (Kidston and Gwynne-Vaughan, 1907–1914). The Permian fossils referred to this Family have been described above. It is true that the fructifications have not been found in relation to such early stems as *Zalesskya* or *Thamnopteris*, and so the final proof of their correct reference to the Osmundaceae is wanting. But, as noted in the introductory chapter (p. 6), numerous sporangia showing in section a structure closely resembling those of living Osmundaceae existed in the Coal period, though possibly they may not have corresponded in detail to the fully elaborated Osmundaceous sporangia of the present day. The remarkable similarity existing between the Permian fossils and living Osmundaceae as regards the whole structure of the shoot (though the stelar state of the Palaeozoic fossils is simpler than in the later types) leaves little doubt of the general correctness of the reference. The Osmundaceous type may therefore be held to date from Palaeozoic times. It is incumbent on those who withhold their assent to produce evidence of the existence of Ferns having this construction of the shoot, but bearing fructifications that indicate some other affinity. Till this is done their objection does not appear decisive, and the conclusion may be accepted meanwhile as the probable truth.

It is open to us to support this probability by the wealth of comparative fact and argument which we now possess. It will be found that the evidence from such sources is quite consistent. The importance of this question to the general presentment of the phylesis of Ferns is so great that a full statement of the grounds for an opinion upon it seems desirable. The data will be arranged under the headings of the twelve criteria laid down in Vol. I, and summarised in the Introduction to this volume:

(1) The upright radial shoot, and the equal distal dichotomy of the Osmundaceae indicate a primitive state, as well as the winged leaf-base and commissure (*Todea*) comparable with those seen among the Coenopterids, Pteridosperms, and Cycads.

(2) The unstable structure of the apical meristems, the three-sided initial of the leaf-apex, and the massive structure of its wings: the fluctuation of the root-tip between the type with a single initial and that with four: and the hovering between the Eusporangiate and Leptosporangiate segmentation in the young sporangium all indicate a place between the Eusporangiate and the Leptosporangiate Ferns as regards cellular construction.

(3) The progression in the sporeling-leaves from equal dichotomy to dichopodial branching of the veins, and uniformly open venation are also primitive characters.

(4) The simple stelar structure, related to that of the protostele, and most markedly so in the Permian fossils, together with the undivided trace, and the mode of its origin in the early fossils, all indicate a primitive character.

(5) The simple uniseriate hairs form a feature common in primitive Ferns.

(6) The non-soral character, and the marginal insertion of the sporangia in *Osmunda* on narrow segments of the sporophyll, agree with what is seen in many Coenopterids; while the superficial insertion in *Todea*, corresponding to that of the Marattiaceae, may be interpreted by abnormal sporophylls of *Osmunda*, and is probably a derivative state consequent on expansion of the leaf-surface.

(7) The absence of indusial protections in the Osmundaceae is also shared by the Coenopterids.

(8) The relatively bulky sporangium, opening by a median slit, with a non-specialised mechanism of the annulus, corresponds to what is seen in the Coenopterids, and also in the Schizaeaceae and Gleicheniaceae.

(9) The spore-output per sporangium (512–128) is intermediate between the large figures for Eusporangiate and the smaller for Leptosporangiate Ferns.

(10) The relatively bulky prothallus is comparable with that of the Marattiaceae.

(11) The sexual organs project from the prothallus as in Leptosporangiatae, but the antheridia have a larger number of spermatocytes than in these Ferns, excepting the Gleicheniaceae; this feature runs parallel with their large spore-output.

(12) The embryogeny is prone as in Leptosporangiate Ferns, and the segmentation is on the same plan, but with differences in detail which

suggest a position between Polypodiaceae and Eusporangiate Ferns (Campbell).

The uniformity with which comparison in respect of each of these twelve criteria points to a primitive state for the Osmundaceae is impressive. All the characters indicate a position for the Family intermediate between the Eusporangiate and the Leptosporangiate Ferns. When the further fact is taken into account, that shoots characteristic of the Family in leaf-arrangement and stelar structure, and especially in the features of the leaf-trace, are recorded from the Permian Period onwards, the conclusion appears fully justified that the stock is a very ancient one. Its antiquity harmonises with the position thus assigned to it by comparison. The Osmundaceae may therefore be held to be a synthetic type between Eusporangiate and Leptosporangiate Ferns.

Nevertheless, the type is an isolated one. There is no Family of living Ferns closely related to them: but it is quite possible that as knowledge increases, affinities with other Palaeozoic fossils may be drawn closer than at present recognised. The Palaeontological history indicates an origin from a protostelic ancestry, and the nearest reference would be to the Botryopterideae, which the similarity of the position and structure of the sporangia would fully support. Starting from the small and protostelic state seen in *Botryopteris* or *Grammatopteris*, we may by comparison trace the probable steps of stelar elaboration, and mark at the same time its relation to size as conveyed by the diameter of the stele appended to each name. The Permian *Zalesskias* (7–13 mm.), and *Thamnopteris* (11 mm.), were of a fair size: though they retain a solid xylem-core this is differentiated into inner storage xylem, and outer conducting xylem, with protoxylem just as it has been found to be in *Diplolabis Römeri* by Gordon. The succeeding Jurassic fossils, *O. Gibbeana* (4·5 mm.), and *O. Dunlopi* (5 mm.) were smaller, showing dimensions about equal to those of living Osmundaceae, which also they resembled in their structure, being dictyoxylic: but *O. Kolbei* (20 mm.), referred to the Neocomian, was not only very much larger, but it is shown to have possessed a dictyoxylic stele with "mixed pith." Unfortunately the state of preservation of the previously named fossils did not allow of examination of their pith-structure. The stelar condition of the Jurassic Osmundaceae corresponds to that of most of the living Osmundaceae, while the state of *O. Kolbei* is matched by the abnormal state of *O. regalis* shown in Vol. I, Fig. 120. The next step is seen in *O. skidegatensis* (24 mm.) of Lower Cretaceous time, a beautifully preserved fossil, showing separate xylic strands, with internal phloem, and incipient dictyostely. This is matched, but without the complete foliar gaps, by the living *O. cinnamomea* (4·5 mm.). The difference in size between this species and the large fossil probably accounts for their absence: for in the largest of the fossils,

O. Carnieri (33 mm.), full dictyostely of an Osmundaceous stele with regular foliar gaps is reached, a fact which accompanies, and is perhaps causally related to, its large size. Its stele is seven times the diameter of that in *O. cinnamomea* (see Fig. 431, p. 142). The march of these successive steps in geological time, together with the progression in size, leaves little doubt as to the stelar story which they convey. The ancient family started from Palaeozoic protostely: it reached its structural and dimensional climax in late Mesozoic time: the living representatives appear to have stood still both in size and in structure at the state attained in Jurassic times, excepting *O. cinnamomea* and occasionally *Todea hymenophylloides*, which reflect structurally the state of the Cretaceous fossils. The facts, apart from all theory, do not suggest any degradation of structure, such as has been suggested on grounds of purely anatomical comparison without any relation to other lines of enquiry. The living Osmundaceae, when examined in relation to all their features, appear to record in their anatomy steps in an upgrade elaboration of the stele, though the most advanced of those steps were already attained in the Mesozoic Period. Such a conclusion accords with the position assigned to them on general grounds of comparison, as taking an intermediate place between the Eusporangiate and the Leptosporangiate Ferns.

The two (or three) surviving genera of the Osmundaceae suggest in their sporophylls the origin of two different states which have figured conspicuously in the later evolution of the Ferns. *Osmunda* retains the marginal or approximately marginal position of the sporangia upon its narrow pinnules, which was characteristic of the Botryopterideae: but *Todea* bears its sporangia superficially upon the lower surface of its broader pinnules. Intermediate states are seen abnormally in *Osmunda* (Fig. 420), and these suggest that the superficial position is derivative, and consequent upon an increase in surface of a primitively narrow sporophyll (Vol. I, Chapter XII). The difference thus seen between the two living genera may be held to prefigure two conditions which become stereotyped in distinct phyletic lines. These will subsequently be designated the Marginales and the Superficiales. The former is represented among early Ferns by the Schizaeaceae, and the latter by the Gleicheniaceae. But the Osmundaceae, consistently with the position which they hold both in the fossil history and as shown by a comparative study of the Class as a whole, include both types. In this again they appear to be a synthetic Family.

It will not be necessary to dwell upon the evidence from spore-numbers per sporangium in the Osmundaceae. It puts into actual figures the intermediate character of these Ferns between the Eusporangiate and the Leptosporangiate types, with an indication of a more primitive state than in any living Schizaeaceae; but there is an equivalence with some Hymenophyllaceae and Gleicheniaceae, though the *Mertensia*-types of the latter may show an even larger output. The figures speak for themselves when compared with those of the whole range of the Filicales.

There remains, however, the outstanding question of the sporangia of those early fossils which have been ranked with the Osmundaceae on anatomical grounds. Their leaves and sporangia are mostly unknown as yet, so far as actual connection with the stocks is concerned. But it is a significant fact that a number of fertile leaves are known, from the Coal Period onwards, bearing sporangia the structure of which is such as to accord generally with that of the modern Osmundaceae. Examples are seen in the Carboniferous fossil *Kidstonia* (Vol. I, Fig. 249, *B*), described by Zeiller (*Bull. Soc. bot. de France*, xliv, p. 195, 1897), and in *Boweria* from the Barnsley Coal, fully investigated and figured by Kidston (see Fig. 310, Introduction, p. 7)[1]. The former has a massive annulus composed of several rows of cells, and is clearly a relatively primitive type of sporangium. In *Boweria* the sporangia are isolated and marginal, being seated on the ends of the veinlets of the pinnule-segments. The annulus consists of two slightly irregular rows of cells, which pass as a band across the apex of the sporangium, and extend a very short distance down the sides. Kidston compares these sporangia with those styled *Pteridotheca* by Scott (*Progressus*, Vol. i, p. 183, 1906: also *Studies*, 3rd. Edn. Vol. i, p. 265): and he remarks, "The plants included in the genus may possibly be referable to Ferns." The alternative would be that they are Pteridosperms. If the two genera last named were actually Ferns, their sporangia and sporophylls would supply a prototype from which by condensation of the pinnule-segments, and simplification of the annulus, the type seen in the modern Osmundaceae would readily follow. Indeed Zeiller has concluded that "the sporangia of *Kidstonia heracleensis* show the nearest affinity to those of the Osmundaceae": but they are not without analogies with those of *Senftenbergia* and *Lygodium*. Thus *Kidstonia* would be grouped with the Osmundaceae, but at the outer fringe of the family, forming a link between them and the Schizaeaceae. A careful re-examination of the sporangial structure in the living Osmundaceae would help in the elucidation of this question, which at the moment must be left as an open one.

It remains to consider the probable relations of the living genera. The first question relates to the validity of *Leptopteris* as a substantive genus. Its filmy habit appears distinctive, though in other features the species fall clearly within the genus *Todea*. A filmy habit is not held elsewhere as a sufficient basis for generic distinction. The filmy Danaeas and Aspleniums are not separated from those substantive genera. Following the practice

[1] "Végét. houil. Hainaut Belge," *Mém. Mus. Roy. d'Hist. nat. de Belgique*, iv, p. 51; and *Fossil Plants of the Carboniferous Rocks of Great Britain*, iv, p. 291, Plates LXXI, LXXII, 1923.

adopted in Hooker's *Synopsis Filicum*, it appears best to recognise *Leptopteris* as a sub-genus of *Todea*, leaving only two substantive living genera in the Family. Neither of these appears decisively the more primitive as regards form or anatomical structure: but the marginal placentation of *Osmunda* is certainly primitive, while the superficial placentation of *Todea* and the filmy texture of *Leptopteris* are both held to be derivative features. The genera may then be placed thus:

OSMUNDACEAE

I. SPORANGIA TYPICALLY MARGINAL.
  (1) *Osmunda* (Linnaeus, 1753) ... ... 9 species.
II. SPORANGIA TYPICALLY SUPERFICIAL.
  (2) *Todea* (Willdenow, 1802)
    § i. *Eu-Todea*, Baker, 1868 ... ... 1 species.
    § ii. *Leptopteris*, Baker, 1868 ... ... 7 species.

BIBLIOGRAPHY FOR CHAPTER XXI

406. EICHWALD. Lethaea Rossica. Vol. i. Stuttgart. 1860.
407. RUSSOW. Vergleichende Untersuchungen. p. 87. 1872.
408. BOWER. Phil. Trans. Pt. ii, p. 565. 1884.
409. BOWER. Quart. Journ. Micr. Sci. Vol. xxv, p. 75. 1885.
410. GARDINER & ITO. Ann. of Bot. Vol. i, p. 41. 1887.
411. RACIBORSKI. Engler's Jahrb. Vol. xiii, p. 1.
412. CAMPBELL. Bot. Gaz. Vol. xv, p. 1. 1890.
413. BOWER. Ann. of Bot. v. p. 109. 1891.
414. ZENETTI. Bot. Zeit. p. 72. 1895.
415. ZEILLER. Bull. Soc. bot. de France. xliv, p. 195. 1897.
416. FAULL. Bot. Gaz. p. 381. 1901.
417. ENGLER & PRANTL. Natürl. Pflanzenfam. i. 4. p. 372. 1902. Here the systematic literature is fully quoted.
418. JEFFREY. Phil. Trans. cxcv, p. 119. 1902.
419. SEWARD & FORD. Linn. Trans. vi. p. 250. 1903.
420. BOWER. Origin of a Land Flora. p. 530. 1908, where the literature to date was fully quoted.
421. FAULL. Trans. Can. Inst. viii, p. 515. 1909.
422. SEWARD. Fossil Plants. ii, p. 324. 1910.
423. SINNOTT. Foliar Gaps in the Osmundaceae. Ann. of Bot. xxiv, p. 107. 1910.
424. KIDSTON & GWYNNE-VAUGHAN. Fossil Osmundaceae. Trans. Roy. Soc. Edin. No. i, 1907. No. ii, 1908. No. iii, 1909. No. iv, 1910. No. v, 1914.
425. GWYNNE-VAUGHAN. Ann. of Bot. xxv, p. 525. 1911.
426. P. BERTRAND. Progressus Rei Bot. iv, p. 189. 1912.
427. Mme M. IOSSA. L'appareil conducteur des Osmundacées et Gleicheniacées. Genève. 1914.
428. CAMPBELL. Mosses and Ferns. 3rd. Edn. p. 346. 1918. Here the literature is fully quoted.
429. SCOTT. Studies in Fossil Botany. 3rd. Edn. Part i, p. 264. 1920.
430. BOWER. Address on Size. Proc. Roy. Soc. Edin. p. 1. 1920.
431. STOPES. *Osmundites Kidstoni*. Ann. of Bot. xxxv, p. 55. 1921.
432. POSTHUMUS. *Osmundites Kidstoni*. Ann. of Bot. xxxviii, p. 215. 1924

CHAPTER XXII

SCHIZAEACEAE

THIS Family includes, according to Christensen's Index, four living genera, *Schizaea* with 25 species, *Lygodium* with 26 species, *Mohria* with 3, and *Anemia* with 64 species. It is of wide distribution, but chiefly within the tropics. *Schizaea* and *Lygodium* are widely diffused, but *Anemia* is almost confined to tropical America, while *Mohria* is restricted to the Cape and adjoining southern islands. In point of time certain fossil genera, such as *Senftenbergia*, *Klukia*, and perhaps *Kidstonia*, have been referred to this affinity. There may be a question as to the correct inclusion of *Senftenbergia* in this Family, and its consequent reference back to the Carboniferous period: but as the Jurassic *Klukia* cannot be held in doubt, this establishes the existence of the Schizaeaceous type early in Mesozoic time. Whereas in the Marattiaceae and Osmundaceae the radial type of shoot prevails, in the living Schizaeaceae there is a pronounced leaning towards a dorsiventral habit. The upright radial type appears it is true in *Schizaea*, in *Mohria*, and in most species of *Anemia*. Frequently, however, the stock is more or less oblique, while in §*Anemiorrhiza* it has a creeping habit. The extreme is seen in *Lygodium*, which has a horizontal underground rhizome with bifurcate branching, and it bears its leaves alternately upon its upper surface, in two nearly coincident distichous rows (Fig. 436). But where the axis is upright or oblique the leaves are disposed in a dense spiral.

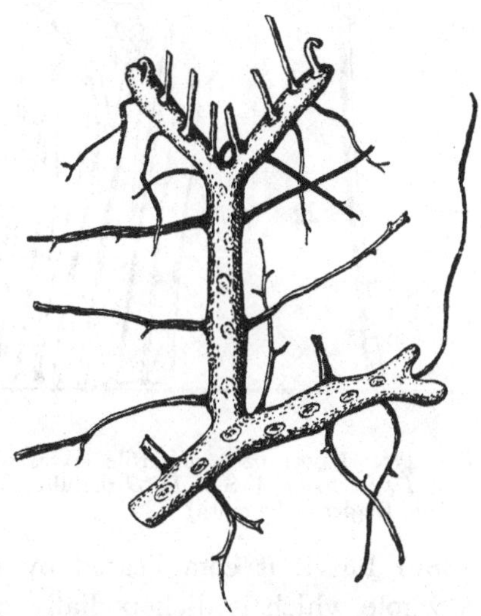

Fig. 436. The dorsiventral and dichotomous rhizome of *Lygodium scandens*, with leaves apparently in a single row. (After Velenovsky.)

The leaves show great diversity of detail in the different genera, but all are referable in origin to distal dichotomy of a long narrow stipe. In *Schizaea* there is very marked and repeated equal forking: the shanks may be more or less completely webbed together below, and they bear the fertile segments on their distal ends (Fig. 437). In *Lygodium* also the leaf-architecture has been traced by Prantl to repeated dichotomy (*Die Schizaeaceen*, Leipzig,

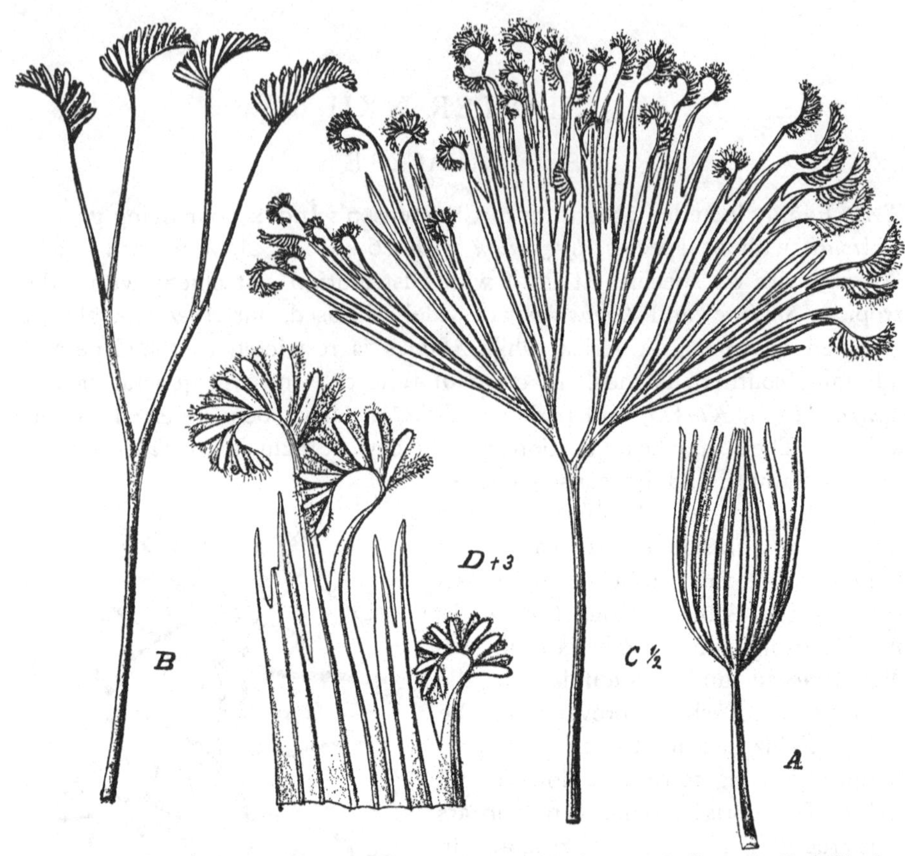

Fig. 437. Upper parts of fertile leaves of *Schizaea*. *A*, *S. pennula* Sw. *B*, *S. bifida* Sw. *C*, *D*, *S. elegans* J. Sm. In *D* the ultimate segments are more highly magnified. (After Diels, from Engler and Prantl.)

1881), but it is complicated by continued apical growth of the leaf as a whole which is dichopodially developed, the lower sterile pinnae being often broad, and the distal fertile pinnae narrower (Fig. 438). The whole leaf may attain a length of 100 feet or more: it acts as a prehensile climber, reaching very considerable heights in scrub and low forest (Vol. I, p. 35, Fig. 43). In *Anemia* and *Mohria* the leaves are of more ordinary type: they are less complex, and the reference to dichotomy is not so clear owing to dichopodial development (Fig 439). The venation is usually open, but some species of *Lygodium* and of *Anemia* show vein-fusions near to the margin. In *Mohria* the sporangia may be distributed over the whole length of the fertile leaf: in *Schizaea* and *Lygodium* the fertile region is distal: but in *Anemia* as a rule only the two lowest pinnae are fertile, thus suggesting Roeper's well-known theory of the origin of the fertile spike in *Ophioglossum* (Fig. 440). Dermal appendages are present in all the genera, and in all except *Mohria* they are simple and filamentous, as they are in the Botryopterideae and Osmundaceae, and many other primitive Ferns.

The sporangia are not grouped in sori, but they are solitary, a number of them being borne on each fertile segment. In *Schizaea* and *Anemia* they

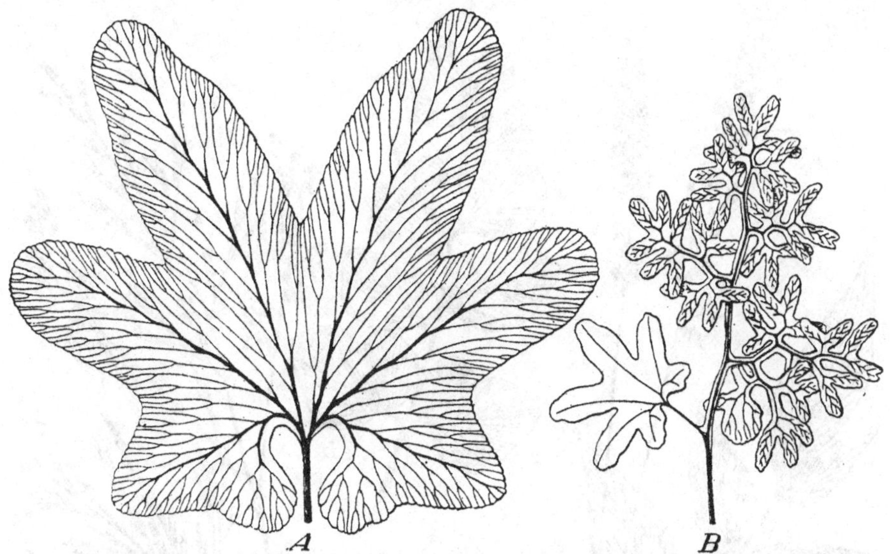

Fig. 438. *Lygodium palmatum* Sw. Leaves, showing their venation. *A*, sterile: *B*, fertile. (After Prantl, from Engler and Prantl.)

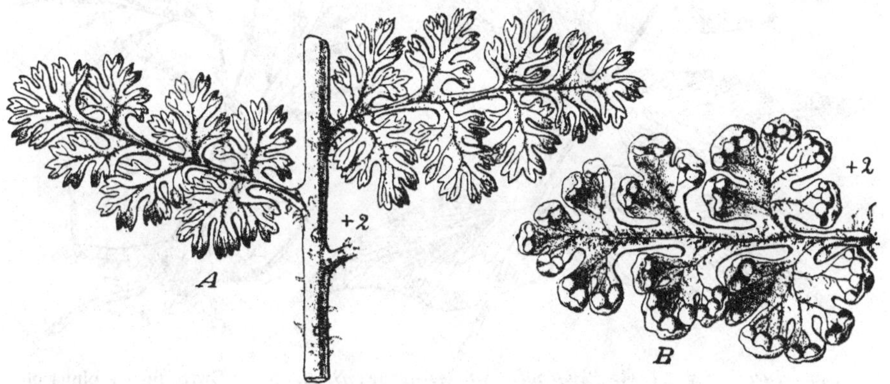

Fig. 439. *Mohria caffrorum* (L.) Desv. *A*, part of a sterile leaf: *B*, a fertile pinna. (After Diels, from Engler and Prantl.)

appear disposed when mature in regular rows, one on each side of the midrib, on the lower surface of the fertile segments. In *Lygodium* and *Mohria* one is seated near each vein-ending (Fig. 441). They may be protected by the curled margin of the pinnule, as in *Mohria* and *Schizaea*: or there may be a special protective growth comparable as regards superficial origin to the indusium of the Hymenophyllaceae, which completely covers each sporangium, as in *Lygodium*. Prantl recognised each solitary sporangium as constituting what he called a "monangial sorus." It will be seen from the

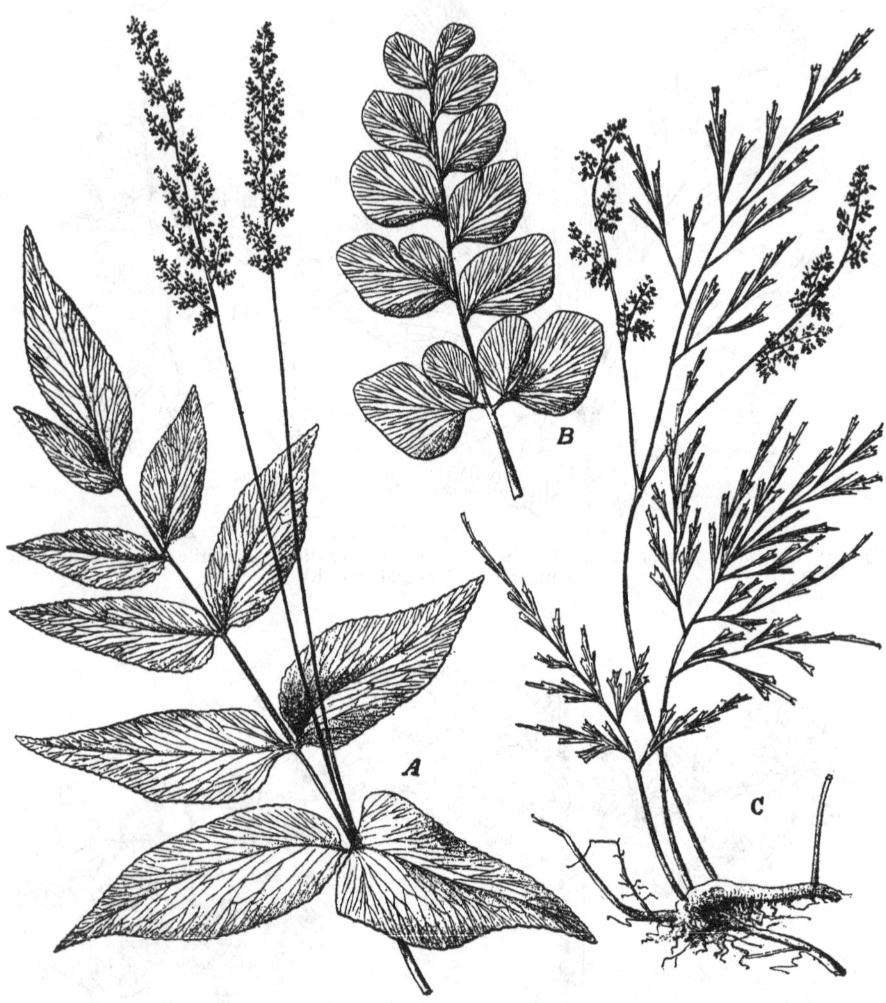

Fig. 440. *Anemia* Sw. *A*, *A. Phyllitidis* Sw., leaf-blade: *B*, *A. aurita* Sw., primary pinna of a sterile leaf: *C*, *A. cuneata* Kze., Habit. (After Diels, from Engler and Prantl.)

development that in all the genera the origin of the sporangia is strictly marginal, while the protective flaps are secondary developments formed after the initiation of the sporangia, by outgrowths not truly marginal but superficial.

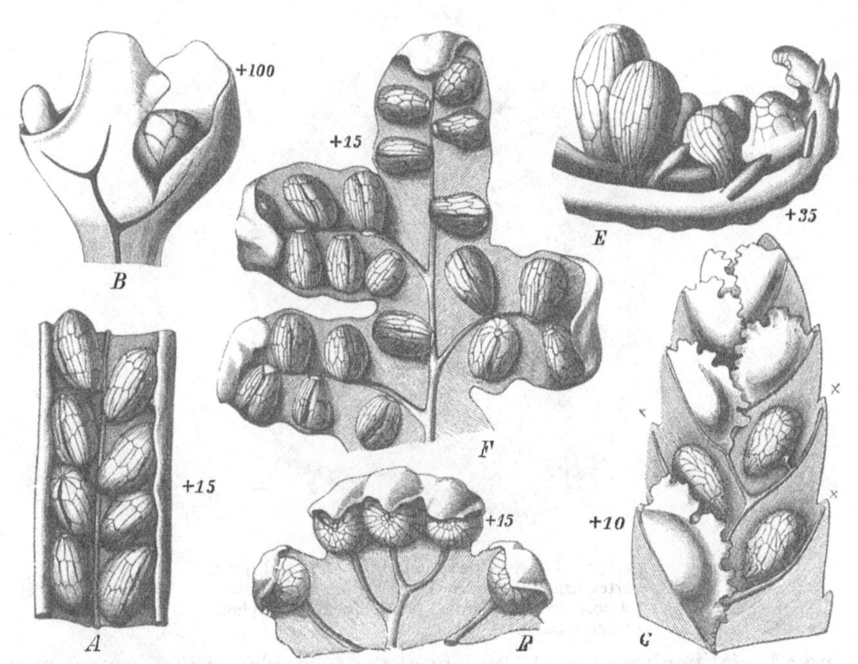

Fig. 441. Disposition of the sporangia of the Schizaeaceae. *A*, *S. dichotoma* J. Sm., part of a fertile segment (sorophore). *B*, *C*, *Lygodium japonicum* Sw. *B*, apex of a young fertile segment. *C*, mature fertile segment, at (×) the flaps have been removed so as to display the sporangia. *D*, *Mohria caffrorum* (L.) Desv., segment of a fertile pinna. *E*, *F*, *Anemia Phyllitidis* Sw. *E*, side view of a young fertile segment. *F*, fertile segment from below. (*A*, *B*, *E*, after Prantl, *C*, *D*, *F*, after Diels, from Engler and Prantl.)

ANATOMY

The Schizaeaceae have great diversity in the habit of the shoot, from the creeping rhizome with its laxly disposed leaves to the ascending or upright stock with leaves crowded and densely spiral. The internal structure also shows marked differences, which follow the external form. In fact the Family is as varied in its stelar structure as that of any in the whole Class of Ferns. The simplest adult structure is found in *Lygodium*, for there the rhizome is traversed by a protostele with solid xylem consisting of tracheides and scattered parenchyma, surrounded by phloem, pericycle, and endodermis. There is no typical protoxylem: the first-formed tracheides are finely scalariform, and are scattered round the periphery of the xylem-core (Fig. 442). (Boodle, *Ann. of Bot.* p. 359, 1901; and p. 511, 1903.) The petiole is traversed by a single strand, which comes off from the protostele with only superficial disturbance of it: in fact the adult plant retains the simple relation between the protostele and the trace which is seen in the sporeling. The foliar strand in the climbing petiole of *Lygodium* is almost cylindrical, with bays of phloem protruding into the xylem (Vol. I, Fig. 165, *g*). There

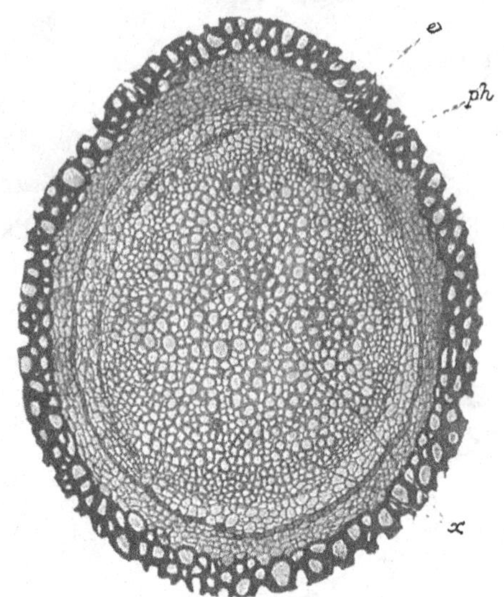

Fig. 442. Transverse section of the stele and inner cortex from the rhizome of *Lygodium dichotomum*. (× 50.) *e* = endodermis; *ph* = phloem; *x* = xylem. (After Boodle.)

are no adaxial hooks, while the xylem of the two sides of the curve is completely fused. This is probably a derivative form from the more usual C-shaped type: in fact it may be held as condensed in relation to the climbing habit (Gwynne-Vaughan, *Ann. of Bot.* p. 495, 1916).

In stems of young plants of *Schizaea* or *Anemia* there is at the very base a small protostele with a solid core of tracheides, and without parenchyma. But sooner or later a mass of parenchymatous cells is found to occupy the centre of a ring of tracheides, and the state may be described as solenoxylic (Fig. 443, iii of *Anemia*, and vi of *Schizaea*). From this ring of xylem a sector may pass to each leaf, opening a xylic gap, and with its phloem it passes off as a foliar trace (Fig. 443 iv of *Anemia*, and v of *Schizaea*), but without any opening of the endodermis. In the adult stem of a large *Schizaea* (*S. dichotoma*), an approach is seen to a more complex state, but without attaining to a settled solenostely. Fig. 444 shows a reconstruction of such a stem by Dr J. M. Thompson. The central column is sclerotic pith, in which occasionally an endodermal island, or a tract showing tracheid-structure, may be seen (near top of figure). Shallow endodermal pockets of very unequal depth occur at the foliar gaps, but do not constitute by their connection any continuous inner endodermis. The stelar structure found in this stem suggests a tentative upgrade advance, in a primitive type of Fern, from a protostelic towards a solenostelic or a dictyostelic structure. Essentially similar details have been observed in other species, showing that this irregular state is normal for the genus. Since tracheides have been seen in the

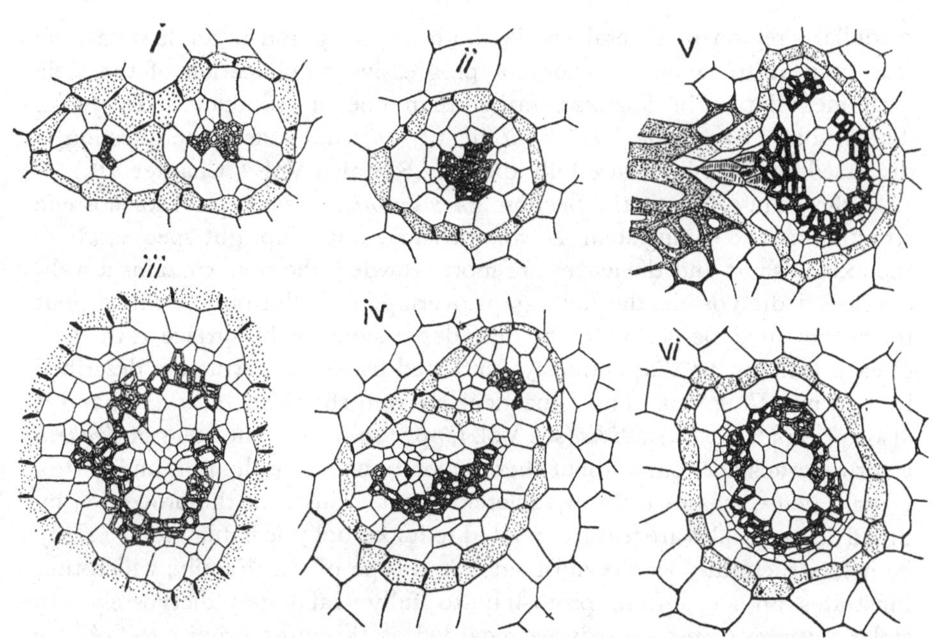

Fig. 443. i–iv. Transverse sections, in succession from below, of the axis of a young plant of *Anemia Phyllitidis*. i, the protostele gives off a leaf-trace without any interruption of the endodermis: ii, protostele higher up: iii, still higher, with central pith: iv, same stele giving off a leaf-trace, again without interruption of the endodermis; note isolated tracheide in the pith.

v, vi. Transverse sections of adult stele of *Schizaea rupestris*. vi shows continuous xylem round central pith: v shows departure of a leaf-trace, to the left a root-trace. The endodermis is dotted. (× 150.)

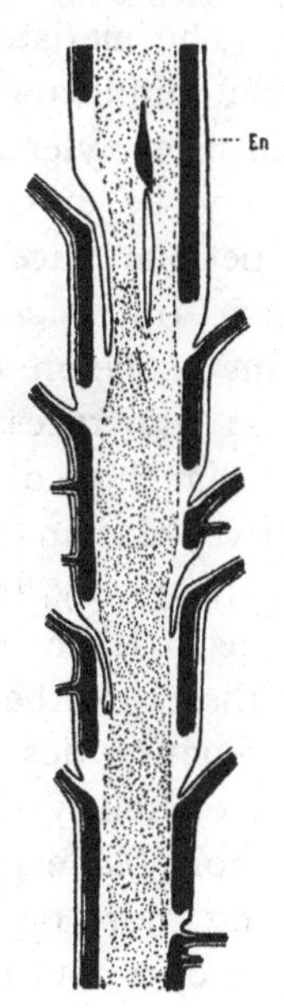

Fig. 444. *Schizaea dichotoma*. Plan of stelar construction in an adult stem. Outer endodermis, *en*; phloem cross-hatched: xylem black: sclerenchyma dotted: parenchyma white. There are endodermal pockets of varying depth, and isolated spindles of inner endodermis and wood in the sclerenchyma. (After J. McL. Thompson.)

medullary region in several species both in young and in adult stems, the facts appear to favour a theory of progressive amplification of the stele with medullation in *Schizaea*, rather than one of reduction. (Dr J. M. Thompson, *Trans. Roy. Soc. Edin.* lii, p. 715, 1920: here the question is discussed, and the literature fully quoted. See also Vol. I, Chapter VII.)

In the genus *Anemia* the Section *Anemiorrhiza* has a typical solenostelic structure in its creeping stem (*A. adiantifolia*): but in upright species, where the axis is short and the leaves are more crowded, the stem contains a well-developed dictyostele, the foliar gaps overlapping. Not only is this so, but their stems may be perforated by deep depressions of the surface at the axil of each leaf, giving that curiously ventilated basket-like structure described by Gwynne-Vaughan. This is particularly well shown in *A. hirsuta*, and it also appears in *A. Phyllitidis* (see Vol. I, pp. 149–150). The stem of *Mohria* is also dictyostelic, but without these involutions, while the pith and cortex contain dense masses of brown sclerenchyma. This with the protection by broad dermal scales are features related to its xerophytic habit. *Mohria* may be held structurally as the most advanced type of the Family, which thus illustrates all steps from protostely to fully elaborated dictyostely. In stelar features *Lygodium* may be regarded as the most primitive, *Schizaea* as holding a middle position, while *Anemia* and *Mohria* are the most advanced.

The stelar advance is carried out with a consistently undivided leaf-trace of a very characteristic structure, as seen in *Anemia* and *Mohria* (Fig. 445). It may be regarded as based upon the underlying C-shaped trace of the Osmundaceae. The meristele has a peculiar saddle-shaped outline, much attenuated in the middle region (compare *Loxsoma*, Vol. I, Fig. 158). There are four exarch protoxylems, and the metaxylem is much enlarged right and left: it may be extended at the margins into adaxial hooks. In *Lygodium* the petiolar trace is on the same general plan, but it is contracted as already noted. In *Schizaea* the trace is peculiarly three-lobed, but with the protoxylems as before on the adaxial face (Fig. 165, *b*, Vol. I). The interest of these leaf-traces lies first in their constantly undivided state: next in the fact that those of *Anemia* and *Mohria* conform closely to those of certain derivative Marginales (*Loxsoma*, *Plagiogyria*, *Davallia speluncae*, Fig. 159, Vol. I): thirdly, in the peculiar modifications to which they are liable in specialised leaves such as *Lygodium* and *Schizaea*, which are still referable nevertheless to the same general scheme.

Lastly, the venation deserves to be noted. In *Schizaea* it is simple, or repeatedly and obviously dichotomous, with blind endings and no vein-fusions. The whole blade presents a near approach to the primitive equal dichotomy, being more or less completely webbed. *Mohria* also has the venation always open, but here the strongly dichopodial branching makes

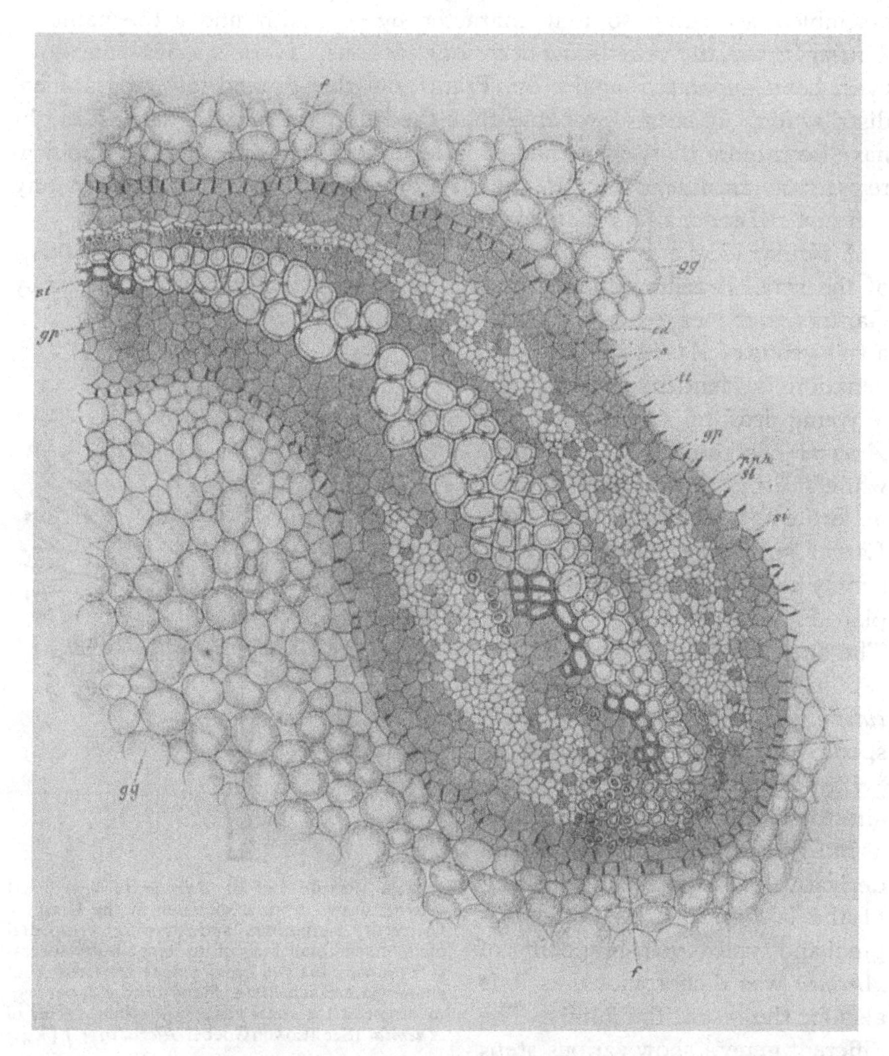

Fig. 445. Half of a meristele from the petiole of *Anemia Phyllitidis*, seen in transverse section. Compare that of *Davallia speluncae*, Vol. I, Fig. 159. *ed*, endodermis: *s*, parenchymatous sheath: *gp*, ground parenchyma: *p.ph*, protophloem: *si*, sieve-tubes: *f*, fibres: *st*, spiral tracheides: *tt*, scalariform tracheides: *gg*, ground tissue. (× 80.) (After Prantl.)

its dichotomous origin less obvious than in *Schizaea*. The other two genera have as a rule open dichotomous venation with varying degrees of dichopodial development. But both run into a low degree of reticulation in certain species with especially wide pinnules. The strongly dichopodial development of the unlimited leaf of *Lygodium* leads to the lateral position of the alternating pinnae and pinnules, and it is only in the final branchings that equal dichotomy asserts itself (Fig. 438, p. 155). In three species of *Lygodium*, assembled according to that character by J. Smith under the name of *Hydroglossum*, the veins show occasional fusions. Those species have, however, been separated again by Prantl on the ground of geographical distribution. It seems probable that the adoption of vein-fusions might have been more than once effected independently in different geographical regions, as an amendment in the structure of a broadening leaf. In any case the existence of a reticulum is regarded as a derivative state.

A similar view may be entertained for *Anemia*. The simplest leaf is that of the small Brazilian species distinguished generically as *Trochopteris* by Gardner, but merged by Prantl as a sub-genus of *Anemia*. The open venation is identical with that of a young leaf of *Osmunda*, or of *Anemia* (compare Fig. 416, *f* and *g*), while the two lowest pinnae may be fertile as in the latter (Fig. 446). Larger species of *Anemia* exhibit simply a further elaboration of that plan (Fig. 440; also Fig. 77, Vol. I). The open venation is retained, except in the broad-leaved *A. Phyllitidis* Sw., and some less-known species grouped by J. Smith as §*Anemidictyon*. Here occasional anastomoses give a coarse reticulum, as in *Lygodium*; probably a late and derivative consequence of widening of the leaf-area. Clearly the original and typical vein-branching of *Anemia* was dichotomous, as it is also for the rest of the family. The different genera show various steps from equal dichotomy to pronounced dichopodial development, and from single-veined segments to advanced webbing. The former is seen very perfectly in *A. millefolia*, where the sterile blade is highly cut so that each lobe contains only a single vein, after the manner of *Sphenopteris*. Webbing appears in various degrees in *Schizaea*, while occasionally it has led in *Lygodium* and *Anemia* to the further step of vein-fusion. It is thus possible to trace in the leaf-architecture indications which lead the eye to recognise, within the family, signs both of the simple beginnings as simple-veined bifurcation, and of the climax as a reticulate expanded blade.

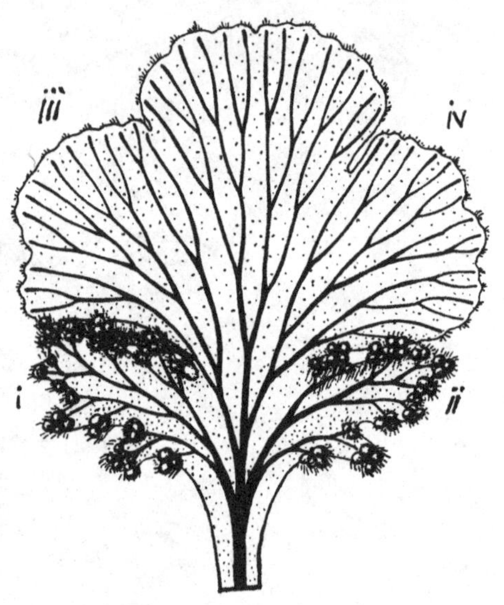

Fig. 446. Fertile leaf of *Anemia* (*Trochopteris*) *elegans*, drawn from a specimen in the Glasgow University herbarium, and showing sympodial dichotomous branching of its veins: i–iv = successive pinnae: the two basal pinnae are fertile, but sporangia are sometimes found on the upper segments or at the distal end, as in other species of *Anemia*. (See also von Goebel, *Flora*, 1915.) (× 4.)

The origin of the pinna-trace is uniformly marginal in the Schizaeaceae, a fact that accords with their primitive position, and the near relation of their leaf-architecture to simple dichotomy.

The dermal appendages are simple hairs in *Lygodium*, *Schizaea*, and *Anemia*, sometimes with glandular terminal cells. But in *Mohria*, which is distinguished by advanced dictyostely, by a type of leaf which corresponds to that seen in many advanced Leptosporangiates, and by a relatively low spore-output from each sporangium, the hairs are replaced by broad scales, each with a distal glandular cell. This state may have been developed in relation to its dry habitat: but it is significant that the more advanced structure of the scale goes along with an advanced state in respect of other features also.

SPORE-PRODUCING MEMBERS

In the Schizaeaceae the sporangia are solitary. Prantl recognised each of them as constituting a "monangial sorus." The purist may regard this as a contradiction in terms: to the student of Fern-phyletics it suggests a primitive state reminiscent of the single distal sporangium, as in *Botrychium* or *Stauropteris* (Vol. I, p. 208). Prantl accurately worked out the development in all the living genera, and found the sporangia to arise in acropetal order on each fertile segment. He ascribes to them an origin from cells of the marginal series, each with a terminal position on a fertile vein. In fact, the position is that which would naturally follow as a consequence of the webbing of the segments which the study of leaf-architecture discloses, so as to form a flat leaf-expanse. Doubt has been cast upon the correctness of Prantl's facts by Diels (*Nat. Pflanzenfam.* i, 4, p. 360): but more recently all the genera have been re-examined, with the result that Prantl's observations have been upheld (*Lygodium*, Binford, *Bot. Gaz.* xliv, p. 214; see Vol. I, Fig. 235: *Schizaea*, Bower, *Ann. of Bot.* xxxii, p. 1, 1918: *Anemia*, Stevens, *Ann. of Bot.* xxv, p. 1059: *Mohria*, Bower, *Ann. of Bot.* xxxii, p. 9). A consequence that follows will be that the indusial protecting flanges or flaps, which differ in detail in the four genera, all appear as accessory surface-growths of the supporting leaf-segment. They arise simultaneously with, or subsequent to, the sporangia which they protect: but growing strongly they force the marginal sporangia into an apparently superficial position. A knowledge of the development is therefore essential for the proper understanding of the morphology of the sporophyll.

The fertile segments of *Lygodium* appear as strobiloid terminals to the sterile region of the leaf (Figs. 438, *B*, and 441, *B*, *C*). Each solitary sporangium is enveloped in a sort of pocket formed on the upper side by what appears to be part of the surface of the blade, on the lower by a membranous flap. It is seated near to the end of a vein which, passing beneath its stalk, terminates in a minute tooth at the margin of the pinnule (Fig. 441, *C*).

The sporangia alternate right and left of the midrib. The development shows that the origin of the sporangium is from a four-angled cell of the marginal series (Fig. 447, *A*). Almost immediately after the identity of the sporangium is established by segmentation the growth of the indusium begins (Fig. 447, *B*, *C*, i). This is at first semi-circular, and the two sides of

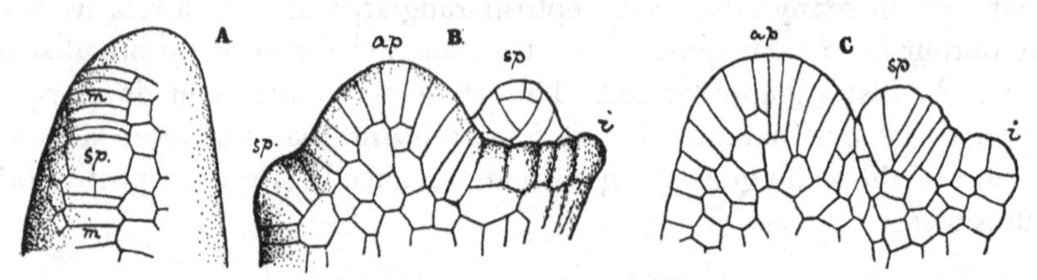

Fig. 447. Origin of the marginal "monangial" sorus of *Lygodium*. *A* = margin of young pinnule in surface view: *m* = marginal cells: *sp* = sporangium. *B* = a similar pinnule in surface view: *ap* = apex: *i* = indusium. (*A*, *B*, after Prantl.) *C* = a similar pinnule in section. (After Binford.)

it differentiate as it develops; the upper more robust lip takes the character of the blade with which it appears to be continuous: the lower remains membranous. The consequence is that the originally marginal sporangium appears as though attached to the lower surface of the adult segment (Fig. 441, *C*).

A similar slide from a marginal to a superficial position appears in *Schizaea*. The fertile leaf is circinate, but with the pinnae reflexed (Fig. 448, *A*, *B*). If

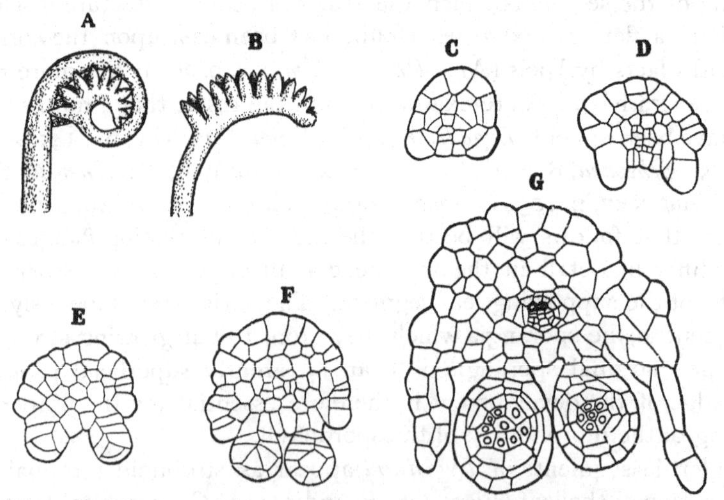

Fig. 448. *Schizaea rupestris*. *A*, *B*, young leaves showing circinate vernation, with pinnae reflexed to the convex (abaxial) side. (× 3.) *C*, *D*, *E*, transverse sections of very young pinnae of *Schizaea rupestris*, showing the marginal origin of the sporangia, which very soon turn inwards towards the lower (abaxial) surface. In *D*, *E*, the indusial flaps are appearing right and left. *F*, *G*, similar sections of older pinnae with sporangia and indusia more advanced. (× 60.)

sections be cut of these when very young the characteristic segmentation of the sporangia is seen in the marginal cells (Fig. 448, *C*, *D*); but very soon strong growth and division spring up in adjoining cells, forming a false margin of indusium (*E*, *F*, *G*), by which the sporangia are forced towards the midrib, and assume a falsely superficial position. One row of them appears on each side of the midrib in §*Eu-Schizaea*, and §*Lophidium* (Fig. 441, *A*). But in §*Actinostachys* there are four rows. This is a consequence of a re-adjustment during development of a single series on each side. The two rows arise from a single marginal row of cells, but as the sporangia grow they are alternately pressed out of line right and left, and consequently two rows of them appear on each side of the adult pinnule (Fig. 449). The origin of

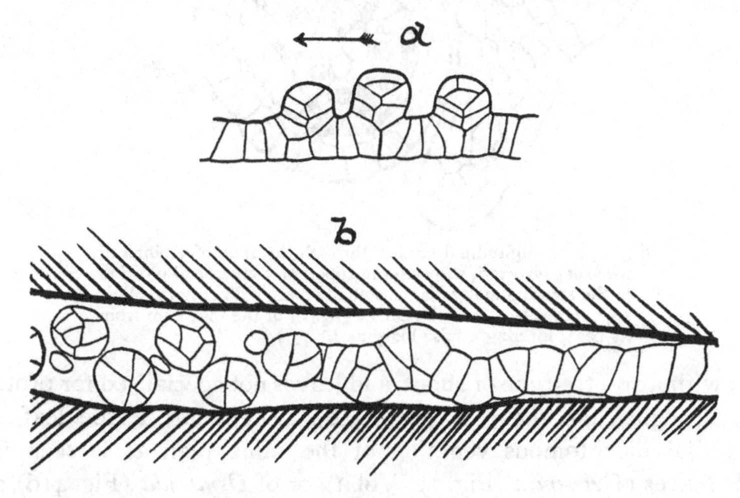

Fig. 449. *a*, *b*. Longitudinal sections following the marginal series of cells of the pinna of *Schizaea digitata*, and the sporangia that arise there. In (*a*) they are cut vertically, in (*b*) horizontally. The latter shows the origin of the two rows of sporangia from a single margin. (×85.)

the sporangia is also proved to be marginal in *Anemia*, but as the pinnule is flatter they are more exposed than in *Schizaea*, appearing falsely superficial as before (Prantl, *l.c.* p. 39; also Stevens, *Ann. of Bot.* xxv, p. 1059). Perhaps the most interesting genus of all is *Mohria*, for here a single sporangium is formed near to the end of each vein, while a marginal flap grows from below it, and covers it like a separate indusium (Fig. 441, *D*). Here again the development shows that the origin of the sporangium is marginal, but it is thrust to an apparently superficial position by a strong growth below the stalk on the upper side. The result is shown in Fig. 450.

Thus it appears for all the genera that the origin of the sporangium is actually marginal. In order to bring the leaves of the Schizaeaceae into relation with those of other primitive Ferns the simplest type of sporophyll may be used, viz., that of the small Brazilian species *Anemia* (*Trochopteris*)

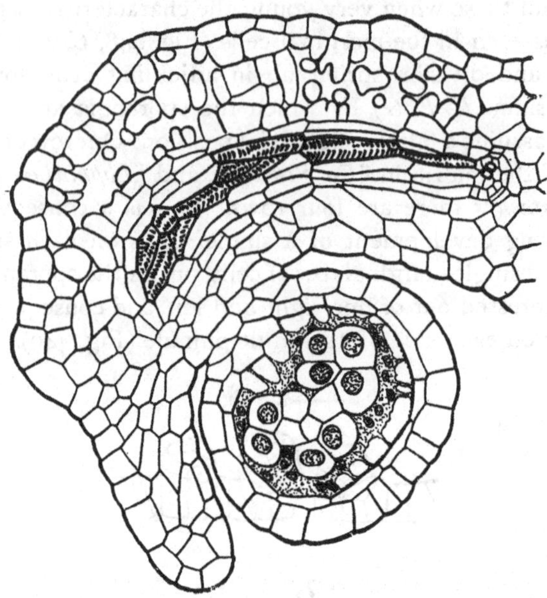

Fig. 450. Longitudinal section through the apex of a pinna of *Mohria caffrorum*, showing the apparently superficial position of the sporangium, which is actually marginal in origin. This appearance is due to the active growth of the indusium from its base, forming a false margin. (× 75.)

elegans, with a rosette-type of shoot, and leaves not specialised for protection against drought, like the rest (Fig. 446). The adult fertile leaf shows a sympodial dichotomous venation of the same plan as is seen in the juvenile leaves of *Anemia* (Fig. 77, Vol. I), or of *Osmunda* (Fig. 416), and it closely resembles also that of *Botrychium* and *Helminthostachys* (Vol. I, Fig. 81). This plainly proves that the architecture of the leaf is based on equal dichotomy, and not very far advanced from that state sympodially. Fig. 446 also shows the sporangia grouped, as in *Anemia*, at the ends of the veins of the two broad basal pinnae. Assuming that their origin is marginal like the rest of the family and of the genus, while they are sometimes solitary as they constantly are in *Mohria*, this simple leaf is itself a near approach to that primitive leaf contemplated in an archetypic Fern (Vol. I, Chapter XVII, p. 339). A lateral webbing of a dichotomous leaf, each branch or vein of which bore a distal sporangium, would give such a leaf as *Mohria*, excepting for the secondary displacement and protection of the solitary sporangia: or it would give that of *Anemia elegans* if the number of the distal sporangia were increased by interpolation along the margin. This is the hypothetical reading of such sporophylls of the Schizaeaceae which naturally follows from the study of their development, together with a comparison of other primitive types of Ferns (Goebel, *Anemia elegans*, *Flora*, Bd. 108, p. 319).

The sporangia are large and sessile, or in *Lygodium* shortly stalked, and they are annulate. The annulus in living Schizaeaceae is usually uniseriate, though occasionally it is more complex. It is contracted towards the apical end of the sporangium, and there is a definite stomium (Fig. 451). However

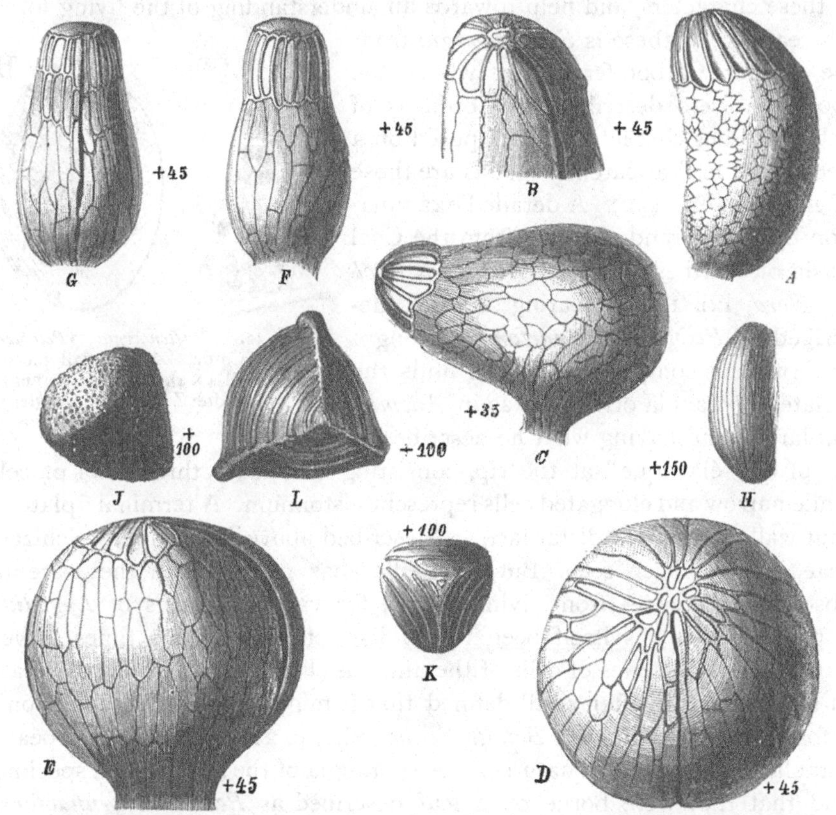

Fig. 451. Sporangia and spores of the Schizaeaceae. *A*, *B*, *Schizaea pennula* Sw. *A*, seen laterally: *B*, the tip seen obliquely from above. *C*, *Lygodium japonicum* Sw., seen laterally. *D*, *E*, *Mohria caffrorum* (L.) Desv. *D*, seen from above: *E*, laterally. *F*, *G*, *Anemia Phyllitidis* Sw. *F*, view from midrib: *G*, from margin of pinnule. *H–L*, spores of the Schizaeaceae. *H*, *Schizaea pennula* Sw. *J*, *Lygodium japonicum* Sw. *K*, *Mohria caffrorum* (L.) Desv. *L*, *Anemia fulva* Sw. (All but *K* after Prantl: *K*, after Diels, from Engler and Prantl.)

contracted the annulus of these Ferns may appear, there is in its centre an apical group of cells designated by Prantl the "plate," which may better be called the "distal face" of the sporangium: it is an important feature for comparison with other Ferns. Sometimes it consists of only one cell (*Lygodium* (*C*), *Schizaea* (*A*, *B*)): sometimes of many (*Anemia* (*F*, *G*), *Mohria* (*D*)). Usually its cells are thin-walled, but in *Mohria* they are thick-walled and irregular (*D*). The rest of the sporangial wall is thin. The sporangia of *Mohria* are radially constructed, the apex and base being opposite (*D*, *E*). In the other genera the sporangia are more or less curved, so as to be dorsiventral: this curvature is slight in *Anemia* and *Schizaea* (*A*, *B*, *F*, *G*), but very marked in *Lygodium* (*C*).

Certain fossils which have been referred to this affinity stamp the antiquity of these characters, and help towards an understanding of the living forms. The earliest of these is *Senftenbergia* from the Upper Carboniferous, of which the sporangia were described by Corda as of Schizaeaceous character, and disposed on a Pecopterid leaf, apparently just as are those of *Anemia* (Fig. 452). A detailed examination by P. Bertrand of a slab from the Coal Basin of Northern France (*Ann. Soc. Geol. du Nord*, lxi, 1912), bearing a fossil described as *Pecopteris pennaeformis* Brongn., confirms this comparison: for he finds the isolated sporangia orientated as in *Anemia*, but larger, and having what he describes as a "plaque élastique" at the tip, consisting of two to three rows of cells, while narrow and elongated cells represent a stomium. A terminal "plate" of thin-walled cells, or "distal face" as described above for the living Schizaeaceae, has not been seen. But, as Zeiller has pointed out, these are not absolute differences from living forms, for various species of *Lygodium*, a genus which has itself been traced back to Cretaceous times, have a partially double series of cells of the annulus (Fig. 454, *A*), while the "plate" in living forms is often so ill defined that it might well escape detection in a fossil (Zeiller, *Bull. Soc. Bot. de France*, xliv, p. 217). The facts appear to establish the Schizaeoid nature of the sporangia of the new French specimen, and that they were borne on a leaf described as *Pecopteris pennaeformis* Brongn. (P. Bertrand, *Ann. Soc. Geol. du Nord*, lxi, p. 222, 1912).

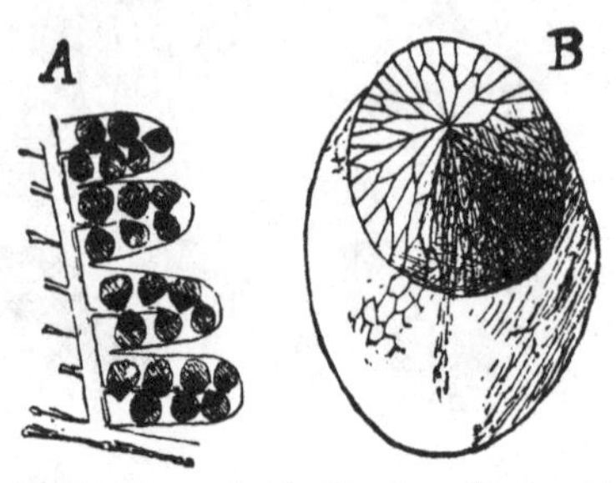

Fig. 452. *Senftenbergia* (*Pecopteris*) *elegans* Corda. *A* = a small piece of sporophyll (× 4). *B* = a sporangium (× 35). (After Zeiller, from Engler and Prantl.)

Whether or not these Carboniferous fossils are really to be included in the Family, there can be little doubt of the correctness of the reference of *Klukia* to the Schizaeaceae. It represents the fructification of a Pecopterid from the Jurassic, and several species have been described by Raciborski (Engler's *Jahrb.* xiii, p. 1, Taf. 1). Here the arrangement of the sporangia, and their structure down to the single series of cells of the annulus and the line of dehiscence, are as in *Schizaea* (Fig. 453). In both of the genera of fossils above mentioned the sporangia appear as though intra-marginal, on the lower surface of the pinnule; but there is no distinction of fertile from sterile pinnules, nor any special indusial protection. From such comparisons it appears that the Schizaeaceous type is an ancient one, and that a Pecopterid foliage in early days bore its characteristic sporangia. Probably in their development these sporangia were also marginal as in the living Schizaeaceae, and had originally a more complex annulus than is usually seen in those of the present day. In this connection the fossil *Kidstonia* from the Coal Measures of Asia Minor has a special interest (Fig. 454). This, as well as other genera, may be com-

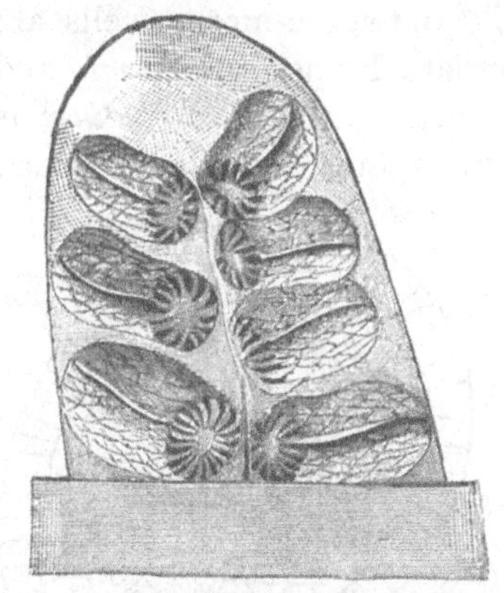

Fig. 453. *Klukia exilis* (Philipps) Raciborski. Fertile pinnule of the last order, seen from below. (× 20.) From the Jurassic of Krakau. (After Raciborski, from Engler and Prantl.)

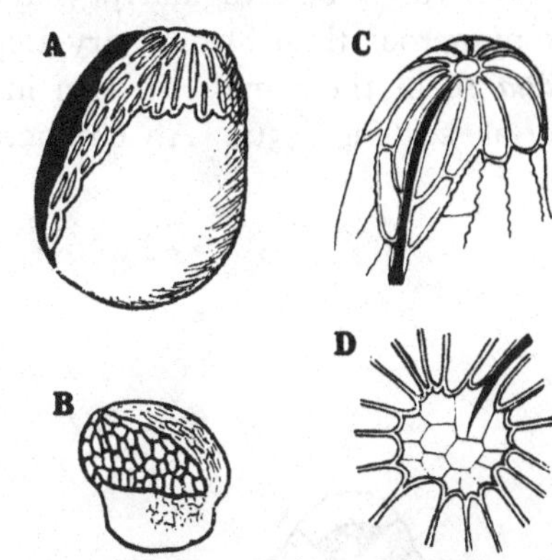

Fig. 454. *A* = a sporangium of *Lygodium lanceolatum* Desv. (× 50), showing an annulus of more than a single row of cells. *B* = *Kidstonia heracliensis* Zeiller, lateral view of sporangium. (× 50.) *C* = *Schizaea*, apex of sporangium with distal wall ("Platte") of many cells. *D* = ditto of *Anemia*. (*A*, *B*, after Zeiller: *C*, *D*, from Engler and Prantl.)

pared also with the Osmundaceae on the ground of sporangial structure (Seward, *Fossil Plants*, ii, pp. 325, 340). They are characterised by having a less specialised, but still an indurated, annulus. Pending a knowledge of further details they may probably be best referred to an indeterminate position among such early Ferns: but their existence appears to strengthen the comparisons of the Schizaeaceae downwards with the Osmundaceous, and ultimately with some Botryopterid or Zygopterid type. (Compare Zeiller, *Bull. Soc. Bot. de France*, xliv.)

On the other hand, certain fossils of later time have been referred tentatively to the Schizaeaceae. For instance, *Tempskya Knowltoni* Seward, from the Lower Cretaceous of Montana. Professor Seward has discussed this in detail, together with various other fossils, such as *Ruffordia* from the Wealden, which may be assigned provisionally to a like affinity (*Ann. of Bot.* p. 485, 1924).

The development of the Schizaeaceous sporangium follows the usual Fern-type in its main features, the essential parts of each, though perhaps not the whole of it, being referable to a single parent-cell with a rectangular base. In all the Schizaeaceae the first segment-wall extends from the outer to the inner periclinal wall of this cell: such a segmentation is characteristic of the more robust types of Fern-sporangium (Fig. 455). This event is followed by two other anticlinal divisions as seen in section, and then comes the periclinal division which separates off the cap-cell from the archesporium. The former gives rise to the greater part of the sporangial wall, while the lower segments complete the wall, and form the short thick stalk. From the archesporium the usual tapetum and sporogenous group arise: the tapetum may sometimes be irregularly 3-layered, while the spore-mother-cells are more numerous than in ordinary Leptosporangiate Ferns. In *Anemia* and *Mohria* 16 of them may be seen in a single section, and in *Lygodium* as many as 20 (Fig. 456). An enumeration of the spores actually produced

Fig. 455. Diagram showing the segmentation in a young sporangium of the Schizaeaceae. The first segment-wall meets the basal (periclinal) wall of the parent-cell: but the second (×, ×) meets the first, and does not extend to the base of the parent-cell.

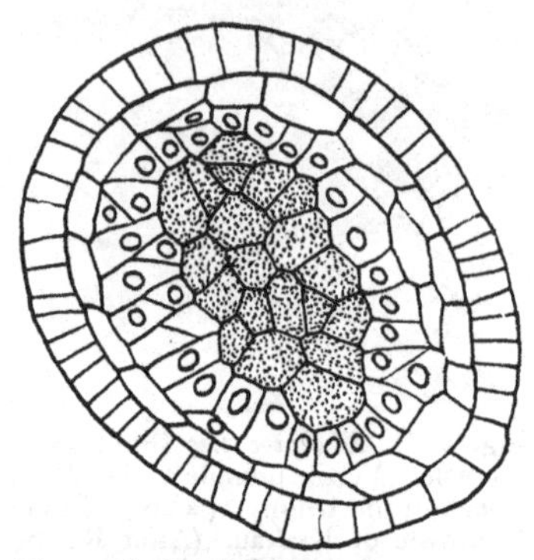

Fig. 456. Section through a sporangium of *Lygodium circinatum*. (After Binford.) 20 spore-mother-cells are cut through; the tapetum is more than doubled. (×480.)

shows that in *L. javanicum* and *dichotomum* the numbers indicate a typical number of 256, while in *L. pinnatifidum* there are only 128. There is thus a difference between species of the same genus, as has already been seen in *Todea*: the lower figure is shared by *Schizaea*, *Mohria*, and *Anemia*. Thus the numbers approach those found in the Osmundaceae: and the largest number appears in *Lygodium*, a genus where the anatomy is more archaic than in the rest of the Family.

THE GAMETOPHYTE

The spores of the Schizaeaceae vary in type. In *Lygodium*, *Anemia*, and *Mohria* they are tetrahedral, but in *Schizaea* they are wedge-shaped. Their walls are marked by more or less projecting ridges, but they are without any deposit of perispore. On germination the same three genera form flattened prothalli of the usual type, but with lop-sided growth, and a lateral cushion (Bauke, *Pringsh. Jahrb.* Bd. xi, 1878). This is not so, however, with *Schizaea* (Britton and Taylor, *Torr. Bot. Club*, xxviii, 1901: Thomas, *Ann of Bot.* xvi, p. 165: Goebel, *Organographie*, Aufl. ii, p. 957: Campbell, *Mosses and Ferns*, 3rd Edn., p. 384). In *S. pusilla*, *rupestris*, and *bifida* it has been seen that the prothallus is filamentous, bearing on short branches peculiar swollen cells, having pale chromatophores, and fungal infection; from these rhizoids spring (Fig. 457). This suggests mycorhizic

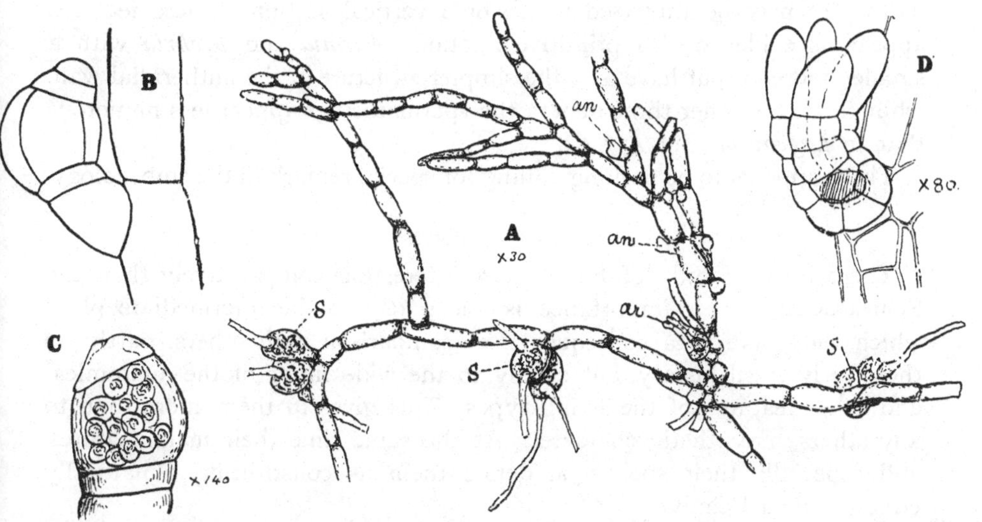

Fig. 457. A=a prothallus of *Schizaea pusilla*, bearing spherical cells (s) with endophytic fungal filaments: ar=archegonia: an=antheridia. B=segmentation of an antheridium of *Schizaea rupestris*; the lid is not always divided. C, D=mature antheridium and archegonium of *Schizaea pusilla*. (A, C, D after Britton and Taylor. B after von Goebel.)

nutrition. The antheridia and archegonia are solitary and distal on special branches. These prothalli, so different from the ordinary type, do not assume the filamentous form as a direct consequence of growth under diminished intensity of light, though this is the effect on many ordinary prothalli. Nor does exposure to stronger light induce them to take a flattened form. As Goebel remarks: "they are blind—they persist in a state which in related forms under favourable conditions exists only as a juvenile stage rapidly passed over." They are in point of fact the simplest prothalli known among the Pteridophyta. They suggest a primitive state, and provoke comparison with green Algae.

The archegonia are of the ordinary Leptosporangiate type, but with straighter necks, a feature of primitive Ferns. The antheridia are peculiar in having their first segmental wall flat instead of funnel-shaped, which accords with the filamentous stalk in *Schizaea*, and it is retained in *Anemia*. There is, however, another feature which differentiates *Lygodium* from the rest of the Family. Schlumberger states that in its antheridium the cap-cell is divided, as it habitually is in the most primitive Ferns: and one of the resulting cells is thrown off on dehiscence (*Flora*, p. 396, 1911). This apparently trivial feature acquires importance as occurring in the antheridium: for its complexity follows that of the sporangium (Vol. I, pp. 290–292). Here *Lygodium*, which has the larger sporangium and spore-output, has also the more complex antheridial wall, and Miss Twiss states that there is a large number of the sperms in each antheridium (156): while 32 spermatocytes may be traversed in a single vertical section. These features appear as evidence of a primitive relation. *Anemia* and *Mohria* with a smaller spore-output have also the simpler structure of the antheridial wall, while in all the other three genera the spermatocytes appear less numerous than in *Lygodium* (Fig. 457, *C*).

There appears to be nothing calling for special remark in the embryology.

COMPARISON

There is no Family of Ferns more interesting comparatively than the Schizaeaceae. Their importance is due partly to the intermediate place which they take as a consequence of comparison with others, partly to their early fossil history, but chiefly to the wide range of the anatomical and other features of the living types. This gives to them, more than to any others, a synthetic character. At the same time their main features, and especially their sporangia, define them as constituting a naturally circumscribed Family.

In their external form they present both the upright radial, and the creeping dorsiventral habit, with distal dichotomy of the axis. In their leaf-architecture they show a wider latitude than any other Family: they pass from equal dichotomy, through all grades of sympodial to extreme dichopodial development, finding their climax in the prehensile leaves of *Lygodium* which may extend to 100 feet in length. But even their most specialised and advanced foliar types are still referable to modifications of dichotomy. From the point of view of anatomy these Ferns comprise the permanent protostele of *Lygodium*: the medullated and otherwise modified protostele of *Schizaea*, the solenostele of *Anemiorrhiza*, the dictyostele of *Mohria*, and even the basket-type of dictyostele of *Anemia Phyllitidis*. Thus all the leading grades of modification of stelar structure are exemplified within the Family. The undivided leaf-trace, based upon the C-shaped trace of the Osmundaceae, is always retained; but it is sometimes greatly modified as in *Lygodium*. The venation of the blade, though open as a rule, is sometimes coarsely reticulate (*Lygodium*, *Anemia*), while the fertile segments show in their various forms obvious evidence of protection against drought. The dermal appendages are as a rule simple hairs, as is usual in primitive Ferns, but *Mohria*, in many respects an advanced type of the Family, bears flattened scales, thereby confirming its relatively advanced position. At the same time it shows how this apparently minor feature follows other more important structural characters, and supports conclusions derived from them. The whole Family is held naturally together by the position and structure of the sporangia. Though the form and detail of the sporophylls may appear to vary in high degree, they all are constructed on the same fundamental plan. The solitary occurrence, and uniformly marginal origin of the sporangia, are regarded as primitive features: nevertheless, the sporangia may in various degrees be thrust down to the lower surface as they mature, a position that is already conspicuous in the early fossils, naturally seen only in the adult state. The protective flaps that invest the individual sporangia in modern forms are absent in Palaeozoic and Mesozoic fossils: but those seen in the Schizaeaceae of the present day offer peculiarly suggestive comparisons with the indusia of the Dicksonieae and Hymenophyllaceae, thus linking with the more advanced Marginales.

The sporangia themselves are uniform in their general features, such as the short thick stalk, and the more or less spherical capsule with longitudinal dehiscence, operated by the distal annulus. This is least specialised in the ancient *Senftenbergia*, which compares on the one hand with the Palaeozoic Ferns of Osmundaceous affinity (Zeiller), and on the other through *Lygodium* with the modern types having a single ring of annular cells. In living forms the segmentation of the young sporangium and its massive stalk link the Family on the one hand with the more primitive types, and on the other with the more advanced Leptosporangiates: a comparison which is further strengthened by the spore-output per sporangium. *Lygodium*, already marked off as relatively primitive by other features, has a typical number of 256 spores, corresponding to the simpler Osmundaceae and Gleicheniaceae: but the Schizaeaceae mostly have a typical number of 128, which is lower than that of most Osmundaceae and other Simplices, though it is in excess of the number in most Leptosporangiate Ferns. These soral and sporangial characters point clearly to a position among the Simplices, but indicate the Schizaeaceae as intermediate between them and Leptosporangiate Ferns.

With such indications drawn from the sporophyte it becomes a question what interpretation is to be put upon the features seen in the gametophyte, and especially upon that filamentous type of it seen in *Schizaea*. Goebel (*l.c.* p. 956) by speaking of the prothallus of *Schizaea* as "blind" to more intense light, and as persisting in the juvenile filamentous state, appears to hold it as primitive, and assent may be given to this view. These are the simplest prothalli among the Pteridophytes, while the distal position of the antheridia and archegonia on short branches suggests at once Algal comparisons.

It is thus seen how interesting is the position of the Schizaeaceae in phyletic comparison. Their undoubted antiquity, and their clear affinity on the one hand with the Simplices and on the other with those Gradate Leptosporangiates that have marginal sori, are wholly in accord with that unusual combination of primitive and advanced features which is disclosed by the study of their living representatives.

The living genus which presents on the whole the most archaic features is *Lygodium*, with its protostelic axis, and its large sporangia with ill-defined annulus sometimes irregularly doubled, and its relatively large spore-output from each. But its leaf is highly elaborated and specialised for climbing, and it occasionally shows reticulate venation. The protection of the sporangia by elaborate indusial growths is doubtless in accord with their exposure at the heights to which the sporophylls climb in the forest. It has been noted above that the output of sperms from each antheridium is large.

Schizaea, on the other hand, has more primitive leaves, with prevalent equal dichotomy and constantly open venation. The stelar structure shows irregular but incomplete advances from the simple protostele, while the sporangia without highly specialised protection have a more perfect annulus, with a diminished output of spores. The gametophyte of *Schizaea* is, however, unique in its simplicity, and the output of sperms from each antheridium is small.

Anemia is distinctly more advanced in the general features of the sporophyte. Its radial construction accords with the dictyostelic skeleton of most of the species, and is led up to by the solenostelic §*Anemiorrhiza*. The leaf is pronouncedly dichopodial, closely resembling that of *Osmunda* or *Botrychium* in its younger stages, but occasionally progressing to reticulation. The sporangia are relatively small and with lower spore-output, but with specialised annulus and enlarged distal face. The output of sperms from the antheridium is relatively low (Kny, Berlin, 1869). The effect of these features is to place *Anemia* in a position of advance.

This position is shared by *Mohria*, which also has a dictyostelic axis with highly dichopodial leaves. The sporangia, however, each with its indusial protection, have a less specialised annulus, but a low spore-output. The dermal protection is by scales in place of the simple hairs which are seen in the rest of the Family.

It is plain that the characters thus summarised do not run parallel: it is therefore difficult to seriate the genera according to them, so as to bring out any single probable line of descent. Taking them cumulatively, however, it may be held that *Lygodium* is on the whole a relatively primitive type, though curiously specialised in accordance with its climbing habit: while *Mohria* is a relatively advanced type, with specialisation in relation to dry conditions: and this conclusion is supported by the detailed, almost trivial-seeming structural fact that in *Anemia* and *Mohria* the cap-cell of the antheridial wall is undivided after the Leptosporangiate type, while in *Lygodium* it is divided, and the output of sperms is large, thus conforming to what is seen in more primitive Ferns. A rough sequence of the living genera may then be given thus:

i. *Lygodium* (Swartz, 1801) ... ... 26 species.
ii. *Schizaea* (Smith, 1795) ... ... 25 species.
iii. *Anemia*[1] (Swartz, 1806) ... ... 64 species.
iv. *Mohria* (Swartz, 1806) ... ... 3 species.

BIBLIOGRAPHY FOR CHAPTER XXII

433. ENGLER & PRANTL. Natürl. Pflanzenfam. i, 4, p. 356, 1902. Here the literature up to date is fully cited.
434. PRANTL. Unters. z. Morph. d. Gefasskrypt. ii, Die Schizaceen. Leipzig, 1881.
435. CAMPBELL. Mosses and Ferns. 3rd. Edn. p. 384, 1918, with full quotation of the literature.
436. SEWARD. Fossil Plants ii, pp. 286, 346, etc., 1910, with full quotation of the literature on fossils.
437. BAUKE. Prothalli. Pringsh. Jahrb. Bd. ix, 1878, also Bot. Zeit. p. 769, 1878, and Bot. Zeit. 1880, Plates 1—6.
438. BRITTON & TAYLOR. Prothalli, Bull. Torr. Bot. Club. xxviii. 1901.
439. HEIM. Flora, p. 329. 1896.
440. THOMAS. Prothallus, Ann. of Bot. xvi, p. 165.
441. SCHLUMBERGER. Flora, p. 396. 1911.
442. GOEBEL. Organographie, Aufl. ii, p. 957.
443. RACIBORSKI. *Klukia*, Engler's Jahrb. xiii, p. 1.
444. ZEILLER. Bull. Soc. Bot. de France, xliv, p. 217.
445. P. BERTRAND. *Pecopteris pennaeformis*, Ann. Soc. Geol. du Nord, lxi, p. 222. 1912.
446. BINFORD. Sporangia of *Lygodium*, Bot. Gaz. xliv, p. 214. 1907.
447. STEVENS. Sporangia of *Anemia*, Ann. of Bot. xxv, p. 1059.
448. TANSLEY & CHICK. Structure of *S. malaccana*, Ann. of Bot. xvii, p. 493. 1903.
449. BOODLE. Anatomy, Ann. of Bot. p. 359, 1901, and p. 511. 1903.
450. JEFFREY. Anatomy, Phil. Trans. B, p. 128. 1902.
451. GWYNNE-VAUGHAN. Lattice-work stem, New Phyt. p. 211. 1905.
452. TWISS. Proth. of *Anemia* and *Lygodium*, Bot. Gaz. p. 178. 1910.
453. GWYNNE-VAUGHAN. Petiole of *Lygodium*, Ann. of Bot. p. 495. 1916.
454. J. M. THOMPSON. New Stelar Facts, Trans. Roy. Soc. Edin. Vol. lii, No. 28. 1920.
455. GOEBEL. *Anemia elegans*, Flora, Bd. 108, p. 319.
456. BOWER. Studies, vii. Ann. of Bot. xxxii, p. 1. 1918.
457. ROGERS. Prothallium of *Lygodium palmatum*, Bot. Gaz. p. 168. 1910.
458. SEWARD. Ann. of Bot. Vol. xxxviii, p. 485, (1924), where the literature on fossils provisionally ascribed to the Schizaeaceae is fully quoted.

[1] There has been some uncertainty as to the correct form of the name *Anemia* (*Aneimia*). I therefore asked the opinion of Dr Daydon Jackson, who replied: "Swartz published the Fern as *Anemia* in his *Syn. Filic.* p. 155, 1806, which name was altered by Kaulfuss in 1824 to *Aneimia* in his *Enum. Filic.* 51." Pfeiffer remarks "ἀνείμων, non vestitus, quare bene correctum per *Anemia* Swartz (*l.c.* 183)." Notwithstanding the incorrectness by derivation, it is better to adopt this than to take the responsibility of correcting and editing botanical nomenclature. A name may be retained as a symbol, however faulty its derivation.

CHAPTER XXIII

MARSILEACEAE[1]

THIS Family comprises three genera of Ferns of semi-aquatic habit, characterised by rather unusual modifications of the vegetative system, and particularly by their heterospory. They are *Marsilea*, which is cosmopolitan, with 56 species: *Regnellidium* from Brazil, with one species: and *Pilularia* with 6 species, inhabiting Europe, America, and Australasia. These plants have commonly been associated with the Salviniaceae under the heading of the Hydropterideae, on the ground of their adaptation to aquatic life, and the heterosporous character which they share with them. But such grounds appear insufficient, since the similarities on which that assumed relation is based may well have been of homoplastic origin. There are reasons, which will appear as the description proceeds, for ranking the Marsileaceae apart from the Salviniaceae, and relating them with the homosporous Schizaeaceae. They may be held as a series derivative from these, but modified in accordance with a semi-aquatic habit.

The sporophyte of *Pilularia* is a creeping rhizomatous plant, rooted in mud, or even floating with pendent roots on the surface of water (Fig. 458). From its rhizome spring leaves which are narrow and subulate with circinate vernation (*a*). Close to the base of each is a bud which repeats the characters of the main shoot, and in fruiting specimens the spherical, short-stalked sporocarp is attached on the adaxial side of the leaf-base. It resembles a pill, hence the name of the plant. The extreme tip of the rhizome is curved upwards: the leaves (*l*) arise alternately from its upper surface, and associated with each are a bud (*b*) and a root (*r*): their relative positions and sequence are shown in Figs. 458, *b*, *c*, *d*, and the probable conclusion is that the buds are really dichotomous branches of the rhizome thrust alternately aside, right and left (Vol. I, p, 78). Simple hairs (*h*) form a protection for the young parts. *Marsilea* is similar in general habit, but it differs in the form of the leaves. These may be floating when the plant grows in water: but if on dry land they rise erect. Each of the aerial leaves bears four equal pinnae at the end of a long stipe, giving the appearance of a four-leaved Clover (Fig. 459). This state is best understood by comparing the juvenile leaves of which the earliest are simple and at first subulate, as in *Pilularia*. Then successively there follow spathulate, then

[1] The name *Marsilea* dates from *Linn. Gen.* ed. i, 326 (1737). Many authors have since written it *Marsilia*; but following the accepted rules the correct form should be the original one, which is accordingly adopted here. The name was derived from Count Luigi Ferdinand's *Massigli* (1656—1730).

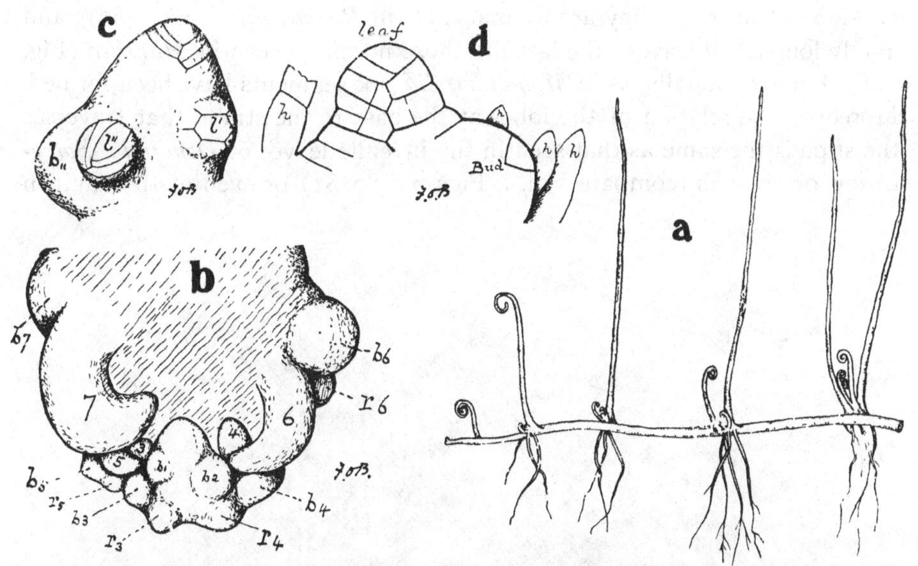

Fig. 458. *Pilularia globulifera.* a = rhizome with alternate leaves, and a bud associated with each of them: vernation circinate. (After Velenovsky.) b = apical bud, with hairs removed, seen from above: 3–7, successive leaves: b_1–b_7, successive buds, one at the base of each leaf: r_3–r_6, the corresponding roots. c = the extreme up-turned tip of a similar apex seen from above: l', l'', the youngest alternate leaves: $b_{,,}$, the bud corresponding to l''. d = section vertically through a corresponding leaf and bud, showing the apical cell of each: h, h, hairs. (a is about natural size: $b \times 70$: $c \times 175$: $d \times 325$.)

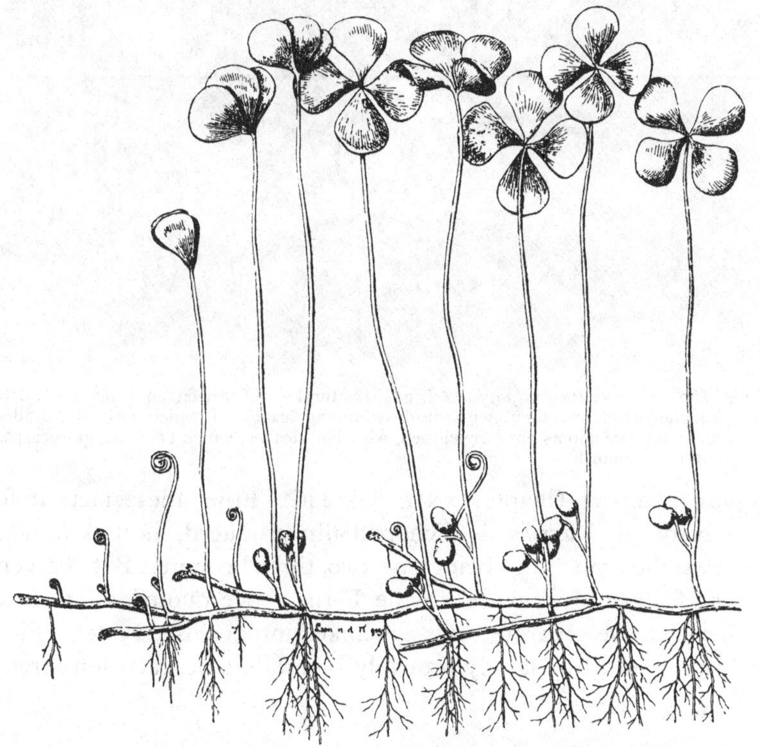

Fig. 459. *Marsilea quadrifolia* L. A small but strong, and richly fructifying plant. (Of natural size. After Luerssen.)

two-lobed leaves (as they are permanently in *Regnellidium* (Fig. 460)), and finally four-lobed leaves; the last are those normally seen in *Marsilea* (Fig. 461): but occasionally, as in *M. quadrifolia*, six segments have been formed. Moreover, the relation of the lobes at the base to the strand that traverses the stipe is the same as that seen in the juvenile leaves of *Osmunda*, *Botrychium*, or *Anemia* (compare Vol. I, Figs. 77, 80, 81), or even of the adult in

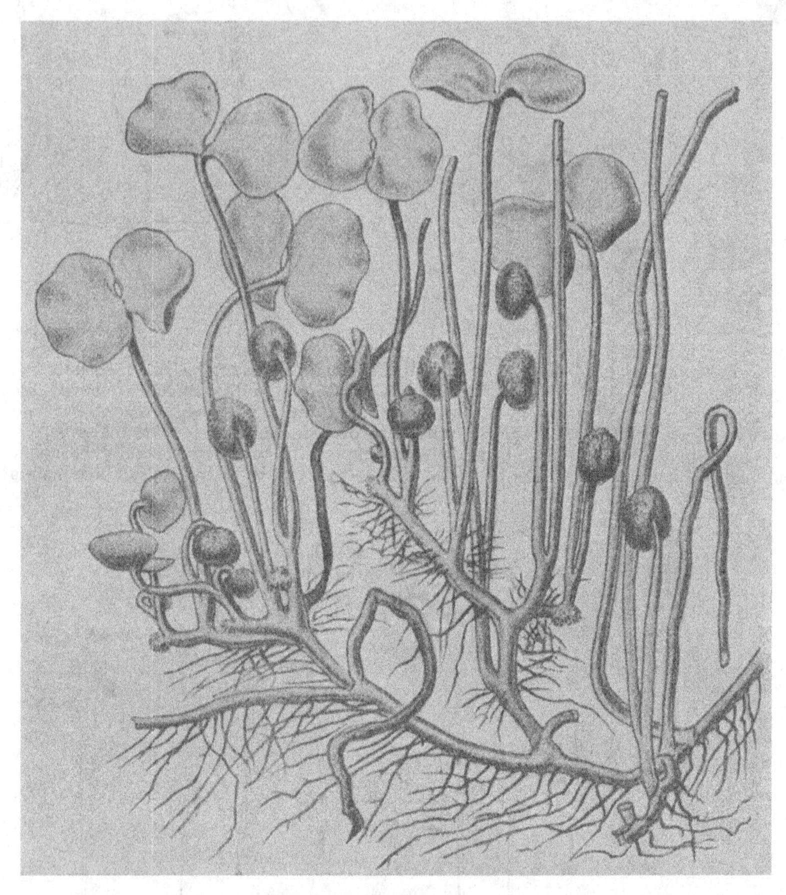

Fig. 460. *Regnellidium diphyllum* Lind. (Natural size.) Specimen from S. Brazil. From an elongated stem, with dead swimming leaves, of which only the stalks remain; aerial shoots have developed, with land-leaves, which bear the sporocarps. (After Lindman.)

A. elegans (compare Chapter XXII, Fig. 446). From these facts it follows that the blade of *Marsilea* is dichopodially produced, as it is in so many early Ferns, the branchings being one, two, three or more. But the venation shows one feature of advance on the Ferns above quoted, in the marginal loops by which the separate veins are linked into a closed system (Fig. 461). The adult leaves of the family probably illustrate a degradation series from

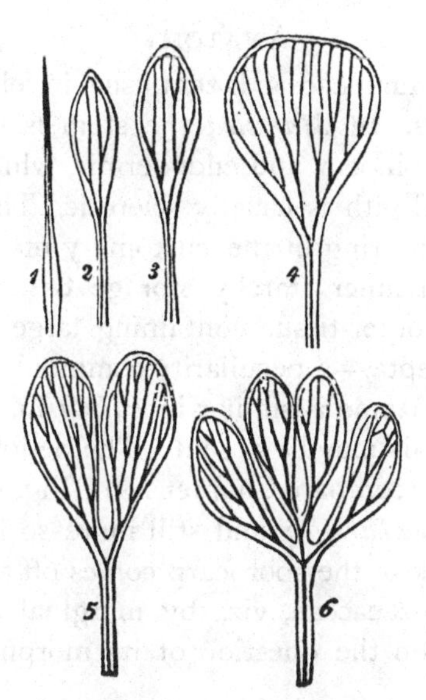

Fig. 461. Successive types of juvenile leaves of *Marsilea*. (After Braun.)

a more highly branched leaf of the type of *Anemia* through 6, 4, and 2 lobes, and finally to the unbranched type of *Pilularia*, which in its adult state simply repeats the cotyledonary stage. *Marsilea*, like *Pilularia*, bears a sporocarp on each leaf, sometimes several, or even many as in *M. polycarpa*. Their relation to the stipe is lateral. These fruiting bodies are hard and woody when ripe, with a hairy surface, and each is borne upon a stalk of variable length which sometimes branches, as in *M. quadrifolia* (Fig. 459). The shoot of *Marsilea* is more or less frequently branched, and the branches are extra-axillary, arising on the flanks of the main rhizome. This again suggests a dichopodial origin, as in *Pilularia*. Though the genus is typically semi-aquatic it includes xerophytic species, such as the Australian *M. hirsuta*, which retains its vitality under long drought, It has, like so many Australian xerophytes, an underground rhizome with tuberous buds which persist after the rest of the plant dies, and germinate as soon as the conditions are favourable, producing fertile leaves.

The newly discovered *Regnellidium* (Lindman, *Ark. f. Bot.* iii, 1904), from South America, is represented by one species, *R. diphyllum*, so called from the distinctive fact that it has constantly only two leaf-lobes. The veins are repeatedly dichotomous, and are united as in *Marsilea* by marginal loops. The veins of the fruit fork also, and the shanks of the same vein may unite at the margin, as well as intramarginally: but the primary veins remain distinct and separate. The new genus thus appears to take a natural place between *Marsilea* and *Pilularia*.

ANATOMY

The anatomy of the Marsileaceae is that usual in relatively primitive Ferns with creeping rhizomes. In *Marsilea* the stem is traversed by a normal solenostele, with inner phloem and endodermis, while in plants grown on solid ground the central pith is usually sclerotic. The undivided leaf-trace separates a sector of the ring in the customary manner. The cortex may be differentiated as an inner starchy storage tissue, separated by a thin sclerotic band from an outer tissue containing large radiating lacunae with thin parenchymatous septa,—a peculiarity commonly seen in aquatic plants. The structure of the rhizome is similar in *Pilularia*, but on a smaller scale, and the inner endodermis may be absent. The petiolar trace is of the same type as that of *Anemia* (compare Chapter XXII, Fig. 10), but it is reduced in size and complexity in *Marsilea* and still more so in *Pilularia*. The vascular supply to the stalk of the sporocarp comes off after the manner of the pinna-trace of the Schizaeaceae, viz., by marginal abstriction, a point of importance as bearing on the question of its morphology (Fig. 462). The

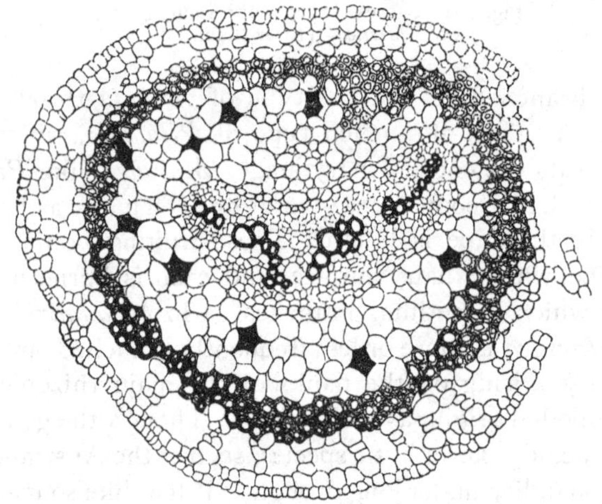

Fig. 462. *Marsilea polycarpa*. Transverse section of the petiole, below the insertion of the lowest sporocarp, and showing on the right the sporocarp-trace, given off marginally, as in the sterile pinnae. (After Miss Allison.) (× 175.)

petiolar trace is marked off by endodermis from the parenchyma outside, which is again differentiated into an inner coherent tissue with tannin-sacs, and an outer lacunar tissue with peripheral epidermis, and sclerotic tissue between. The roots are diarch, with a conical apical cell, and regular lacunar cortex. In *Pilularia* peculiar spirally coiled cells occupy the lacunae, acting probably as mechanical props, as in the Nymphaeaceae.

The chief interest in the sporophyte of the Marsileaceae lies in the sporocarp which, from the evidence of anatomy as well as of the ontogeny, is a lateral offshoot of the leaf, in fact a pinna. As it is a complex body, and difficult to understand, the simpler type seen in *Pilularia* will be described first (Fig. 463). The sporocarp is here spherical: cut transversely four

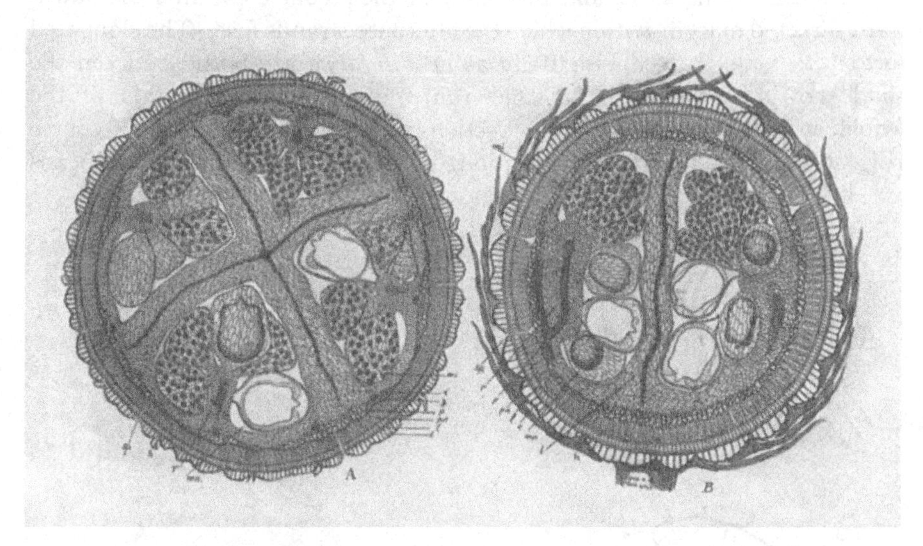

Fig. 463. *Pilularia globulifera* L. *A*, transverse section of the ripe but still closed sporocarp, about half-way up, at the limit between mega- and microsporangia. *B*, approximately median longitudinal section of the same. (Both × 20.) The firm and resistant outer wall is formed of prismatic cells, and has a hairy covering. It is lined, and the cavity partitioned by mucilaginous tissue styled the "indusium," which swells with water when ripe and evacuates the sporangia: these are produced in four sori, attached to marginal receptacles. (From Luerssen in *Rab. Krypt. Flora*, Vol. iii.)

loculi are seen, containing numerous sporangia borne on four peripheral receptacles, the whole being enclosed by a firm rind (Fig. 463, *A*). The sporangia are of two sorts: megasporangia, each containing one megaspore visible with the naked eye; and microsporangia each containing many minute microspores. Cut longitudinally (*B*) the loculi are seen to extend from base to apex of the sporocarp, the elongated receptacle usually bearing microsporangia distally and megasporangia towards the base. Each loculus is enclosed peripherally by part of the massive wall of the sporocarp, while centrally it is enveloped by a mantle of soft tissue often called the "indusium." The internal tissues become mucilaginous when ripe, and are liable to swell on access of water. The external wall is composed of an outer epidermis, with hairs and stomata, lined by two deep bands of hard prismatic cells. Internally to this shell is softer parenchyma enclosing the vascular strands, one of which supplies each receptacle, while two others run longitudinally in each of the spaces between (Fig. 463, *A*). When the inner tissues swell on access of water the shell breaks into quarters by longitudinal slits, and the sporangia and spores are evacuated in a mass of mucilage.

The sporocarp of *Marsilea* is constructed on the same plan, but with a larger number of receptacles. The outline of it is seen from Fig. 464, *A*, where the midrib, or "dorsal bundles" of Johnson, is directed upwards: from it alternating veins arise, and they fork in their course towards the downward-directed margin, within which the alternate strands fuse. The elongated receptacles, constructed essentially as in *Pilularia*, are borne between the shanks of these lateral veins, and run transversely to the axis of the whole sporocarp. A transverse section would then follow their course (Fig. 465, *A*), and show the position of the sporangia with their heads

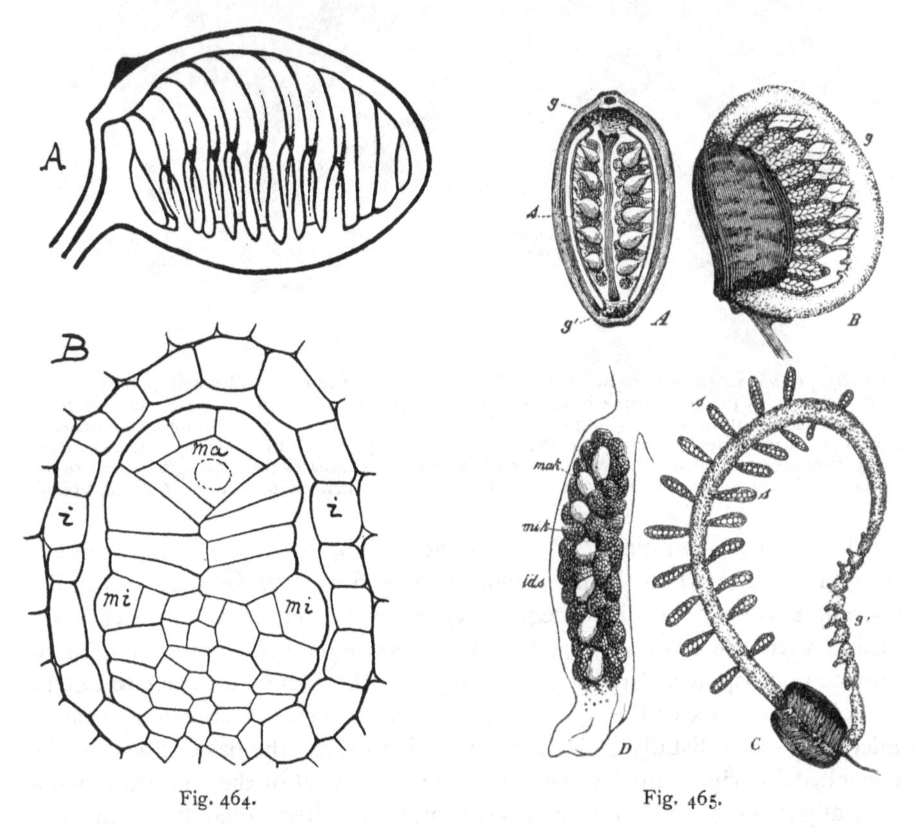

Fig. 464. Fig. 465.

Fig. 464. *Marsilea quadrifolia*. (After Duncan Johnson.) *A*, a view of the inner side of one of the valves of a nearly mature sporocarp, slightly simplified. *B*, part of a horizontal section of a receptacle, showing the relations of a megasporangium (*ma*), and the microsporangia (*mi*), and of the indusium (*i*).

Fig. 465. *Marsilea salvatrix* Hanst. *A*, transverse section of the ripe but still closed sporocarp. (× about 4.) *g*, *g'*, gelatinous ring: *s*, sorus. *B*, swollen and opened spore-fruit, with the swelling gelatinous ring, *g*, which draws out with it the attached sori. (× about $2\frac{1}{2}$.) *C*, fully opened spore-fruit, with the extended gelatinous ring, its ends still attached to the fruit-wall. The sori now free are borne upon its dorsal side, while mammiform processes from which they have broken away are seen on the ventral side. (Natural size.) *D*, a sorus magnified six-fold, with microsporangia (*mik*), megasporangia (*mak*), and "indusium" (*ids*). (After Hanstein from Luerssen.)

pointing inwards. Each receptacle, as in *Pilularia*, is closely invested by an "indusium," and the structure is shown in section in Fig. 466, *G*, *H*

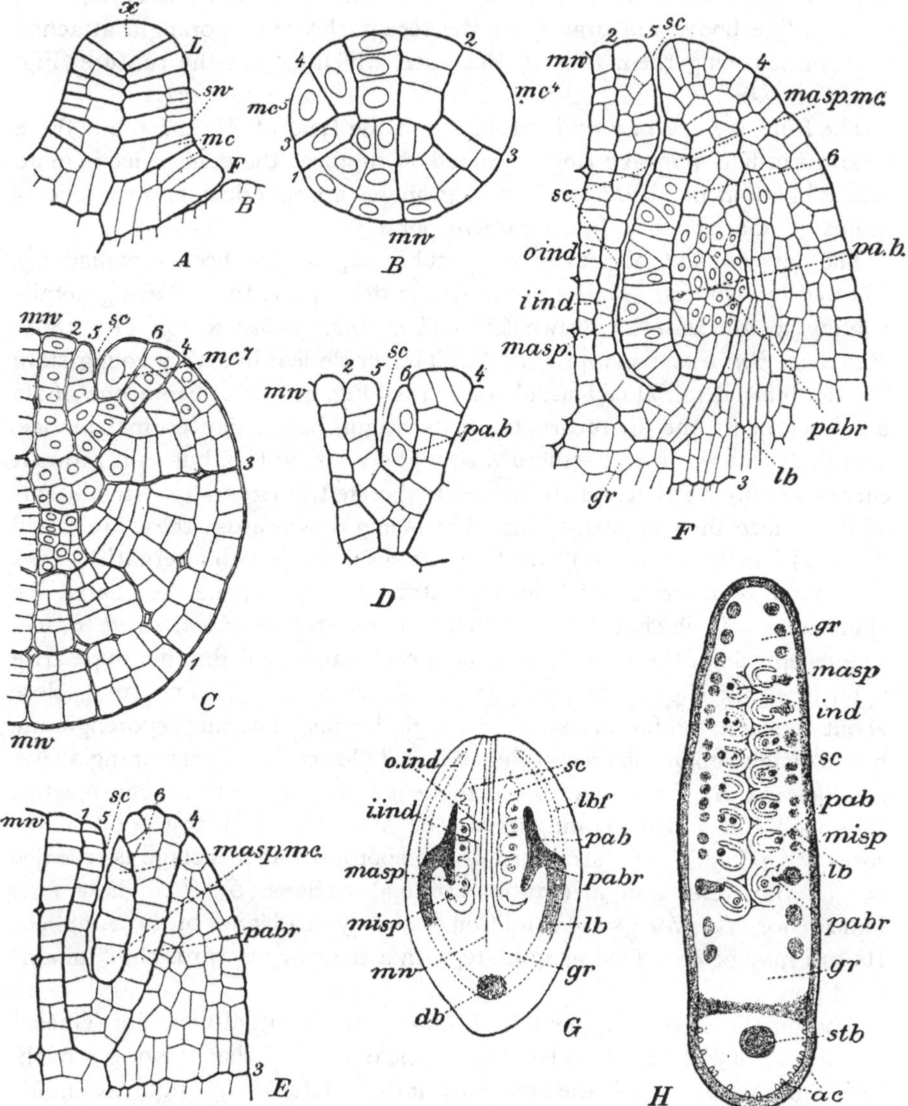

Fig. 466. *Marsilea quadrifolia* L. *A*, young leaf. *F*, the mother cell or apical cell of a sporocarp: *x*, the apical cell of the leaf: *mc*, *sw*, divisions in the segments from the apical cell. *B*, trans. sec. of a young sporocarp, whose growth by marginal cells is already established: *mw*, the median wall: 1–4, the successive anticlines in each half: *mc*⁴ and *mc*⁵, the marginal cells. *C*, development of one half of the leaf: *mc*, mother-cell of sorus: *sc*, the soral canal. *D–F*, stages of development of the sorus, and soral canal. *G*, trans. sec. of flaps of a sporocarp. *H*, long. sect. of same: *ac*, air-canal: *db*, dorsal bundle: *gr*, gelatinous ring: *i.ind*, inner indusium: *lb*, lateral branches of dorsal bundle: *lbf*, twig of a lateral branch of dorsal bundle: *masp*, megasporangia: *masp*, mother cells of megasporangia: *misp*, microsporangia: *mw*, median wall: *o.ind*, outer indusium: *pab*, placental-bundle: *pabr*, branch of same: *sc*, soral canal: *stb*, bundle of stalk of sporocarp. (*A–E* × about 250: *G*, *H* × 45.) (After Duncan Johnson, from Engler and Prantl.)

At maturity the whole sporocarp splits like a bivalve shell on access of water: the internal tissues behave as in *Pilularia*, but the whole tract of tissue at the ventral margin remains coherent; as it swells it emerges as a worm-like body, and draws out the receptacles and sporangia attached to them, exposing them fully to the water which caused the rupture (Fig. 465, *B*, *C*, *D*).

The fruit of *Regnellidium* is generally of the type of *Marsilea*, but there is no worm-like mucilage-ring extended on rupture: the dehiscence is more nearly like that in *Pilularia*. The megaspores are spherical, and the microspores number about 60 in each microsporangium.

The development of these very peculiar organs has been exhaustively followed by Duncan Johnson, from whose description the following details relating to *Marsilea* are drawn (*Marsilea*, *Ann. of Bot.* xii, p. 119, 1898; *Pilularia*, *Bot. Gaz.* xxvi, p. 1, 1898). The whole leaf in a very young state has the bifacial initial cell usual for Ferns (Fig. 466, *A*). From one of its basal segments the sporocarp takes its origin, being, like normal pinnae, lateral (*F*). It assumes a structure very like a young leaf, but as it grows it curves sharply so that the distal end is directed horizontally. Involutions of its surface then appear on the side facing downwards: these are called the soral canals (*sc*, *C—F*), and they are disposed in two alternating rows (*H*). Each of these is lined on its central face by a tissue designated the "indusium," of which two layers may be recognised (*i.ind*, *a.ind*, *F*, *G*). The receptacle of the sorus faces this in each canal, and the sporangia arise upon it with a slightly marked acropetal succession (*F*). The canals close about the time the formation of sporangia begins. The megasporangia are borne distally upon the receptacle, and are followed by microsporangia (*ma*, *mi*, Fig. 464, *B*). In a strict sense there is thus a gradate sequence, while Campbell states that sporangia may be interpolated at any point in the sorus (*l.c.* p. 438). The structure of the sporocarp at this stage is revealed by the transverse and median-longitudinal sections (*G*, *H*). These facts indicate for *Marsilea* a soral condition distinctly in advance of the Simplices. Its sori may be described as gradate, with a tendency towards the "mixed" condition.

The details of development in *Pilularia* are essentially similar (Goebel, *Bot. Zeit.* p. 45, 1882). Von Goebel has clearly shown that the origin of the sori is marginal: so that the sporocarp may be referred in origin to a modification of a typical Leptosporangiate sporophyll (*Organographie*, 1918, ii, 2, p. 1136).

The development of the sporangia follows the ordinary plan for Leptosporangiate Ferns, but Campbell states that it corresponds most nearly to that of the Schizaeaceae. There is no highly elaborated opening mechanism, the sporangia appearing as hyaline sacs. But Campbell has noted in

Pilularia americana certain elongated but not indurated cells, corresponding in position to the annulus of *Schizaea*, and these he regards as a vestigial representation of it (Vol. I, Fig. 253). The number of the spore-mother-cells is usually eight, but some or all of these may divide again, so that the whole number ranges from eight to sixteen (Campbell). The extreme number of the microspores is 64. In the megasporangium only one of the spore-mother-cells develops beyond the tetrad stage. One of its daughter-cells enlarges greatly at the expense of the others, and forms a single megaspore, with a thick, highly differentiated mucilaginous wall.

The germination of the spores of *Marsilea* proceeds very rapidly after evacuation from the sporocarp. Divisions within the microspore separate two basal vegetative cells of the prothallus, while the rest of the contents form two groups of spermatocytes together with tabular cells as their protective walls (compare Campbell, *l.c.* Fig. 245). It appears that here there are probably two antheridia, each with 16 spermatocytes: this small number resembles that seen in the antheridia of *Schizaea* or *Anemia*. The spermatozoids are peculiar in the large number of their spiral coils. The megaspores are essentially tetrahedral like the microspores, but oval in form; they may be ½ mm. in length, and are covered by a thick differentiated mucilaginous coat. The apex, which corresponds to the centre of the original tetrad, is marked by a hemispherical projection covered by a thin brown wall. Here the female prothallus is formed, while the rest of the spore is stored with nutriment (Fig. 468). The prothallus is minute, consisting only of a single archegonium embedded in a cap of tissue fitting over the apex of the large storage cavity within the spore. Seen from without at the time of fertilisation it is as in Fig. 467.

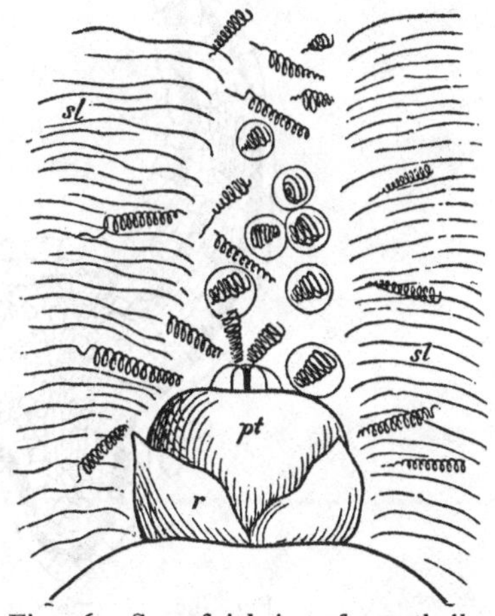

Fig. 467. Superficial view of a prothallus of *Marsilea salvatrix* Hanst. *pt*, the prothallus: at its apex the neck of the archegonium. *sl*, the mucilage-layers of the epispore forming a funnel, with many spermatozoids. *r*, the torn coat of the megaspore. (× about 150.) (From Engler and Prantl.)

The embryo undergoes segmentation after the scheme of the Leptosporangiate Ferns, and the stem, leaf, root, and foot have the normal relations. As the large oval spore will as a rule lie on its side the axis of the archegonium will be approximately horizontal. The first or basal wall includes that axis: but, as Leitgeb showed, it can adjust itself in relation to gravity so that there will be an upward-directed, epibasal, shoot-forming half, and a downward-directed, hypobasal, root-forming half. The relations of the growing embryo will therefore be as in Fig. 468. The prothallus grows for

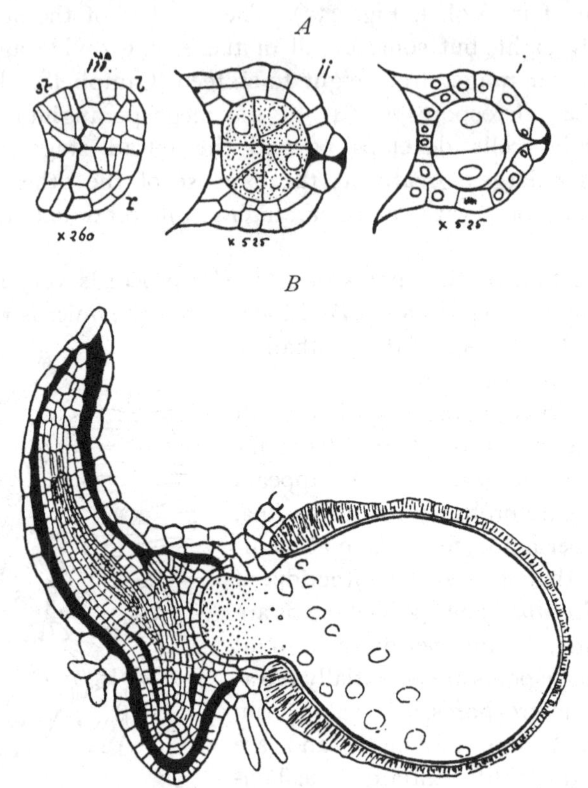

Fig. 468. *A*, embryos of *Marsilea vestita*. (After Campbell.) They are orientated in their probable natural position, since the oval megaspore will normally lie flat on its side, and the prothallus is apical. *B*, median longitudinal section of a young sporophyte and megaspore of *Pilularia*. (After Campbell, but orientated as it would be in nature.) (× 70.)

a time and protects the sporeling, while the nourishment is derived from the large storage-spore till the sporeling is established.

COMPARISON

The relation of the Marsileaceae to the Schizaeaceae has been definitely suggested and supported, chiefly by the observations of Campbell. The essential facts are well known for the two families, and the most important of them have been stated in a condensed form in the preceding pages. Numerous details less direct in their bearing have been omitted, but are readily accessible in the works referred to at the end of this chapter. Systematic comparison of the two Families, taking up the several criteria in use here should lead to some definite conclusion whether or not the suggested relationship is a real one.

(1) The rhizomatous shoot of the Marsileaceae, with alternate leaves and lateral branchings which suggest a reference to a dichopodial branching, find their correlative in *Lygodium* and in the creeping species of *Anemia* (§*Anemiorrhiza*). But it would also compare with rhizomes of many other creeping Ferns.

(2) The apical meristems conform to what is usual in Leptosporangiate Ferns: even the segmentation of the young leaf is the same with its marginal series of cells.

(3) A comparison of the leaves within the Family shows successive steps from a simple unbranched subulate leaf to those with two, four, or even six pinnae. The venation of these pinnae is clearly dichotomous, and the equal forking is obvious in the last branchings. But all of them are found on exact analysis to be referable to dichotomy developed sympodially (compare Bower (471), Figs. 8, 9). This conclusion coincides with the result of analysis of the primordial leaves of *Osmunda*, *Botrychium*, *Anemia* and *Lygodium*, and may be regarded as a general feature in relatively primitive Ferns. *Schizaea* itself illustrates the more primitive equal dichotomy in its sterile region: but in the fertile region of its leaf the branching is dichopodial (*l.c.* Fig. 6 bis). A feature of advance in the leaves of *Marsilea* and *Regnellidium* is the marginal anastomosis of the veins: but in advanced species of *Lygodium* and of *Anemia* vein-fusions are present. The leaves of the Marsileaceae may then be accepted as being originally constructed on a plan not unusual in relatively early Ferns: that plan is seen most obviously in the adult leaves of *Marsilea*, with four or sometimes six pinnae: but it is simplified in *Regnellidium* to only two pinnae, while in *Pilularia* the leaf is unbranched. This state is seen in the juvenile leaves of *Marsilea* itself, and is exemplified also in the sterile blades of many species of *Schizaea*, and particularly in *S. rupestris*.

(4) The solenostele of the rhizome, and the undivided leaf-trace of the Marsileaceae find their correlatives in various primitive Ferns, in particular the details show some near similarity in the creeping species of *Anemia* included in §*Anemiorrhiza*. In *Pilularia* the inner endodermis may be absent: and thus its stele approaches that simple state seen in some of the smaller species of *Schizaea* (Chapter XXII, Fig. 443). Lastly, the marginal origin of the pinna-trace, which may itself be held as indicating the pinna-nature of the sporocarp, corresponds to its marginal origin in the Schizaeaceae: it is found to apply in *M. polycarpa* for each of the numerous sporocarps (Fig. 462).

(5) The dermal appendages of the Marsileaceae are simple hairs, as is commonly the case in primitive Ferns, and in particular in the Schizaeaceae,

excepting *Mohria*, which appears to be an advanced xerophytic type, with protective scales.

(6, 7) The spore-bearing organs present the most exacting problem. Where the sporocarp is solitary the origin, position, and lateral vascular connections are sufficient evidence of its pinna-nature. The numerous sporocarps all on one side of the petiole in *M. polycarpa* suggest a succession of pinnae on an anadromic helicoid plan, such as is seen in the leaf of *Pteris semipinnata* (Vol. I, p. 88, Fig. 82, *C, D, E*). Further, the branching of the pedicels in *M. quadrifolia* (Fig. 459) has been referred on developmental grounds to a production of pinnules, themselves developed as sporocarps (Johnson, *l.c.* p. 139). Accordingly the whole series of structures seen in the Marsileaceae are referable in origin to parts of a normally branching leaf, and the sporocarps may be regarded as pinnae or as pinnules (compare Von Goebel, *Organographie*, ii, 2, p. 1136). Turning now to the Schizaeaceae, the spore-production in them is not limited to any definite part of the sporophyll. In *Mohria* and *Lygodium* it is generally spread throughout its length: in *Anemia*, though it is often restricted to the lowest pair of pinnae, this is not constant for the genus (Goebel, *Flora*, Bd. 108, p. 319). In *Schizaea* it is located on distal branchings of the dichotomising sporophyll. Thus the variety seen in the Marsileaceae is matched by that in the Schizaeaceae: in either family the fertile region may be whole pinnae or pinnules.

The inner constitution of the sporocarp itself presents a further problem. The alternate position of the elongated receptacles, and the superficial origin of the sporangia facing on to the soral canals, which are surface-involutions, are facts that suggest that the whole body is of the nature of a condensed sporophyll, or rather part of it. Such a view involves fusion of the parts together, and an additional tissue-development to produce what is styled the "indusium." Lastly certain curvatures in the whole body will demand consideration. Duncan Johnson remarks (*l.c.* p. 129) for *Marsilea* that "there is never any curling in the extreme tip of the capsule suggesting the circinate coiling of the leaf." He describes how the young sporocarp "bends ventrally upon itself." But there is no mention of its rotation. A natural structural indication of the orientation of the whole body when mature is supplied by the outline of the vascular strand of the stalk, especially at the point where it enters the sporocarp. A section through the so-called "raphe" shows the vascular strand of the type of the Marsileaceae and Schizaeaceae, orientated so as to suggest that the raphe is dorsal (abaxial), and the sori ventral (adaxial) (Fig. 469, *A*). To attain this position there must have been some slight rotation, about $\frac{1}{4}$ of a circle: this will be clear from Fig. 469, *B, C*. In view of these points the hypothesis seems to be tenable that the sporocarp consists of a rachis bearing two rows of pinnules: this is indicated by the venation, for both families have con-

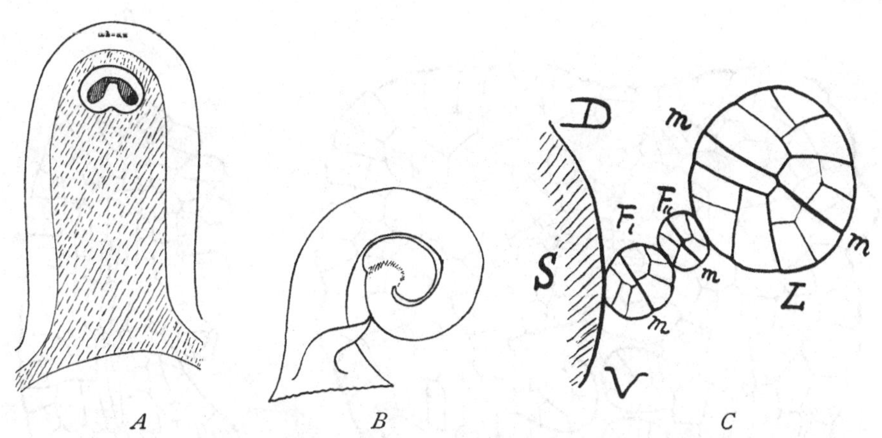

Fig. 469. *Marsilea quadrifolia*. *A*, section through the "raphe" of the sporocarp, showing by the orientation of the meristele that the raphe marks the abaxial side. *B*, young leaf, showing the lateral insertion of the young sporocarp. *C*, relations of stem (*S*), leaf (*L*) with its median plane (*m*, *m*), and two sporocarps (*F*,, *F*,,). *D*, dorsal, *V*, ventral directions. (*B* and *C* after Duncan Johnson.)

stantly an undivided leaf-trace and pinna-trace: that these pinnules are folded towards the ventral side, which is directed downwards, and there fuse laterally: while the space between is filled by an upgrowth of tissue forming the "indusium." In support of this view a comparison with one of the simplest of the Schizaeaceae, viz. *S. rupestris*, may be made: its sterile leaf is flattened but simple, as it is in *Pilularia*. The fertile leaf bears 6–10 slender, spreading, serrated spikes on each side. These are shown in Fig. 470, and are recognised

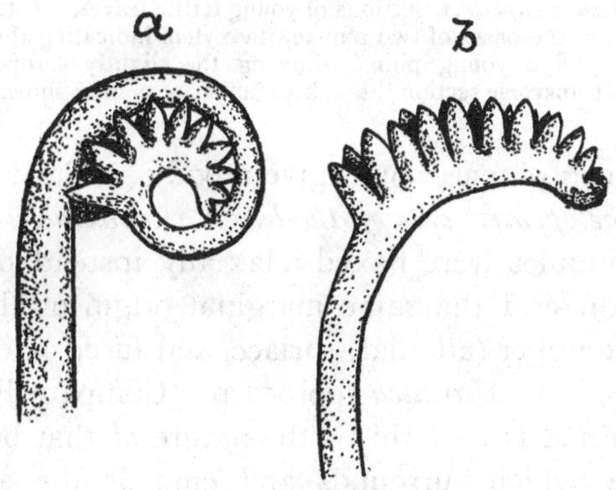

Fig. 470. Young leaves of *Schizaea rupestris*, showing circinate vernation with the pinnae reflexed to the convex (abaxial) side. (× 6.)

as pinnae reflexed so that their lower surfaces, to which the originally marginal sporangia are directed, face inwards. A transverse section through them is shown in Fig. 471, *A*, and a longitudinal section in Fig. 471, *B*: while their natural position relatively to one another is shown in Fig. 471, *C*.

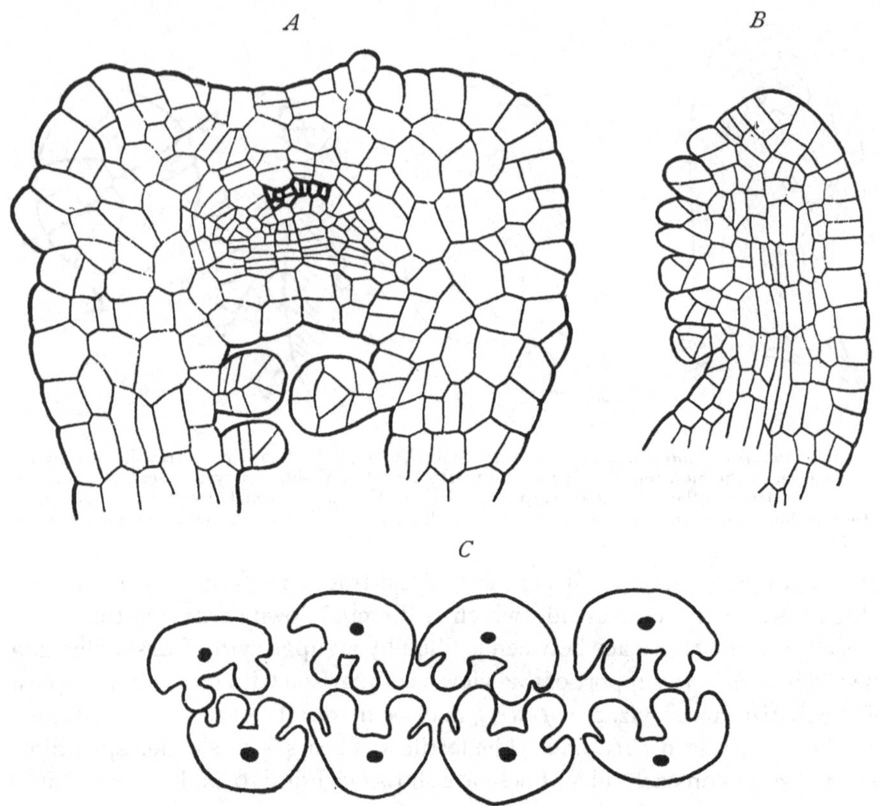

Fig. 471. *A*, *B*, *C*. *Schizaea rupestris*, sections of young fertile leaves. *A*, transverse section of the rachis and longitudinal of the bases of two pinnae, the xylem indicating the adaxial face. (× 85.) *B*, longitudinal section of a young pinna, showing the slightly acropetal succession of the sporangia. (× 85.) *C*, transverse section through a number of pinnae, showing their relation to one another. (× 35.)

Lateral fusion of such pinnae would give a body similar to that postulated, *but orientated on the opposite side of the leaf*, i.e., *abaxially*. If, however, two similar series of pinnules were flexed adaxially instead of abaxially, with the same alternation, and the same marginal origin of the sporangia, but inclined now to the upper (adaxial) surface, and fused laterally, they would give the structure of the *Marsilea* sporocarp. (Compare Fig. 466, *H*.) The hypothesis entertained is that this is the nature of that body, consolidated by indusial tissue, which surrounds and embeds the adaxially-directed sporangia. But before it can be considered as demonstrated a critical re-examination of the details of development will be necessary, and especially of the vascular tracts.

(8) The sporangial segmentation is of the Schizaeaceous type, but the adult features are chiefly negative, though Campbell recognises in the abortive annulus and oblique form a similarity to *Schizaea*.

(9) The spore-output is smaller (at most 64 against 128 or 256 of the Schizaeaceae), while heterospory introduces entirely different and derivative relations.

(10, 11) The very reduced prothallus presents negative features, though it may be remarked that a very simple prothallus, such as that of *Schizaea*, would lend itself readily to reduction. The archegonia are standardised as usual, but the antheridia present in the low number of the spermatocytes a point of similarity with *Schizaea* and *Anemia*.

(12) The embryology corresponds so exactly with that of Leptosporangiate Ferns that it confirms these otherwise peculiar plants in their Filical relationship, while their general features point to a relatively primitive position.

The result of this comparative analysis is to render support to Campbell's recognition of a real relationship of the Marsileaceae to the Schizaeaceae, the similarity extending along the whole line of the comparison. Perhaps the most divergent point is between the abaxially-flexed pinnae of *Schizaea* and the (hypothetically) adaxially-flexed pinnules of the sporocarp of *Marsilea*. But on this discrepancy it is fair to remark (i) that the sporangia being assumed to be marginal, as in fact they are in the Schizaeaceae, they would be as easily displaced to the upper (adaxial) as to the lower (abaxial) surface: (ii) that the hypothesis demands a folding of the pinnules of the sporocarp in the same way as the pinnae of the sterile leaf of *Marsilea*, that is adaxially: and (iii) that the indusial developments of the Schizaeaceae are themselves so various that it is no great demand on probability that the tissue filling in the sporocarp and protecting the sporangia in the Marsileaceae should not comform to any one of the Schizaeaceous types. The hypothesis is not definitely proved, but it can at least be claimed that there is high probability of its truth. In any case the relationship to the Schizaeaceae seems beyond doubt. The Marsileaceae may be held to be a family of that same fundamental type, though specialised for aquatic life, and heterosporous. But it does not seem possible to link them definitely with any single genus of the Schizaeaceae: the similarities are, however, closest to *Schizaea* and *Anemiorrhiza*. In support of the comparison with *Schizaea* it may be noted that *S. rupestris* is to be found growing in very moist conditions, under dripping rocks, in the Blue Mountains, New South Wales, as a veritable Hydropterid in habit.

The early existence of the Marsileaceae in geological time has not been satisfactorily demonstrated. Suggestive but not very convincing fossils have been recorded from the Rhaetic, Jurassic, and Lower Cretaceous floras. But *Sagenopteris*, which was formerly compared with the Marsileaceae, is now believed to be the foliage of a primitive Angiosperm (H. H. Thomas, "Caytoniales," *Phil. Trans.* Vol. ccxiii, B, p. 199).

The three living genera may be best seriated according to the complexity of their leaves, viz:

(i) *Marsilea* (Linnaeus, 1753) ... 56 species. Leaves 4-lobed.
(ii) *Regnellidium* (Lindman, 1904) ... 1 species. Leaves 2-lobed.
(iii) *Pilularia* (Linnaeus, 1753) ... 6 species. Leaves unbranched.

The features of advance which the sporocarp of the Marsileaceae possesses over the sporophyll of the Schizaeaceae do not consist only in Heterospory. The sori also show indications of gradate succession and even of mixed interpolation of sporangia. There is, however, no evidence of the Marsileaceae having led to any higher type of organisation, notwithstanding these advanced features.

BIBLIOGRAPHY FOR CHAPTER XXIII

459. HOFMEISTER. Higher Cryptogamia, Engl. Edn. Ray Soc. 1862.
460. HANSTEIN. Pringsh.-Jahrb. iv. 1865.
461. RUSSOW. Vergleichende Untersuchungen. St Pétersbourg. 1872.
462. LEITGEB. Sitz. d. k. Akad. der Wiss. Bd. lxxvii. 1878.
463. BAKER. Fern Allies. London. 1887. Here the systematic treatment up to its date is given.
464. LUERSSEN. Rab. Krypt. Flora, iii, p. 606, 1889: with full references to the systematic literature.
465. DUNCAN JOHNSON. *Marsilea*, Ann. of Bot. xii. p. 119. 1898.
466. DUNCAN JOHNSON. *Pilularia*, Bot. Gaz. xxvi, p. 1. 1898.
467. ENGLER & PRANTL. Natürl. Pflanzenfam. i, 4, p. 403, 1902. Here the literature is fully quoted.
468. LINDMAN. Ark. f. Bot. iii, 1904. *Regnellidium*.
469. SEWARD. Fossil Plants, ii, p. 473. 1910.
470. ALLISON. Sporocarp in *M. polycarpa*, New Phyt. x, p. 204. 1911.
471. BOWER. Leaf-Architecture, Trans. Roy. Soc. Edin. li. 1916.
472. GOEBEL. Organographie, 2te. Aufl. ii, Teil, p. 1134. 1918.
473. CAMPBELL. Mosses and Ferns, 3rd. Edn. p. 417, 1918. Here the literature is fully quoted.
474. BOWER. Studies VII, Ann. of Bot. xxxii, p. 1. 1918.
475. BOWER. Primitive Spindle, Proc. Roy. Soc. Edin. Oct. 1922.

CHAPTER XXIV

GLEICHENIACEAE

THIS Family is represented by 80 living species, referred in Christensen's Index to two genera: *Stromatopteris* which is monotypic, its single species being native in New Caledonia: and *Gleichenia* with 79 species distributed throughout the Tropics, whence they extend far southwards, but only in less degree north, and they are absent from the northern temperate zone. This genus has been divided into three sub-genera: (i) *Dicranopteris* (= *Mertensia*), including flabellate forms which there is reason to regard as primitive: (ii) *Eu-Gleichenia*, where the leaflets are contracted, probably in relation to xerophytic conditions, and (iii) *Platyzoma*, with leaflets also contracted, which is represented by a single, very peculiar species, native in North East Australia. The Family is a very natural one, showing general uniformity of character, though the details allow of the species being so arranged as to illustrate important evolutionary steps.

Among these Ferns a creeping habit is constant, with a rhizome which is deeply buried in *Stromatopteris*, but in the rest it spreads at or near the surface of the soil, while it may sometimes take an ascending position. Upon it the leaves are usually solitary, often with long internodes, but sometimes more closely grouped (*Platyzoma*). The axis dichotomises frequently, the forking being independent of the leaf-insertions (Fig. 472). The leaves, in which the vein-endings are always free, are sometimes simply pinnate (*Platyzoma* and *Gleichenia* in the sporeling state): but usually they show high degrees of branching, together with a peculiar straggling habit. Owing to the continued apical growth and widely divaricating branching of the leaves, these Ferns often form dense impenetrable thickets at the margins of forests and on open savannahs.

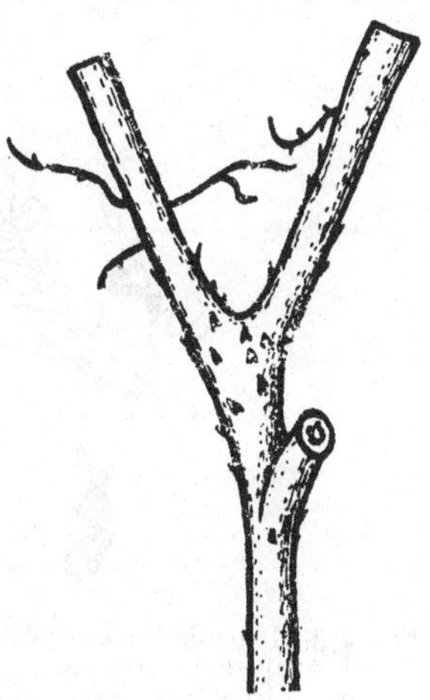

Fig. 472. *Gleichenia linearis* (Burm.) Clarke, showing dichotomy of the rhizome. (Natural size.)

The branching of the leaf has often been erroneously described as dichotomous, but according to Goebel no species of *Gleichenia* has a dichotomous leaf (*Organography*, Eng. Edn., ii, p. 319; also II Aufl., p. 1043. 1918), though it is certainly the fact that equal forking (as in a *forma furcata*) does occur in *Platyzoma* (Thompson, *Trans. Roy. Soc. Edin.*, p. 635. 1916), and in *Stromatopteris* (Christ, *Geog. d. Farne*, Fig. 89). The error has arisen by mis-interpreting the arrested leaf-apex in *Gleichenia* as an adventitious bud. A correct understanding of the leaf-architecture is obtained by study of the ontogeny. Observations have been made on young plants of *Gleichenia* (*Dicranopteris*) *linearis*, in which the juvenile leaf may at first be constructed on a dichotomous system, as is usual in Ferns, but showing even in the cotyledon advanced sympodial development (Fig. 473, *a*, *d*, *cot.*).

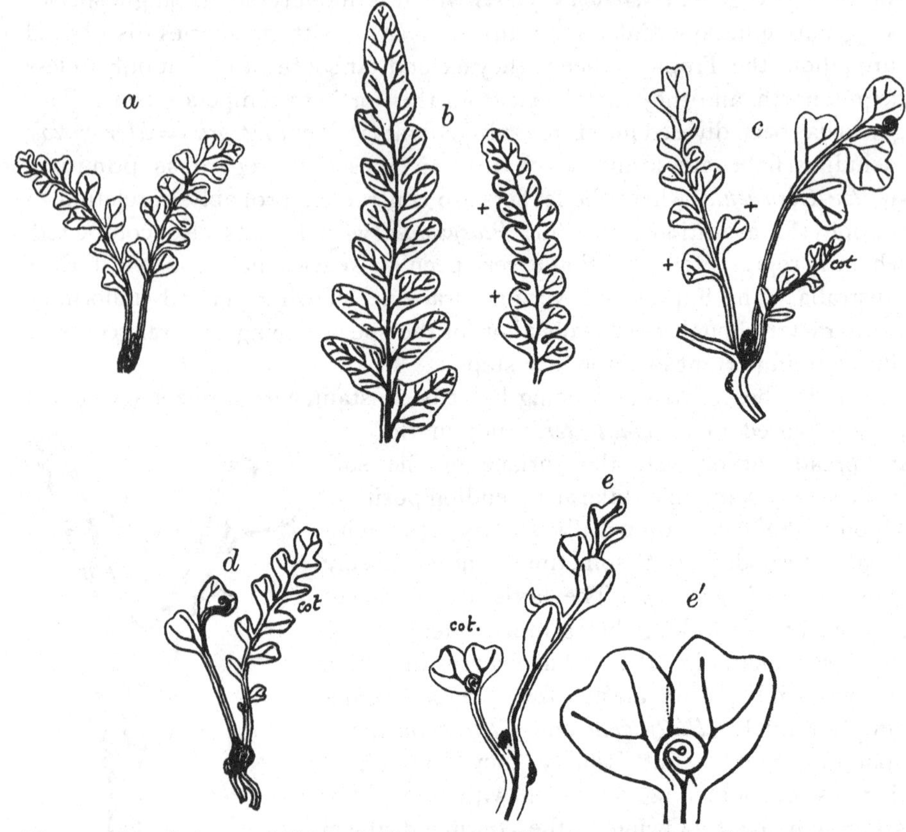

Fig. 473. Juvenile leaves of *Gleichenia* (*Dicranopteris*) *linearis*. *a*, *b*, show ordinary pinnate types, with sympodial development. *c*, *d*, *e*, show early stages of arrest of the apex, such as lead to the characteristic habit of the genus: *e'* is a cotyledon with its apex arrested after two pinnae have been produced. (*a*, *c*, *d* × 5; *b* × 4; *e* × 10; *e'* × 30.)

The later leaves are wholly so constructed, as appears from their apical regions, the leaflets often showing distally a single forking of the veins: but passing downwards there are clear steps to a scorpioid sympodium in their venation (*b*). The leaf-construction thus initiated may be continued indefinitely, though in some simple Gleichenias and in *Platyzoma* the elaboration does not proceed further than a primary pinnation. But the young plants of *G. linearis* and of some others illustrate that the "false dichotomy" arises in such a scheme, by temporary arrest of the apical growth. The tip of the leaf (*c*, *d*), and even of the cotyledon itself (*e*, *e*′) may retain its circinate coil, while the pinnae below it expand. It is clear in Fig. 473, *e*′, that the two equal pinnae are not themselves the result of equal dichotomy, but pinnae sympodially produced according to the scheme illustrated in (*a*). The whole apical region is delayed in its development: but though dormant it still possesses a power of indefinite further growth. This scheme may be repeated in each pinna or pinnule, giving the key to the indefinite foliar development so often realised in the genus when forming thickets on savannahs. The degrees of branching of the leaves have been made the basis of systematic division of the section *Dicranopteris* (*Mertensia*) into four sections (Fig. 474). A further elaboration is described by Goebel in a mode of protection of the resting bud, found in some species (Goebel, *Organography*, Engl. Edn., p. 318). The pinnules that stand nearest to the apex develope

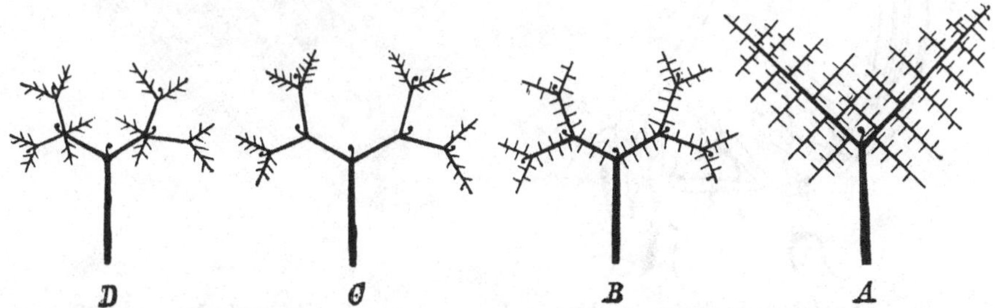

Fig. 474. Scheme of branching of the leaf in the our sections ot *Mertensia* Willd. = (*Dicranopteris*). They illustrate different results of the interruption of apical growth in the pinnate leaf. (After Diels, from Engler and Prantl.)

as scales covering it; they have sometimes been mistaken for adventitious or aphlebioid growths. As a matter of fact the whole structure of the leaf, including its protections, can be referred to a normal pinnate development altered by temporary arrest of the apex, and by precocious development of certain pinnae or pinnules. As regards the pinnules themselves two broadly distinct types are seen: that with relatively long flat pinnules (*Dicranopteris* = *Mertensia*), and that having short and semicircular pinnules with convex upper surface (*Eu-Gleichenia* and *Platyzoma*). The latter is probably a derivative xerophytic type.

Dermal appendages of various forms are found in the Family. They vary from simple uniseriate hairs to flattened scales, and they are sometimes seated on emergences of considerable bulk. In *Eu-Gleichenia* broad scales protect the dormant leaf-apices and appear also on the leaf-bases and rhizomes: but simple hairs are associated with them: both fall off early in many species. The same holds for most species of § *Dicranopteris*; but in *G. linearis* and *pectinata*, which we shall find to be distinguished strongly from the rest of the genus by marked characters of stele and sorus, scales are absent, and tufted hairs are found (Fig. 475). In *Stromatopteris* the dermal appendages are very various, ranging from simple hairs to scales, and they may be seated on emergences of various sizes. In *Platyzoma* there are only simple hairs. In this it compares with *G. linearis* and *pectinata*. It will be seen later how these divergent facts bear upon the phyletic comparison.

The sori of the Gleicheniaceae are always superficial, being disposed in a single row on either side of the midrib of the fertile segment. These segments may be relatively long, with long rows of sori, as in the *Dicranopteris* type (Fig. 476); or short and concave, so that there is space only for

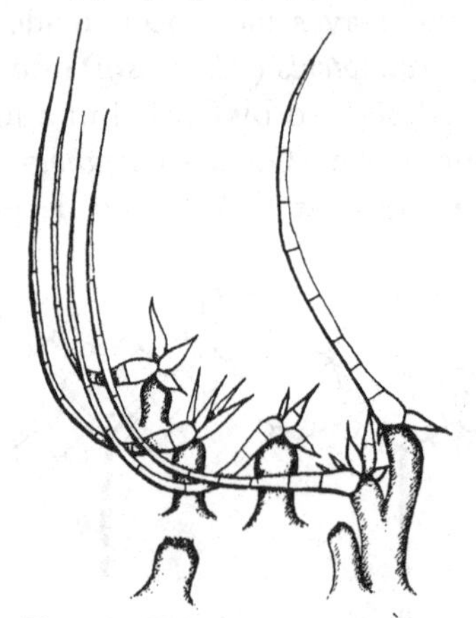

Fig. 475. Hairs from the base of the leaf of *Gleichenia pectinata*. Each is seated on a massive emergence, from which it is easily detached, the emergence remaining, as in the Cyatheaceae. (× 50.)

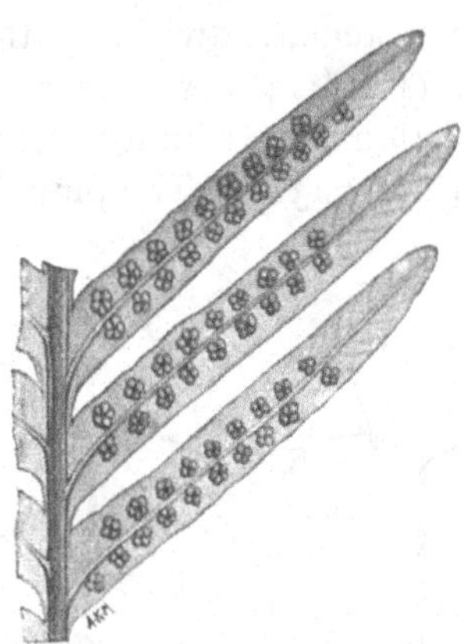

Fig. 476. *Gleichenia flabellata* R. Br. Midrib and three pinnules, showing the arrangement and constitution of the sori, with a variable number of sporangia in each.

few sori or even a single one, as in the *Eu-Gleichenia* type, and in *Platyzoma*, and *Stromatopteris*. The sorus is as a rule typically radiate-uniseriate, the sporangia being attached in a simple ring round a central receptacle. There is no indusium. The number of the sporangia varies in different species, 2 to 5 or 6 being common numbers: but in the Cretaceous fossils from Greenland Professor Seward found a larger number.

ANATOMY

The axis of those sporelings of *Gleichenia* which have been examined contain a small protostele. In the great majority of species this structure is maintained throughout life. But as the creeping rhizome increases the protostele also enlarges, and finally settles down to a cylindrical form, of approximately constant size, as it is seen to be in *G. flabellata* (Fig. 477). There is a central core of xylem consisting in the adult of tracheides interspersed with parenchymatous cells. Mesarch protoxylem-strands are present near to the periphery, and those towards the upper side are continuous outwards into the petiolar trace. The xylem, which is sometimes fluted (*G. circinata*), is surrounded by a relatively narrow band of phloem, with a broad band of pericycle, and a continuous endodermis delimits the

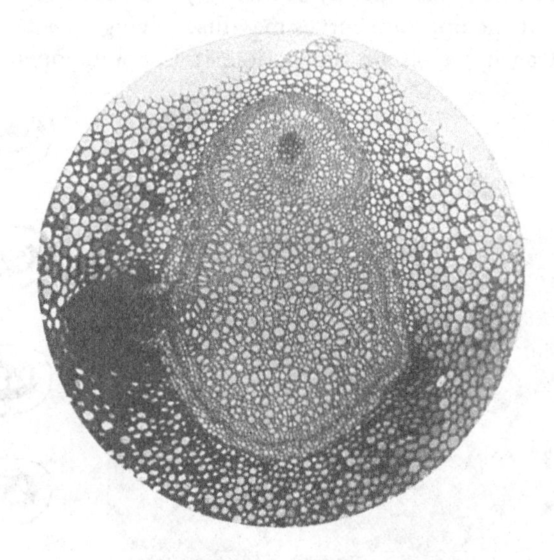

Fig. 477. Transverse section of the protostele in the rhizome of *Gleichenia flabellata* R. Br., from a section by Gwynne-Vaughan, showing a leaf-trace being given off. (× 30.)

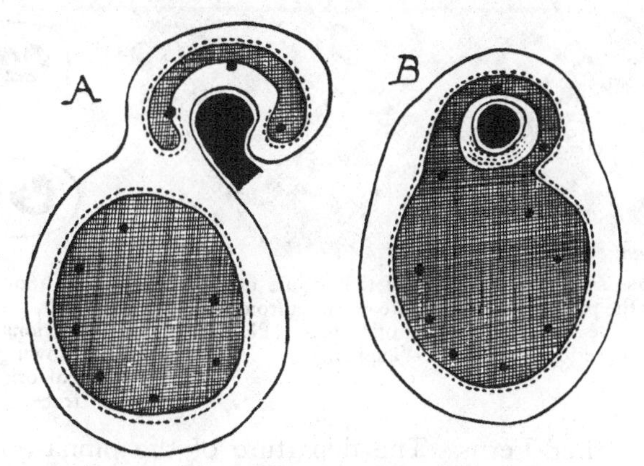

Fig. 478. Transverse sections of the rhizome of *Gleichenia flabellata*, showing the detachment of the leaf-trace from the protostele, with its included protoxylems. (After Tansley.)

stele (Fig. 478). The vascular supply to the roots is thrown off superficially, and mostly from the lower side of the stele. The leaf-trace, which is a deeply in-rolled horse-shoe, with 3 protoxylems subsequently increased to

4, passes off with a shallow ad-axial pocket, as a **C**-shaped tract of xylem, with internal phloem, and endodermis un-interrupted at the departure. Its curve may enclose a mass of sclerenchyma.

The **C**-shaped trace is frequently so strongly contracted in the lower part of the petiole that the opposite curves may fuse, giving a pseudo-stelar effect (Fig. 479). But as it passes up the petiole it is apt to open out with free

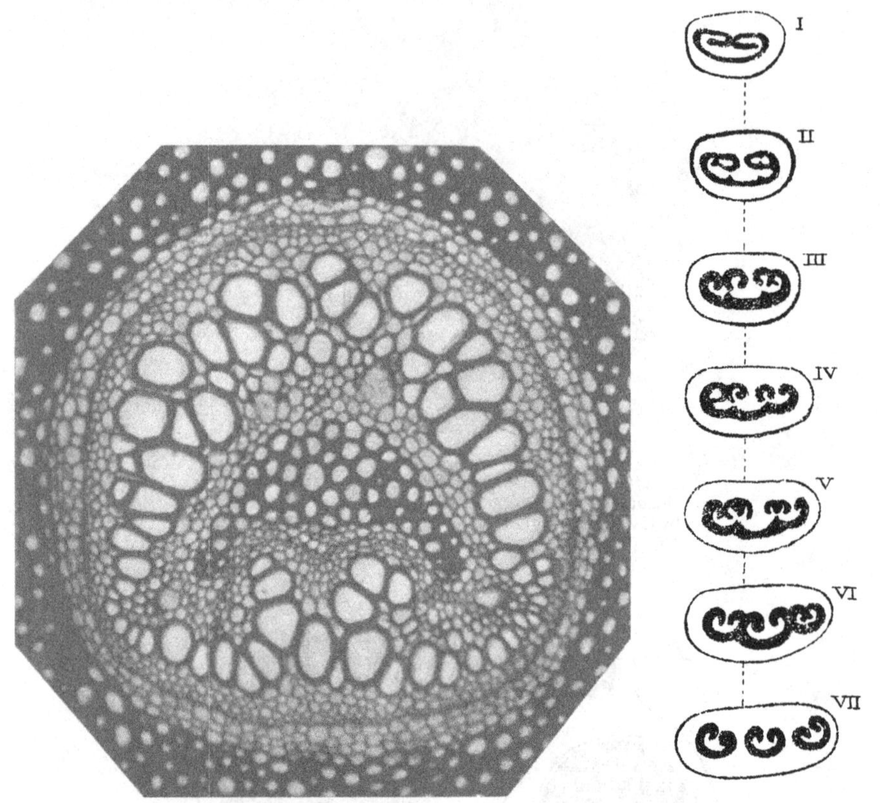

Fig. 479. Transverse section of the base of the petiole of *Gleichenia dicarpa*, showing the pseudo-stelar structure resulting from contraction of the horse-shoe-like xylem, till its margins fuse. (Photograph by Dr Kidston from a section by Gwynne-Vaughan.)

Fig. 480. *Gleichenia linearis*, series of transverse sections of the rachis, showing the extra-marginal origin of the pinna-trace. ($\times 4$.)

margins, as in other Ferns. The departure of the pinna-traces is extra-marginal, as it usually is where the margins of the meristele are strongly incurved (Fig. 480). Consequently the arrangement of the vascular tissues at the base of the first pinnae may appeared involved: but comparison with what is seen in *Lophosoria* (Vol. I, p. 174, Fig. 170) shows that apart from the lateral involutions of the trace, which are absent in *Gleichenia*, the method is the same as that common for the Cyatheaceae.

Certain species of § *Dicranopteris*, viz. *G. linearis* and *pectinata*, differ from the rest in soral construction, and the latter species stands apart also in the absence of scales, stiff hairs being present. Nevertheless *G. linearis* conforms to the rest of the genus in being protostelic: but *G. pectinata* shows a structural advance on any other species, in being solenostelic in the adult state. Its structure is illustrated diagrammatically by Fig. 481, which demonstrates that it is a solenostelic *Gleichenia*, with slightly fluted xylem. The shallow pocket seen in *G. flabellata* if extended from node to node would give the structure present in *G. pectinata*. It becomes therefore a point of some importance to see how the transition from protostely to

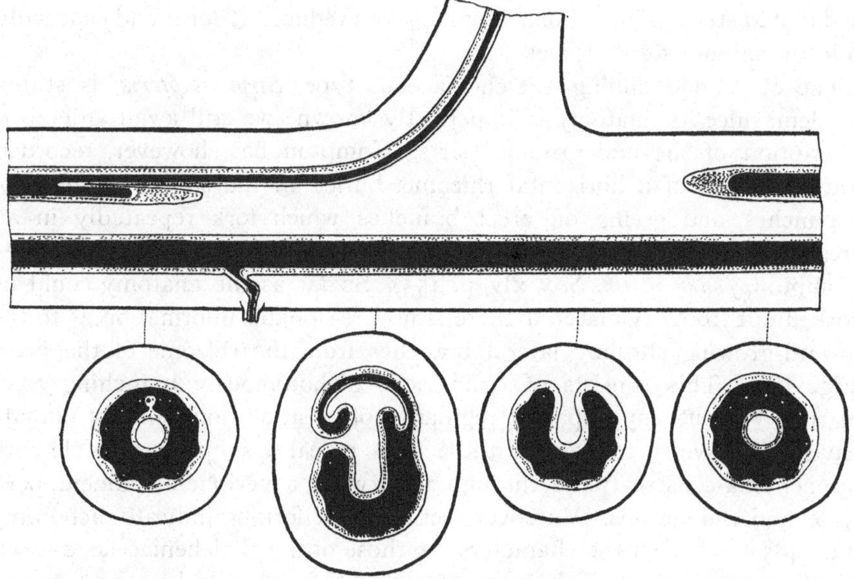

Fig. 481. Diagram illustrating the solenostelic structure and attachment of the leaf-trace in *Gleichenia pectinata*. The transverse sections show the structure corresponding to the several points indicated. (After Boodle.)

solenostely is brought about in the individual life. A full account of the change has been given in Vol. I, pp. 142–4, where Fig. 134 suggests the ontogenetic steps which may be held as illustrating not only an individual advance, but probably also a phyletic departure from the primitive protostely of the Family.

On the other hand, it has been shown in Vol. I, p. 131 how *Platyzoma* has made a similar advance, but less complete. Here the whole habit of the plant is different. The leaves of this small xerophyte are closely crowded on the shortened rhizome, and are of very different sizes: the pinnate fertile leaves far exceeding the simpler sterile leaves, with which they alternate in successive zones. These facts alter the balance of the requirements for conduction, and the vascular system differs accordingly, though still the underlying Gleicheniaceous scheme may be traced (Fig. 482). The general stelar structure of this Fern has been already described, with its internal sclerotic pith which fluctuates in bulk, and its inner endodermis which is of inconstant occurrence, and connects at no point with the outer endodermis (Vol. I, p. 132, Fig. 125). A transverse section of the stem is seen in Fig. 483 *A*, while the origin of the leaf-trace of one of the larger leaves is shown in Fig. 483 *B*, *C*, to be typically that of protostelic Ferns. The natural interpretation of this structure is that *Platyzoma* is a Gleicheniaceous Fern of rather advanced type, but still imperfectly solenostelic, in which the departure of the relatively small leaf-traces causes only a local disturbance of the medullated stele. The petiolar bundle is of a reduced **C**-form and especially so in the minute sterile leaves.

The other outstanding Gleicheniaceous type, *Stromatopteris*, is still a problem, since its anatomy is imperfectly known: we still await a detailed description of the underground parts. Compton has, however, recorded "the existence of a horizontal rhizome, buried in the soil to a depth of 3–4 inches, and giving off erect branches which fork repeatedly in an irregular fashion. Roots are borne sparsely on this horizontal rhizome" (Compton, *Journ. Linn. Soc.*, xlv, p. 453). So far as the anatomy could be worked out from dry material, there is now reasonable information as to the upward-growing shrubby lateral branches from the rhizome of the plant (Fig. 484). This consists of cylindrical, dichotomously branching axes bearing, without any apparent regularity of position, long simply pinnate leaves. Transverse sections of these axes reveal a simple protostele with parenchyma-cells scattered through the xylem, a peripheral phloem, pericycle, and endodermis. Moreover sections at a forking indicate dichotomy (Fig. 485 *a*). In fact the characters are those of the Gleicheniaceae, excepting the erect position. The apices of these axes appear to be soon arrested. The petiolar supply at its base may resemble very nearly that of the axis: but higher up it is clearly of the **C**-type, and as it pursues its course into the pinnate region the minute pinna-traces are given off in extra-marginal manner, as in *Gleichenia* (Fig. 485 *b*). Pending more detailed information these shrubby upward-growing regions may be held to be erect forking rhizomes bearing simply pinnate leaves.

On these anatomical facts two views are possible. Either that the protostelic state of most Gleichenias is primitive, and that the solenostelic type has been derived from it: or that the protostelic types are reduced from a solenostelic ancestry. The former view appears the more probable in so primitive a family: and especially in view of the fact that *G. pectinata*, which is solenostelic, has sori and sporangia of a character advanced in the direction of the simpler representatives of the Cyatheaceae, which are solenostelic or

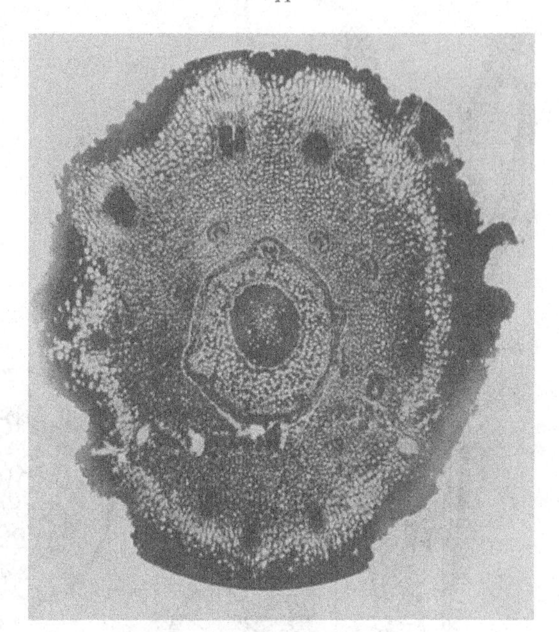

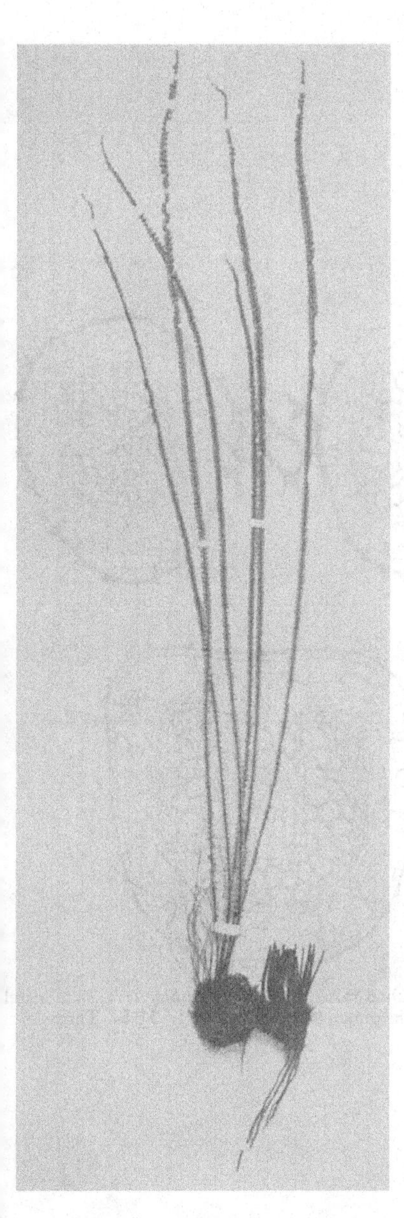

Fig. 482. Habit of *Platyzoma microphylla* R. Br., showing heterophylly, and the hairy investment of the abbreviated rhizome. The small leaves near the apex are sterile, the longer leaves fertile. (From specimen coll. Prof. Baldwin Spencer. Boorrooloo. N.T., Australia. 1902. ½ Natural size.)

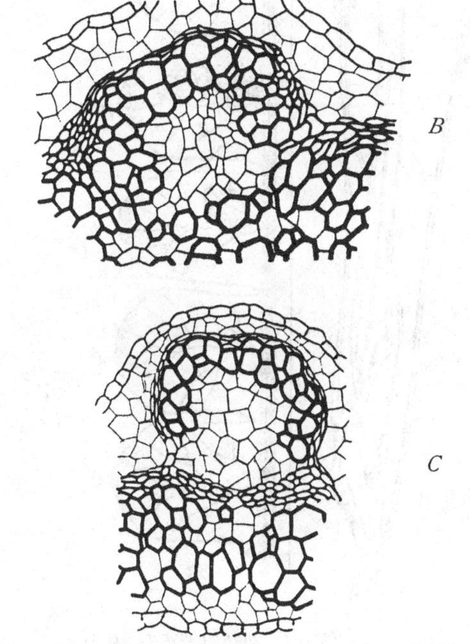

Fig. 483. *Platyzoma*. Transverse section of rhizome, *A*, showing soleno-xylic stele: *B*, *C*, showing the protostelic origin of the small leaf-traces. (After J. M. L. Thompson.)

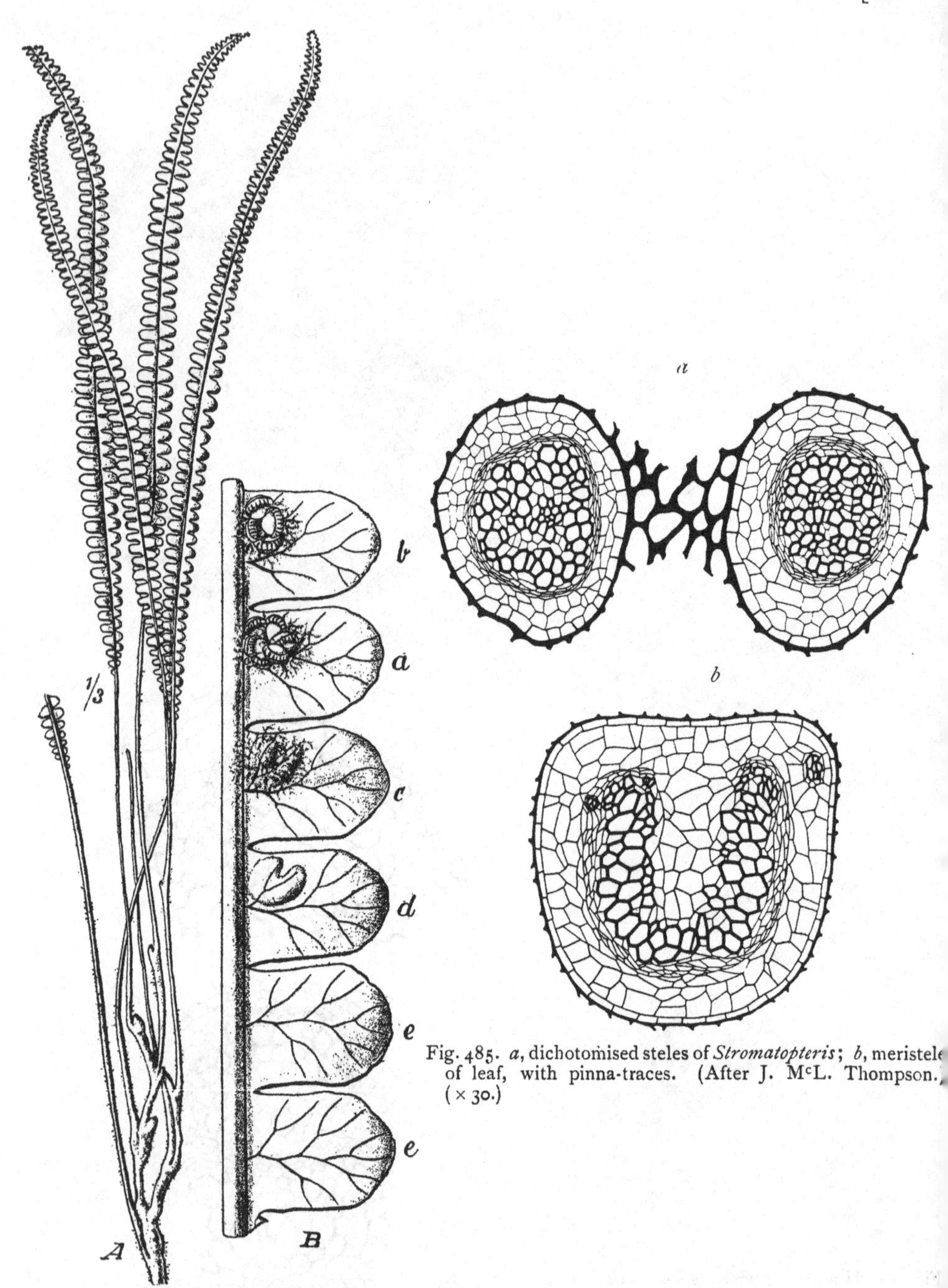

Fig. 484. *Stromatopteris moniliformis* Mett. *A* = habit of one of the upright branches that rise from the horizontal rhizome. *B* = portion of a leaf: *a*, *b*, with sori intact: *c*, sporangia removed, but hairs remaining: *d*, hairs also removed, to show the receptacle: *e*, sterile segments. (After Diels, from Engler and Prantl.)

Fig. 485. *a*, dichotomised steles of *Stromatopteris*; *b*, meristele of leaf, with pinna-traces. (After J. McL. Thompson.) (× 30.)

dictyostelic. Moreover the ontogeny of *D. pectinata* points to a similar conclusion

SPORE-PRODUCING MEMBERS

The naked sorus of the Gleicheniaceae consists of a low circular receptacle bearing a variable number of sporangia, which are quite separate from one another normally, though instances of synangia which resemble a fusion of two sporangia occur occasionally. The sporangia usually form a single row round the receptacle, and their orientation in this case is constant, the longitudinal slit of dehiscence facing directly towards the centre of the rosette-like sorus. *Gleichenia flabellata* may be held as a central type of the genus (Fig. 476). It has two parallel rows of leaf-segments of the Pecopteris-type so common among Palaeozoic Ferns, and each segment bears a row of sori on either side of its midrib. The number of sporangia in each sorus is commonly 4 to 6. From such central numbers deviations may be found within the genus, on the one hand to fewer sporangia, on the other to a greater number. The former is seen in *G. flabellata* itself, for the distal sori may bear three, two, or even only one: these are evidently results of reduction below the normal, and are probably consequent on defective nutrition. But in *Eu-Gleichenia* the number is commonly 3 to 4, while in *G. dicarpa* two is the usual number. These species all have small rounded, usually concave pinnules, each bearing a single sorus on the lowest acroscopic vein: their sporangia are commonly protected by "ferrugineous wool," or sometimes by "paleaceous hairs." They inhabit sub-tropical or temperate regions, sometimes at high levels (*G. dicarpa*). The conclusion is that these are exiguous, xerophytic members of a tropical family, with sori reduced in number, and containing fewer sporangia than the central types; and that protection is afforded by the concave pinnules, fluffy hairs, and even by scales. These species are far from deserving the title of "*Eu-Gleichenia*," being apparently derivative, and specialised in relation to climate. The same probably holds for *Stromatopteris* and *Platyzoma*, both of which are clearly xerophytic, and also show reduction in number of sori and of the sporangia composing them. These reduced types have secured themselves by their concave pinnules, and protective scales and hairs.

On the other hand there are species included in §*Dicranopteris* which diverge from the *G. flabellata*-type of sorus both in the number and in the position of the sporangia. This is so in *G. linearis* (Burm.) Bedd. (= *G. dichotoma*, Willd.), a large and variable species, probably the commonest of the genus, spread through the tropics of both hemispheres. The number of capsules may here be 6 to 10 (Fig. 486). These are arranged as in the radiate-uniseriate type: but the centre of the sorus, vacant elsewhere, is often occupied by one or more sporangia (Fig. 487: compare also Fig. 489).

The extreme state for this Family is seen in *G. pectinata*, in which the number of sporangia in a sorus may frequently be 10 or 12 (Fig. 488). These are densely crowded together on the central receptacle. Where the number of the sporangia in the sorus is more than 5, as in *G. linearis* and *pectinata*, single sporangia may be displaced perhaps by lateral pressure, so that their dehiscence is not strictly radial to the centre of the sorus, but oblique: and this irregularity naturally appears where the centre of the sorus is occupied by one or more sporangia. In such examples we

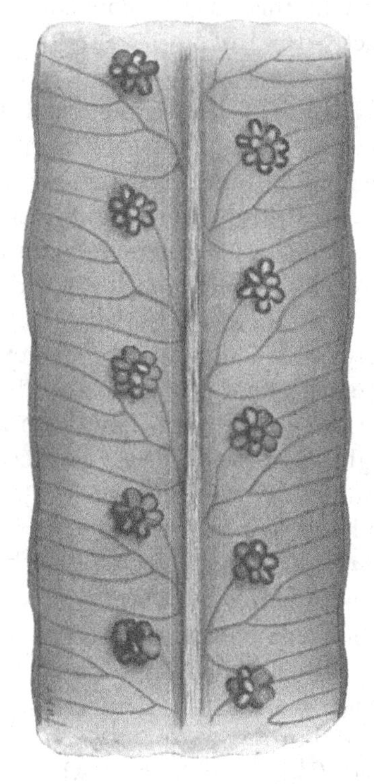

Fig. 486. Pinnule of *Gleichenia linearis* showing the venation, and mature sori. Some of the sporangia are seen to have dehisced. (Enlarged. Drawn by Mr Maxwell.)

Fig. 487. A single sorus of *Gleichenia linearis* on a larger scale, showing the disposition of the sporangia, several of which have opened by a median slit. (Drawn by Mr Maxwell.)

see the extreme development of the Gleicheniaceous sorus. In none of them is an indusium present, though a dense flocculent growth of hairs may surround the sorus, especially in the xerophytic types. It will be realised that by such species as those last named the gap is bridged between two well-marked types of sorus. On the one hand is the type of the Marattiaceae, which is shared by most of the Gleicheniaceae, where a single series of sporangia surrounds the periphery of the low receptacle: on the other is the type of the Cyatheaceae, and other Gradatae, where the receptacle is more or less elongated, and its apex covered by numerous sporangia: and it is important to note that the first steps are taken within the genus *Gleichenia*.

The sporangia of the Gleicheniaceae are pear-shaped with a very short stalk, and an annulus consisting typically of a single row of cells: it is complete round the sporangial head except at the line of dehiscence, which is median and directed away from the lower side of the leaf (Fig. 489, *i–n*). Its oblique course defines two thin-walled areas: the central or basal which faces obliquely to the centre of the sorus, and the peripheral or distal which

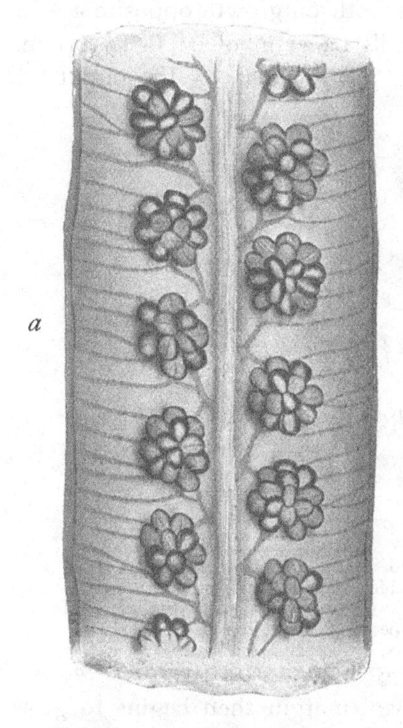

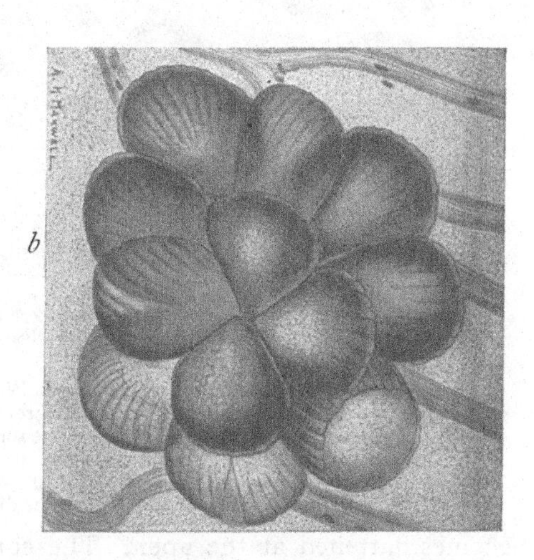

Fig. 488. *a*, *b*, *Gleichenia pectinata* Pr. *a*, shows the venation and the mature sori, some of which have dehisced. *b*, shows a single sorus, with its sporangia so closely packed as to be flattened against one another, one has already dehisced, one sporangium on the right is inverted. (Drawn by Mr Maxwell.)

faces obliquely towards the leaf-surface. These correspond to the basal and distal surfaces in the Schizaeaceae, both being constructed on essentially the same plan. (See Vol. I, p. 254. Fig. 250.)

There is considerable difference in size of the sporangia in the genus *Gleichenia*. Those species which have a small number of sporangia in the sorus, such as *G. rupestris* and *circinata*, have relatively large sporangia, and their spores are large (Fig. 489, *i*, *j*, *k*): those which have more numerous sporangia have them of smaller size, e.g. *G. linearis* (Fig. 489, *l*, *m*, *n*). In *G. circinata* the form is almost that of a kettle-drum: the stalk is short, and consists of a central group of cells surrounded by a peripheral series: it is thus thicker than in ordinary Leptosporangiate Ferns. The sporangium of *G. linearis* is more elongated, and smaller, and the stalk is thinner, and has no central group of cells. *G. flabellata* holds a middle position between these two types as regards size and shape, but it will be seen that the number of the spores produced in each sporangium is much higher than in either of the above species, though the spores are individually smaller than in *Eu-Gleichenia*.

The sorus of *G. flabellata* arises as a smooth outgrowth opposite a vein (Fig. 490, *a*), a considerable number of cells being involved in its origin. Having grown to a height almost equal to the thickness of the pinnule it

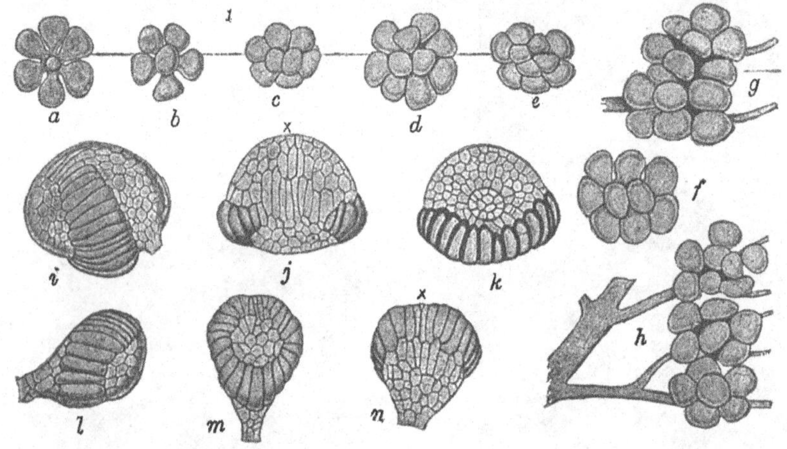

Fig. 489. *a–h* = sori of *Gleichenia dichotoma* Willd. *a–c*, show sori of radiate type, but with one or more sporangia in the centre, which is usually in this genus vacant. *f, g, h*, show states of apparent fission of the sorus. (*a–h* × about 14.) *i, j, k*, sporangia of *Gleichenia circinata* Sw., seen from different aspects: *x*, is the line of dehiscence: in *k* the broken stalk faces the observer. (× 50.) *l, m, n*, sporangia of *Gleichenia dichotoma* Willd., seen from different aspects. Note the difference in size from *G. circinata*. (× 50.)

becomes flattened at the apex. The convex margin then begins to grow out as rounded processes which develope into sporangia (*b*). There is some variety in the segmentation according to the size of the future sporangium. In the larger type of *G. flabellata*, or in *G. circinata*, a conical sporangial cell (×) is formed: in it successive oblique divisions follow without exact rule of sequence, the earlier of which go to form the massive stalk (*b, c*): the later define the central cell laterally, when a periclinal wall follows cutting off the cap-cell. This last appears at a time when the sporangial head projects but slightly from the surface of the receptacle, a state reminiscent of the Eusporangiate Ferns. Thus the whole sporangium is relatively massive, and results from segmentations more numerous than those seen in ordinary Leptosporangiate Ferns, though the last segmentations follow a sequence usual for them.

In the more attenuated *G. linearis* the sporangium is from the first more elongated, and its stalk less massive (Fig. 490, *g*). The cap-cell appears at a time when the sporangial head is more clearly in advance of the adjoining tissue, and the central cell is never actually immersed in the tissue of the receptacle. In fact the development is as in Leptosporangiate Ferns. Sometimes the central sporangia are absent (*g*): but when present they arise simultaneously with the rest, and occupy from the first the central area of the receptacle (*h*, *i*). Their presence cannot then be accounted for simply by displacement due to pressure: it is to be ascribed rather to interpolation

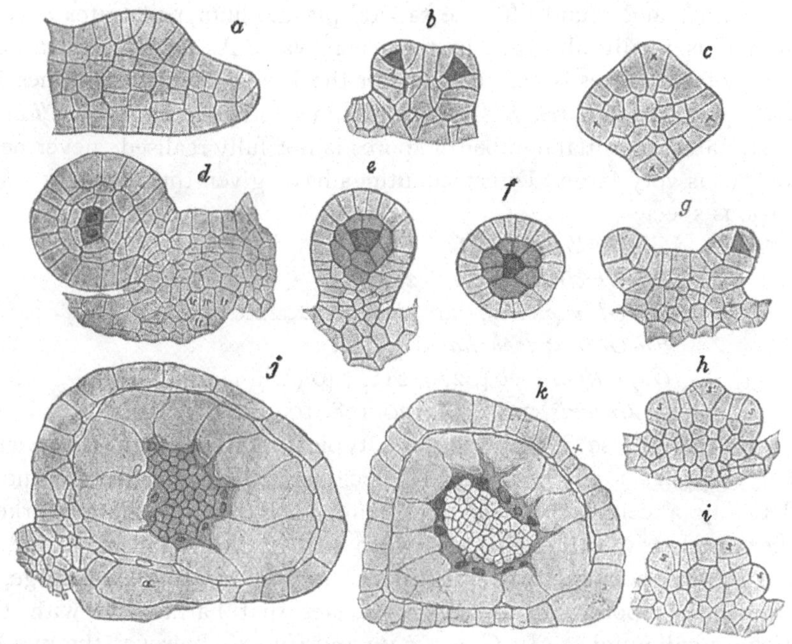

Fig. 490. *a*, *b*, *c*, sori of *Gleichenia flabellata*. *a*, *b*, in vertical, *c*, in horizontal section. *d*, *e*, *f*, sporangia of *G. circinata*, showing central cell and tapetum. *g*, *h*, *i*, sori of *G. dichotoma* (*linearis*): in *g* the centre is vacant, in *h* and *i* young sporangia appear in the vacant space. *j*, *k*, sporangia of *G. flabellata* with spore-mother-cells formed, and very numerous. (*a*–*i* × 200; *j*, *k* × 100.)

of accessory sporangia not normally present in the *flabellata*-type. It is important to follow these differences within the genus in detail, for they appear to provide within a near circle of affinity some interesting transitions between the Eusporangiate and the Leptosporangiate types of sporangial formation.

The divisions in the sporangial head follow in the main the Leptosporangiate type. The annulus soon makes its appearance, only part of it originating from the cap-cell. The tapetum divides as usual into two layers, of which the outer remains narrow and tabular, and traces of it may be found in the mature sporangium. The cells of the inner layer enlarge greatly, and become polynucleate: the cytoplasm and nuclei aggregate closely round the sporogenous cells, and the walls are finally absorbed (Fig. 490, *j*, *k*). The definitive sporogenous cell in *G. flabellata* undergoes successive divisions (Fig. 490, *d*, *j*), but these are continued beyond the limited number usual in Leptosporangiate Ferns, producing a very considerable mass of spore-mother-cells, of which 45 is a mean number exposed in a single vertical section (Fig. 490, *j*): their total number is probably about 360. If each of these on tetrad-division produced four spores the potential output would be 1440. As tetrad-division approaches the spore-mother-cells separate and round off: the tapetal plasmodium penetrates between them, and is finally absorbed in the usual way. As in the Schizaeaceae the form of the spores is not constant for the Family, being sometimes two-sided (*G. pedalis* and *Stromatopteris*), sometimes tetrahedral (*Eu-Gl. dicarpa*). The estimated potential number of spores is not fully realised: nevertheless the output is very large. Direct countings have given the following results for various species:

G. flabellata, 794, 695, 838, 634.
Eu-Gl. circinata, 241, 242.
Eu-Gl. rupestris, var. *glaucescens*, 220, 232, 244.
Eu-Gl. hecistophylla, 265, 272.
G. pectinata, 204, 229, 218, 240 (Williams).
G. linearis, 251, 319, 120, 128, 194, 228 (Williams).

The largest of these figures suggest a typical number of 512–1024, which probably applies for *G. flabellata*, a species regarded on other grounds as phyletically a central type for the genus. All the rest show markedly smaller numbers, mostly falling between such typical numbers as 128 and 512. It has been noted that in *Eu-Gleichenia* the sporangia are large, but the individual spores are also large, a fact that harmonises with their relatively small number. In *G. pectinata* and *linearis*, however, the numbers are relatively small and the sporangium is small also. An examination of it in the young state has shown that the number of the spore-mother-cells in a single section is about 15 to 20, and their total number may be computed as approximating to 64. This would give a probable number of spores per sporangium approximating to 256, though the actual countings are sometimes much lower (Fig. 491). It thus appears that within the genus there is considerable variety in sporangial spore-output. *G. flabellata* compares more nearly with the Eusporangiate Ferns than with the Leptosporangiates, while *Eu-Gleichenia* and particularly *G. pectinata* and *linearis*, with smaller numbers, correspond to the Osmundaceae and some Schizaeaceae in approaching the simpler Leptosporangiate state.

Stromatopteris is somewhat variable in the structure of the sporangium: but it is clearly of a Gleicheniaceous type, though sometimes the annulus is almost horizontal, and the distal face (the plate) is small as in *Schizaea*, while occasionally it is partially doubled. The spore-output is large, counts of "about 480" and "at least 416" having been made (Thompson). It thus falls in this character between *G. flabellata* and other species observed. *Platyzoma* is also peculiar in the variability of structure of the sporangium, as well as in the inconstancy of size and number of the spores (Fig. 492). The annulus is

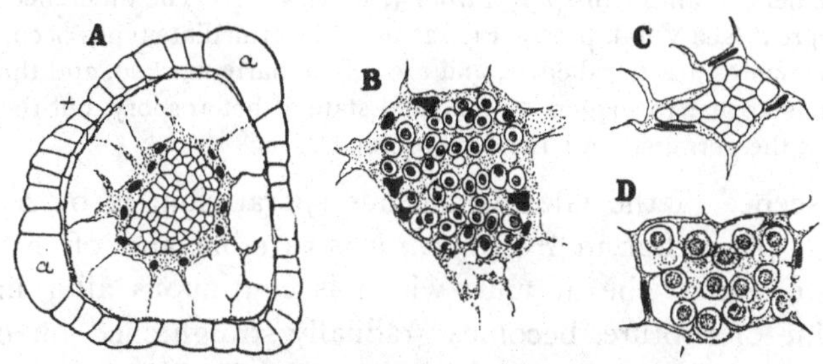

Fig. 491. *A* = section of a sporangium of *Gleichenia flabellata*, showing over 60 spore-mother-cells in section. (× 100.) *B* = spore-mother-cells of the same, older and separated from one another in the tapetal plasmodium. (× 165.) *C* = spore-mother-cells and tapetal plasmodium of *Gleichenia dichotoma*: only 20 are seen in section. (× 100.) *D* = cells separated in tapetal plasmodium. (× 165.)

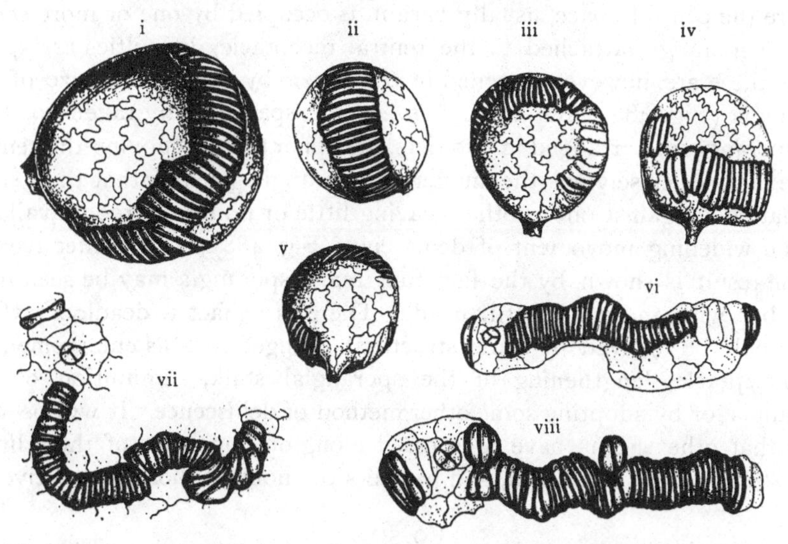

Fig. 492. Sporangia of *Platyzoma*, all of same magnification (× 24), showing variability in size and structure. i, ii, are extremes of difference in size, with oblique annulus, as in *Gleichenia*: iii, annulus irregular: iv, annulus almost transverse: v, annulus nearly vertical, interrupted at the stalk: vi, vii, ruptured sporangia, with 3-rowed stalk: viii, ditto with 4-rowed stalk. The stomium is not strictly defined, but the dehiscence is lateral. (After J. McL. Thompson.)

sometimes nearly horizontal, sometimes nearly vertical, and the sequence of its cells is very irregular. The stalk is typically three-rowed, but four-rowed stalks occur. The insertion of the stalk frequently interrupts the annulus, as it does in many Leptosporangiate Ferns, while the point of dehiscence, which is defined by a thin-walled, but not highly differentiated stomium, is clearly lateral. It thus appears that the sporangial structure of

Platyzoma links together by its irregularities the Gleicheniaceous ype of sporangium with that of the Leptosporangiate Ferns (Compare *Ceratopteris*, Thompson, *Trans. R. S. E.* Vol. liii, ii, Pl. vii). The sporangia of this Fern could be ranged after two types, large and small, the latter being in the majority. The spore-number is far lower than in any other Gleichenioid Fern. The large sporangia contain large spores numbering from 16 down wards, the smaller contain small spores, from 32 downwards. The difference in size of the spores is very great (see Vol. I, p. 265, Fig. 258), but intermediate types occur between the extremes. The spores are tetrahedral, and are all similarly marked, and they have thick walls. The facts appear to suggest an incipient state of heterospory, but there can be no proof of this till the germination has been observed

The dehiscence of the Gleicheniaceous sporangium is by a slit in the median radial plane, but in *Platyzoma* it is variable, and often lateral with a definite stomium. The annulus which is continuous all round, except along the line of rupture, becomes gradually straightened on drying, the whole sporangium thus widening laterally. Plainly this mode of dehiscence requires lateral space, and it is thus ill-suited for a crowded sorus, though it will serve well enough for lax sori, such as those of *Eu-Gleichenia*, or possibly for those of *Dicranopteris* where the centre of the sorus is unoccupied. Where the central space, usually vacant, is occupied by one or more short stalked sporangia attached to the central receptacle, difficulties are apt to arise: these are, however, lessened in *G. linearis* by the smaller size of the sporangia (Fig. 486). But in *G. pectinata* the space is fully taken up: here the sporangia are arranged in two tiers, the upper fully occupying the central space, and so closely are the numerous sporangia packed that their sides are flattened against one another, leaving little or no lateral room available for the widening movement of dehiscence (Fig. 488). The ineffectiveness of the result is shown by the fact that many sporangia may be seen fully ripe, but with the spores still inside. There is in fact a deadlock, which could only be relieved by some structural change: such as enlargement of the receptacle, lengthening of the sporangial stalk, diminution of the sporangia, or by adopting some other method of dehiscence. It will be seen later that other Ferns have progressed along one or other of these lines: but *Gleichenia* has stood still, and has adopted none of those alternatives.

FOSSILS

The *flabellata*-type of foliage is essentially Pecopterid (Fig. 476), but that does not define affinity: nor do the continued apical growth and peculiar branching of the leaf. Both of these features are such as might well be shared with plants of other relation, such as the Pteridosperms. The identification must depend upon the conjunction of such characteristics with a soral structure clearly Gleicheniaceous. The genus *Oligocarpia*, founded on Fern-like fronds from the Coal Measures bearing circular sori with few sporangia, has been quoted as proving the existence of Gleicheniaceae in Palaeozoic times (Fig. 493). Seward, however does not consider the evidence as conclusive (Hooker Lecture, 1922). The sorus is certainly of the Gleicheniaceous type, and the sporangia closely resemble those of *Dicranopteris*, but it is a question whether the annulus consisted of a single row of cells. Still this in itself is not distinctive, for *Stromatopteris* occasionally has an irregularly double annulus, and the like is sometimes seen in *Gleichenia*. Moreover, if the Zygopterid sporangia, with their broad annular band, represent a prototype of later Fern-sporangia, we should expect intermediate forms to show irregularities such as *Oligocarpia* appears to present. (See Kidston, ii, 4, p. 284.) But further, a specimen of *Oligocarpia* shown me by Dr Kidston seemed to suggest that the sorus was synangial. If it were so, which is doubtful since each sporangium is annulate, this again does not rule out a Gleicheniaceous affinity, for in the Marattiaceae synangial sori exist (*Danaea*, *Marattia*), while others have separate sporangia (*Angiopteris*, etc.). Moreover occasional synangia occur in modern Gleicheniaceae. The conclusion thus appears to be justified that fossils, having characteristic features showing a general resemblance to those of the Gleicheniaceae, certainly existed in the Coal Measures. But it is only in more recent geological time that the clear proof of the occurrence of Ferns definitely referable to the Family is forthcoming.

Fig. 493. Sorus of *Oligocarpia Gutbieri* Göppert, from a photograph by the late Dr R. Kidston. The annulus had the appearance of being a single series of cells. It was difficult on examination of the specimen to be sure there was no synangial fusion, but the presence of the annulus would make fusion appear to be improbable, as the annulus would then be functionless.

The oldest examples of authentic Gleicheniaceae are from the Keuper of Switzerland. From this period onwards, through the Jurassic to the Chalk, frequent records show that the Gleicheniaceae were widely distributed: but it is in the Cretaceous floras that the genus was especially abundant, extending even to West Greenland. In the Wealden of Belgium the recognition of the type is further supported by the discovery of rhizomes having the characteristic stelar structure (Bommer, Seward). Though it must therefore be admitted that the recognition of the Gleicheniaceae as definitely a Palaeozoic Family is doubtful, still it is certain that they already existed in early Mesozoic times, and flourished in the Chalk Period. Their record is very like that of the Schizaeaceae. Both families had already acquired in early Mesozoic time those characteristics by which they are recognised today, while less certain indications of both types reach back to the Palaeozoic Period.

The Gametophyte

The prothalli of *Stromatopteris* and of *Platyzoma* are unknown: observations are therefore restricted to the sections *Dicranopteris* and *Eu-Gleichenia*, but it will be seen that they present features of phyletic significance. The prothallus was first described by Rauwenhoff (*Arch-Néer*, xxiv, 1891, p. 157), but his observations have been extended into further detail by Campbell (*Bull. Jard. Bot. Buit.* viii, 1908, p. 88). It corresponds in development to that of the Polypodiaceae, though it may have a midrib like that of *Osmunda*, and has like it abundant adventitious growths formed upon the margin or

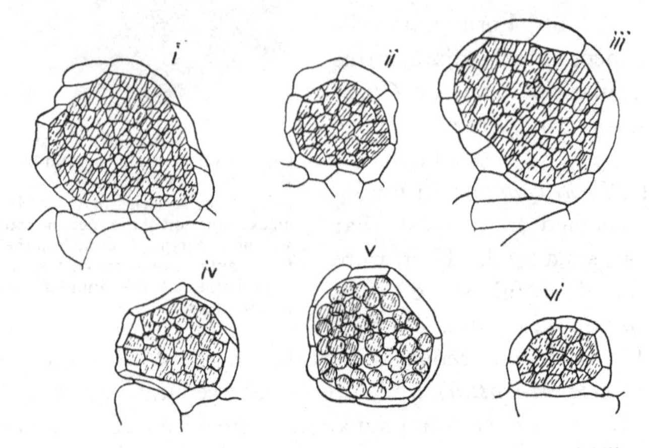

Fig. 494. Antheridia of the genus *Gleichenia*, showing their variability, and the large size of some of them: i, ii, *G. laevigata* (102, 35 spermatocytes in sections): iii, *G. pectinata* (66): iv, v, *G. dichotoma* (33, 45): vi, *G. polypodioides* (19). (All after Campbell.)

the ventral surface, and developing into normal prothalli (compare Campbell, *Mosses and Ferns*, 3rd. Edn. Fig. 208). The habit is almost that of *Fossombronia*: old prothalli contain an endophytic fungus. The archegonia have a long and straight neck, as in the Osmundaceae, and are scattered over the whole thickened midrib. The antheridia present more important features: they have been examined in several species. They are mostly large, and resemble those of *Osmunda*. In *G. laevigata* the diameter may reach 100μ, and each may contain several hundred sperm-cells: in fact those of *G. laevigata* and *pectinata* are larger than those of any Leptosporangiate Fern yet described (Fig. 494, i, iii). But this size is not constant for the species: when crowded, smaller antheridia may be formed (Fig. 494, ii), and these may contain fewer spermatocytes.

Reference has already been made to the parallel between spore-number and spermatocyte-number in various Ferns (Vol. I, p. 291). In no genus of

Ferns does this come out in so interesting a way as in *Gleichenia.* It has already been seen that the genus is variable in spore-output, which is larger in *Dicranopteris* than in *Eu-Gleichenia.* The spore-number in *G. flabellata* is higher than in any other recorded Leptosporangiate Fern. For the purpose of comparing this with the number of spermatocytes, a rough measure may be obtained by counting the number of these traversed in a single section, though this naturally falls far below the total: but it will serve for a rough comparison with similar sections through the mass of spore-mother-cells. The result is shown in the following table:

Name	Spermatocytes in section	S. M. cells in section	Spore-output
D. (*Gl.*) *laevigata*	(large) 102, (small) 35	—	—
D. (*Gl.*) *flabellata*	—	66	512–1024
D. (*Gl.*) *pectinata*	66	—	256
D. (*Gl.*) *linearis*	45, 33	20, 15	128–256
Eu-Gl. polypodioides	19	—	—
Eu-Gl. circinata	—	—	256
Eu-Gl. rupestris	—	—	256

From these data, which should be extended by other observers, it appears that, speaking generally, *Dicranopteris* has a larger spore-output per sporangium than *Eu-Gleichenia.* So far as the available evidence goes it has also a larger output of sperms per antheridium. These facts point directly to a primitive relationship for *Dicranopteris.* On the other hand, the spore-output for *Eu-Gleichenia,* though considerable, rarely exceeds 256 per sporangium. The only record to hand gives a very low figure for the spermatocytes per antheridium, in *Eu-Gl. polypodioides.* Both these features suggest for *Eu-Gleichenia* a derivative position compared with *Dicranopteris.* It need not be expected that the numbers should run exactly parallel: it may be held as remarkable that the parallel should come out as clearly as it does.

Gleichenia pectinata is thus seen to possess relatively large antheridia and many sperms, together with a relatively large spore-output: but *G. linearis* has fewer sperms per antheridium, while its spore-mother-cells (20, 15 in sec.) and spores actually counted (120–228) are fewer than in other species of *Dicranopteris.* These two species have been seen to stand apart in other respects from their Section, and this is borne out by the figures, especially for *G. linearis.* They confirm these species as holding a peculiar and an aloof position: they have hitherto been ranked with *Dicranopteris*, but are not typical of that sub-genus.

COMPARISON

While the Schizaeaceae with their four living genera are marked by the wide range of their characteristics, both formal and anatomical, the Gleicheniaceae are essentially a monotypic family: indeed at one time or another all the living representatives have been ranked under the single genus *Gleichenia* Nevertheless the various living species show variations both in their anatomical and their soral characters that provoke phyletic comparison: and this is all the more pointed since those differences appear in plants so clearly akin to one another. At the same time there can be no doubt of the relatively primitive character of the Family as a whole. This follows not only from the soral features, but also from the general morphology and particularly from the anatomy: with this also the geological evidence is in full accord.

The prevalence of equal dichotomy in the usually extended rhizome is significant for the genus *Gleichenia*, and it appears also to dominate the development of the upward-growing shrubby branches of *Stromatopteris*, though it is still uncertain what is the relation of these branches to the underground rhizome. In *Platyzoma* a dichotomous branching of the main axis appears occasionally. A general comparison of the shoot of *Gleichenia* with the dichotomous rhizome of *Lygodium* appears justified, while the form of the shoot of *Platyzoma* with its crowded leaves has features in common with the xerophytic *Mohria*. The leaves throughout the Family are based upon sympodial dichotomy: but it may run in *Gleichenia* to an unusual and very characteristic development, which depends upon a periodic interruption of the apical growth. Branching may be rare or absent, as it is in *Stromatopteris* and *Platyzoma*, and a like state is seen in the small Andean species *G. simplex*, which appears to retain permanently the type of leaf characteristic of the sporeling. Such plants serve to link the peculiar leaves of the larger Gleichenias with ordinary leaf-development as seen in primitive Ferns. The larger types are comparable when fully analysed with what is seen in *Lygodium*, allowance being made for difference of proportion of the parts, for the interruptions of apical growth in *Gleichenia*, and for the cessation of apical growth in the pinnae, which arrives earlier in *Lygodium* than in *Gleichenia*. There are no basal or stipular developments in either of these Families, such as characterise so many of the most primitive Ferns.

The most peculiar member of the Family as regards its leaves is the xerophytic *Platyzoma*, for it is heterophyllous (Fig. 482). The leaves are densely crowded, and are inserted all round the creeping axis. The fertile leaves are pinnate, resembling the simple leaves of the *Eu-Gleichenia* type. They appear in successive crowded zones upon the growing plant which alternate with zones bearing narrow linear sterile leaves of much smaller size. These betray their nature as arrested leaves of the same type as the fertile by the fact that abortive pinnation may be found at the apex of some of them. It thus appears that *Platyzoma* is a specialised Gleicheniaceous type.

The prevalence of protostelic structure in the Family is an index of its relatively primitive position, and compares with the like state in *Lygodium* which is also a genus with unlimited growth of the leaf-apex: and like it the **C**-shaped meristele of the petiole is liable to contraction so as to present a pseudo-stelic appearance. This is in accordance with the elongated straggling or climbing habit. The greatest interest, however, centres in the advance to typical solenostely of the axis, as seen in *G. pectinata*, which is a species specially advanced also in the type of its sorus. The anatomical facts provide material of special value for comparison with the Cyatheaceae. The partial advance of the stele of *Platyzoma* towards solenostely has as its special interest the absence of any foliar gaps, a fact to be related to the relatively small size of the leaf-trace, and the slight structural disturbance caused by its departure. This provides a pointed comment on the undue importance attached by some to the foliar gap, and the so-called "phyllosiphonic" state. This now appears to be simply a consequence of the relative size and influence of axis and leaf, and not a characteristic of the Class. *Platyzoma* presents in fact the anomaly of a solenostelic but not a phyllosiphonic type of Fern.

The dermal appendages of the Gleicheniaceae are very various. They may appear as soft, simple or branched hairs, as stiff bristles, or as more or less flattened scales (ramenta). These are often perched upon massive emergences of tissue, which probably correspond to those spines of the Cyatheaceae which persist as their prickly "armature." They are specially prevalent in both Families about the leaf-base. It is less clear in this Family than in any other what are the phyletic relations of these several appendages. But in Vol. I, Chap. XI the conclusion followed inductively from facts relating to many early Families of Ferns that the linear hair is a relatively primitive feature, and the flattened scale derivative. Any difficulty in respect of the evidence yielded by the modern representatives of this essentially Mesozoic Family cannot be accepted as invalidating the conclusion derived from a wide induction from the facts in other Families. Many of its living species are pronounced xerophytes, and this habit is recognised as promoting the formation of protective ramenta.

Hairs only, and no ramenta, are found in the two outstanding species, *G. pectinata*, and *linearis*, as well as in *Platyzoma*. In the first of these the hairs take the form of long stiff bristles, with a tuft of shorter divergent spines radiating from the base, each being seated on a massive emergence

(Fig. 475). It is but a slight step structurally from this to the scales seen in many species, including not only the xerophytic *Eu-Gleichenia*, but also representatives of the less specialised section *Dicranopteris*. For instance in *G. flabellata* the scales have a peltate insertion upon the stalk, with short bristles radiating from the margin and base, suggesting an origin from some branched hairs such as those of *G. pectinata* by broadening the basal region of the main bristle (Fig. 495). In this Family the dermal appendages do not greatly help the phyletic argument: for though *G. pectinata* and *linearis* appear to be relatively primitive in respect of their dermal appendages they are advanced in their anatomy, and in their soral characters. But it is a frequent experience that advance does not necessarily march parallel in all features compared, and this appears to hold for these two species: for though advanced sorally they bear appendages of more primitive type (see Studies II, *Ann. of Bot.* xxvi, p. 272).

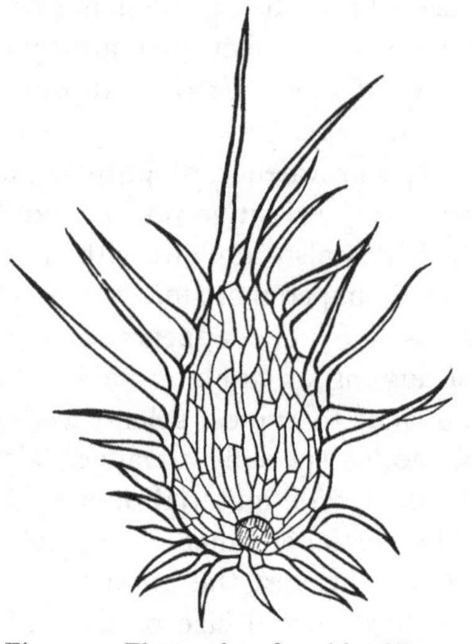

Fig. 495. Flattened scale, with stiff marginal spines, from the rhizome of *Gleichenia flabellata*. (× 20.)

The position and structure of the sorus of the Gleicheniaceae compares on the one hand with those of Marattiaceae and of *Todea*, and on the other with that of the Cyatheaceae and Matonineae. The absence of all indusial structures is a significant fact for comparison on the one hand with the Osmundaceae and other Simplices, on the other with the simpler Cyatheoid Ferns. It contrasts broadly with the presence of indusial growths in certain Schizaeoids and Dicksonioids. The departure of the sorus in *G. linearis* and *pectinata* from the uniseriate type characteristic of most Gleicheniaceae and of all the Marattiaceae is an important fact. It appears in species without scales, but one of them is advanced anatomically to full solenostely in the adult state, though the ontogeny shows how this emerges from protostely, and suggests a phyletic origin of this more advanced state. The same two species bridge over also an important step in sporangial structure. The details have been described above, but the net result appears in the spore-output per sporangium, which amounts in *G. flabellata* to numbers comparable only with those of Eusporangiate Ferns: *G. pectinata* and *Eu-Gleichenia* appear to take a middle position in the family, while *G. linearis* has the smallest output per sporangium, thus leading towards that relatively small number characteristic of the Leptosporangiate Ferns. The low output goes along with a larger number of sporangia in the sorus. Such features clearly indicate *G. pectinata* and *linearis* as holding an intermediate position between the Gleicheniaceae and certain Ferns ranked with the Cyatheaceae.

In respect of the gametophyte the intermediate features are no less significant. Its general form in the Gleicheniaceae is reminiscent of that of the Osmundaceae, a point supported by the long straight neck of the archegonia. But it is from the antheridia that the clearest indications are derived: for they are larger, and the spermatocytes are more numerous in such a type as *G. laevigata* than has been observed in any of the Leptosporangiate Ferns. But the numbers are variable within the species, and even in the individual: in certain species which may be regarded as specialised members of the Family the number of spermatocytes appears to fall to a figure approaching that usual in Leptosporangiate Ferns.

From these comparisons in respect of the several criteria mentioned it appears that the Gleicheniaceae hold a singularly interesting intermediate position between the Eusporangiate and the Leptosporangiate Ferns. This conclusion is in full accord with their palaeontological history. There are indications of the existence of similar Ferns in the Palaeozoic Period. But the existence of true Gleicheniaceae from early Mesozoic times onwards is fully made out, with a climax in the Chalk Period. Such a history might in itself suggest that intermediate position which is so amply proved by detailed comparison of the living representatives. Thus the two lines of evidence mutually support one another. There are probably no other genera of living Ferns that serve so effectively as *Gleichenia* and *Osmunda* to bridge the gap between the Eusporangiate and the Leptosporangiate Ferns. The early existence of the Gleicheniaceae, the Osmundaceae and the Schizaeaceae shows that in early geological time intermediate forms existed between these two branches of the Class. Their survivors of the present day serve to demonstrate the essential unity of the Class of the Filicales, and to mark some at least of those lines of modification which have been involved in the evolution of the more recent Leptosporangiate types.

PHYLETIC GROUPING OF THE FAMILY

The facts upon which a phyletic grouping of the Family may be based are embodied in the preceding pages. Little alteration is required from the systematic sub-division of the Gleicheniaceae now generally accepted. *Stromatopteris* (Mettenius, 1861) is upheld as a substantive genus. The further knowledge of the anatomy, heterophylly, and apparent heterospory of *Platyzoma* (Smith, 1793) justify its reinstatement as another substantive genus. The comprehensive genus *Gleichenia* (Smith, 1793) should on the ground of the new facts relating to its anatomy, its sorus, sexual organs, and dermal appendages be divided into three Sub-genera: viz. (i) *Dicranopteris* (Bernh.), a name already adopted in place of Willdenow's name

Mertensia, and covering the same species excepting as mentioned below: (ii) *Eu-Gleichenia* in the sense of the *Synopsis Filicum*: and (iii) *Eu-Dicranopteris*, under which name I propose to rank those two outstanding species, known as *G. pectinata* (Willd.) Pr., and *G. linearis* (Burm.) Clarke, the latter including *G. nervosa* (Klf.) Spr. Syst., which is held to be a variety of *G. linearis* in the *Synopsis Filicum*. Hooker had already separated these species from the rest of the genus, each in a distinct section, but without any special designation. They have in common the presence of hairs, not of scales as in the other two sections: the sorus has its centre not vacant as in other Gleicheniaceae, but occupied by one or more sporangia. Moreover, *G. pectinata* has solenostelic structure. These characters sufficiently distinguish this section from the other two.

As regards the phyletic relations of the genera and sub-genera thus distinguished, a central position is to be accorded to the sub-genus *Dicranopteris*, and particularly as represented by such a species as *G. flabellata*, on the ground of the Pecopterid-type of the leaves, with flat pinnules, and of the consistently protostelic structure of the axis: of the numerous uniseriate sori, with 4–6 sporangia, having relatively large spore-output from each, and sometimes with large antheridia and numerous spermatocytes. These features are all held to be relatively primitive, but they may be associated with scaly ramenta, and contracted petiolar meristeles, which are features of relative advance.

The sub-genus *Eu-Gleichenia* is held to comprise xerophytic derivatives from the central type, as shown by the contracted and often concave pinnules, usually with only a single sorus of very few sporangia, each having a relatively low spore-output. Ramenta are present. The same may be said of *Stromatopteris*, which is also regarded as a specialised xerophyte, its underground rhizome being an important feature in its success. The spore-output per sporangium is, however, larger than in *Eu-Gleichenia*, and the dermal appendages are very variable, including ramenta.

Platyzoma is also a highly modified xerophyte, as shown by the prevalence of sclerenchyma in the rhizome which has signs of advance towards solenostely, and is densely covered with hairs. The exiguous leaves resembling those of the *Eu-Gleichenia*-type and its isolated sori confirm this. The spore-output is the smallest in the Family, and it goes along with the peculiar heterospory. Thus *Eu-Gleichenia*, *Stromatopteris*, and *Platyzoma* may all be regarded as xerophytic derivatives from a type like *Dicranopteris*.

It will not be possible to arrange the Family in a simple linear sequence, so as to show correctly the phyletic relations of the genera and sub-genera. It must therefore suffice to enumerate them in catalogue form, giving in a few words the substance of the foregoing comparisons. *Gleichenia* itself undoubtedly holds a central position, and the genus is sub-divided thus:

I. GLEICHENIA, Smith. 1793. 79 species.

i. Sub-genus *Dicranopteris*, Bernh. 1806.

Includes probably the most primitive types, as shown by their large output of spores per sporangium, and of spermatocytes per antheridium. Protostelic axis. Protective scales.

ii. Sub-genus *Eu-Gleichenia*, Hooker. 1846.

Specialised xerophytes, with reduced leaf-surface, fewer sori, and sporangia, and lower spore-output from each. Protostelic axis. Protective scales.

iii. Sub-genus *Eu-Dicranopteris*, Bower. 1925.

Habit of *Dicranopteris*, but with hairs only. Sorus crowded, with as many as 10 or 12 sporangia, and a lower spore-output from each. Advanced stelar structure in *G. pectinata*.

II. STROMATOPTERIS Mettenius. 1861. 1 species.

Specialised xerophyte: scales and hairs: spore-output rather high. Protostelic rhizomatous axis.

III. PLATYZOMA, R. Br. 1810. 1 species.

Specialised xerophyte, hairs, not scales: heterophyllous: initial solenostely: annulus horizontal to vertical with lateral dehiscence: very low spore-output per sporangium: initial heterospory.

BIBLIOGRAPHY FOR CHAPTER XXIV

476. RAUWENHOFF. La Gén. Sex. d. Gleichéniacées. Arch. Néerl. xxiv. 1891.

477. BOWER. Studies IV. Phil. Trans. cxcii, p. 29. 1899.

478. BOODLE. Anatomy of the Gleicheniaceae. Ann. of Bot. xv, p. 703. 1901.

479. ENGLER & PRANTL. Natürl. Pflanzenfam. i. 4, p. 350. 1902.

479 *bis*. TANSLEY. Ann. of Bot. xix, p. 477. 1905.

479 *tris*. TANSLEY. New Phyt. vi, p. 135. 1907.

480. UNDERWOOD. Bull. Torrey. Cl. 34, p. 243. 1907.

480 *bis*. CAMPBELL. Prothallus of *Kaulfussia* and *Gleichenia*. Ann. of Jard. Buit. viii, p. 69. 1908.

481. BOWER. Origin of a Land Flora. 1908, p. 553.

482. SEWARD. Fossil Plants. Vol. ii, p. 351. 1910.

483. BOWER. Studies II. Ann. of Bot. xxvi, p. 269. 1912.

484. THOMPSON. Anatomy of *Platyzoma*. Tran. Roy. Soc. Edin. li, Part iii, p. 631. 1916.

485. THOMPSON. A further contribution on *Platyzoma*. Trans. Roy. Soc. Edin. lii, Part i, p. 157.

486. THOMPSON. *Stromatopteris*. Trans. Roy. Soc. Edin. lii, Part i, p. 133. 1917.

487. THOMPSON. Morph. of the stele of *Platyzoma*. Trans. Roy. Soc. Edin. lii, Part iii, p. 571. 1919.

488. GOEBEL. Organographie. II Aufl., p. 1043. 1918. Also Eng. Edn., p. 318.

489. CAMPBELL. Mosses and Ferns. 3rd edn., pp. 366, 635. 1918.

490. SCOTT. Studies in Fossil Botany. 1920, p. 263.

491. COMPTON. Plants collected in New Caledonia. Linn. Journ. xlv, p. 453. 1922.

492. SEWARD. Hooker Lecture. Linn. Journ. xlvi, p. 219. 1922.

CHAPTER XXV

MATONIACEAE

THIS family is represented by three species of living Ferns; *Matonia pectinata* and *M. sarmentosa*, are well known, both being of limited distribution in the Malayan region. A third species (*M. Foxworthyi* Copel.) has been described by Copeland from Amboina, with characters closely resembling those of *M. pectinata*. But Ferns referred to this affinity on the characters of leaf and fructification played a prominent part in the vegetation of the Secondary Rocks, and have been traced back as far as the Rhaetic period: this fact accords with the unmistakeable analogies which they show to the Gleicheniaceae.

Fig. 496. Habit of *Matonia pectinata* from a photograph by Mr Tansley of a group of plants in a wood on Gunong Tundok, Mount Ophir. (From Seward.)

The living species differ in habit. *M. pectinata* is a stout, ground-growing species, with elongated creeping rhizome covered by filamentous hairs, and branching in an apparently dichotomous manner. It bears solitary leaves at considerable distances apart, on its upper surface. These grow to a height of 6 to 8 feet, and have a characteristic pedate construction of the lamina, which is referable to a dichotomous system of branching (Fig. 496).

It is in point of fact a very perfect example of a catadromic helicoid dichopodium: even the middle lobe, which often appears to hold a terminal position, has been recognised as the inner branch of the second dichotomy (see Vol. i, Fig. 82 *A*, which is a much reduced photograph of a leaf). The segments themselves are pinnatifid, and the solitary sori are borne on their wings at points near to the midrib. Another species, *M. sarmentosa*, found strictly localised at Sarawak, Borneo, has dichotomous rhizomes, and straggling pendent leaves. At first sight the branching of its leaves appears quite different from that of *M. pectinata*: but this is due partly to the unequal development of the dichotomies, certain of the branches being represented only by arrested buds: partly it is due to their sympodial development: but still the dichopodial branching appears to hold for both. The branching in *M. sarmentosa* as worked out by Prof. Compton (*New Phyt.* viii, p. 204, 1909) is readily related to a primitive dichotomous type. The rachis forks

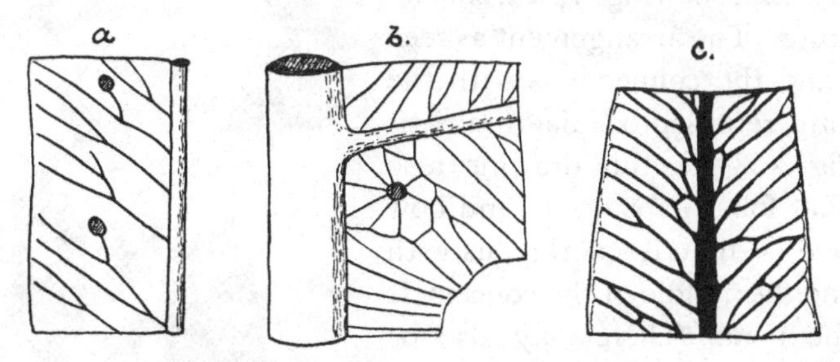

Fig. 497. *Matonia*, portions of leaf, to show the venation and vascular connections of sori. *a*, *Matonia sarmentosa*. *b*, *M. pectinata*. (After Diels.) *c*, *M. pectinata*, sterile region. (After Seward.)

repeatedly: either both branches may develope equally, and elongate, forking again and again: or one of the branches may develope fully while the other after bearing a few pinnae ends in an aborted bud. The latter is the more usual, as is seen in Diels' illustration (Engler and Prantl, i. 4, Fig. 183). In both species the venation is dichotomous, and mostly free (Fig. 497), but in *M. pectinata* there are vein-fusions (*c*), and especially about the insertion of the sori (*b*), though even these are absent in the more attenuated *M. sarmentosa* (*a*). The sori of the former species are large, and are seated solitary on a plexus of veins: those of *M. sarmentosa* are smaller, and more numerous, forming a row on either side of the midrib, as in *Gleichenia*. In both of these closely related species there is an overarching indusium formed from the apex of the receptacle.

The dermal appendages of the genus are simple hairs, often thin-walled at the base, but with the acute tip indurated; in fact they have the same structure and appearance as those of *Gleichenia pectinata*, but without any basal emergence.

The genus *Matonidium* occurs in the Jurassic of Yorkshire (Seward, *Jurassic Catalogue*, Vol. I). The genus *Matonia* has been traced back to the Cretaceous Period. Krasser has stated that *Matonidium Wiesneri* from the Cretaceous of Kronstadt (Moravia) is indistinguishable from *Matonia pectinata* itself.

ANATOMY

The adult rhizome of *Matonia* possesses one of the most complicated solenostelic structures known among Ferns: in the young stem, however, simpler conditions are found which suggest how the final structure was arrived at. In the most complex rhizomes, which may be about 8 mm. in diameter, three concentric vascular rings are seen embedded in parenchyma, each showing typical solenostelic structure. The arrangement as seen at a node, and the connections with the leaf-trace, are represented diagrammatically in Fig. 498, *C*: this drawing also indicates that foliar gaps occur, and how the leaf-trace is directly continuous with the outer and the middle of the concentric rings at the node. There may also be connection with the innermost ring, but this occurs at some little distance from the actual node, and so it is not shown in the drawing. The result is that the whole vascular system is connected, but only at intervals of its whole length; the parenchymatous tracts in which the cylinders are embedded are also connected through the leaf-gaps.

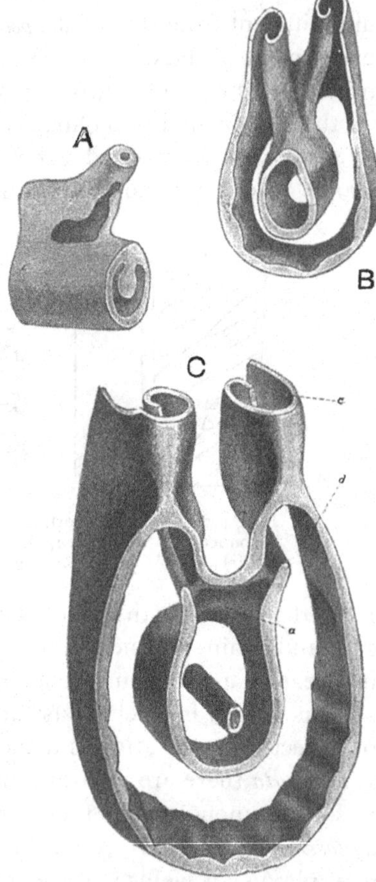

Fig. 498. Drawings from wax models of the stelar system of *Matonia pectinata*. *A*, from a young stem showing a node. *B*, from an older stem, showing a node seen from behind. *C*, still older node, seen from in front. *A* × 25. *B* × 12. *C* × 10. (After Tansley and Lulham.)

The ontogeny suggests how this complicated structure is related to what is seen in other and simpler Ferns. The young axis contains at first a slender protostele: but this soon expands, and a strand of phloem appears in the midst of the xylem. This internal phloem appears to be a phloem-pocket decurrent from the axil of the next higher leaf, but there is as yet no true leaf-gap. The stele soon widens into a solenostele with internal endodermis and central parenchyma. Meanwhile at the nodes a ridge of xylem projects internally, which becomes more prominent at subsequent nodes, and is continued forwards into the internode further and further at successive nodes, till that of one node eventually connects with a similar xylem-dilatation of the next node (Fig. 498, *A*). A continuous central strand is thus produced, which is connected at the nodes with the outer cylinder. The process thus described may then be repeated in that central strand: it thus becomes cylindrical, forming the second vascular ring, which is still connected at the nodes with the foliar system (Fig. 498, *B*), and a fresh strand may originate internally from it: this in its turn becomes cylindrical in the most advanced types, but still maintains its connection with the middle and outer rings in the neighbourhood of the nodes. The whole development is in fact an extreme type of the elaboration of the solenostele described by Gwynne-Vaughan in other Ferns (*Ann. of Bot.* xvii, p. 703). The smaller and more lax species, *M. sarmentosa*, with a rhizome about 4 mm. in diameter, stops short before this extreme complication is reached: it appears to settle down to a condition between (*A*) and (*B*) in Fig. 498, according to the observations of Compton (Fig. 499). All these facts suggest that size has an influence upon the complexity of the vascular structure.

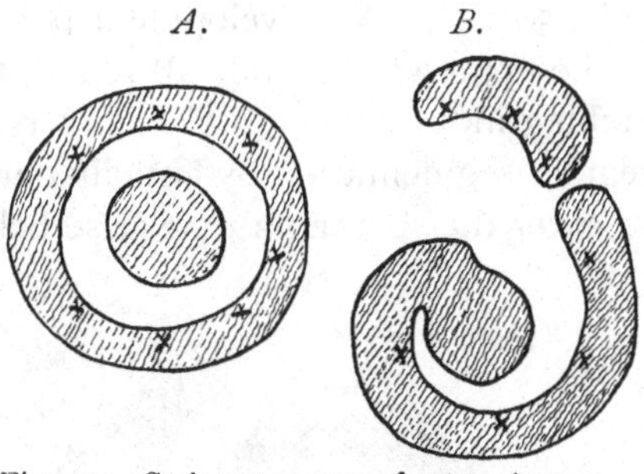

Fig. 499. Stelar structure of *Matonia sarmentosa*, after Compton. *A*, an internode: *B*, structure at a node, with leaf-trace. The protoxylems are indicated by crosses.

The leaf-trace as it passes up the petiole becomes a deeply inrolled meristele in *M. pectinata*, but less so in *M. sarmentosa*. Here the trace comes off as that of *Gleichenia* does (Fig. 499, *B*, compare Fig. 478, p. 197). It possesses three protoxylems, as in that genus, and the upper part of its petiolar course shows features of the meristele strikingly like those of *Gleichenia*, but not so strongly contracted (Compton, *l.c.* Fig. 39). Finally, the origin of the pinna-traces is extra-marginal, as in *Gleichenia*.

The roots are found to be triarch.

Spore-producing Members

The anatomical similarities thus disclosed between *Matonia* and *Gleichenia* provoke comparison of their sori also. It will be found that while the position and general plan of the sori is the same, there are differences in many details. The sporangia of the sorus in *Matonia*, commonly six to nine in number, form a simple ring-like series round the receptacle, and are covered till maturity by a thick and leathery hemispherical indusium, which is ultimately deciduous. The orientation of the sporangia is not uniform: that of the majority is as in *Gleichenia,* but many have the annulus inclined, a consequence probably of crowding. This is seen also in the fossil *Laccopteris* of Rhaetic and Jurassic time. The annulus is incomplete, with an ill-defined stomium, while the rupture is by a ragged lateral slit, opened by the straightening ring (Fig. 500). The sorus originates as a smooth upgrowth from the lower surface of the pinnule, opposite to a vein, a considerable number of cells being involved from the first: no definite mode of segmentation has been recognised (Fig. 501, *F*). As development proceeds the margin of the upgrowth extends all round, to form the over-arching indusium (*i, i*): this, undergoing a somewhat regular segmentation by anticlinal walls, curves so as to cover the sporangia that arise below (*s*, Fig. 501, *F*). The indusium thus

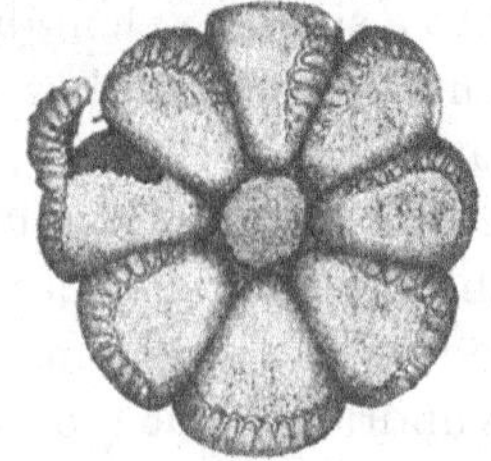

Fig. 500. A sorus of *Matonia pectinata* from which the indusium has been naturally shed. (After Seward.)

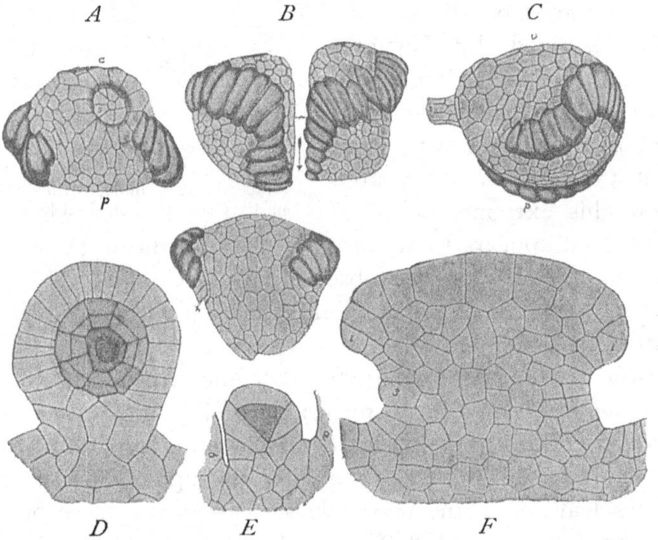

Fig. 501. *Matonia pectinata. A, B, C,* as well as the central figure, represent mature sporangia in various aspects. *F*, young sorus: *i, i*, indusium: *s*, sporangium. *E*, young sporangium with cap-cell formed: *a*, its acroscopic, *b*, its basiscopic side. *D*, sporangium with tapetum doubled. (*A–C* × 50: *D–F* × 200.)

precedes the appearance of the sporangia, as in many other indusiate Ferns. The sporangia originate from single cells which have commonly a square base, though it may be a question whether this is always so. The segmentation is by walls inclined to one another: the segments thus produced surround a central triangular wedge-shaped cell, from which finally the cap-cell is cut off in the usual way (*E*).

The further segmentation of the central cell follows the course usual in Leptosporangiate Ferns: a double tapetum is formed (*D*), of which the inner cells become greatly enlarged, and their nuclei clustering round the sporogenous group of cells and undergoing fragmentation, present an appearance very like that in *Gleichenia*. The central cell divides into 16 spore-mother-cells, and the typical number of spores should be 64: but countings gave figures between 48 and 64 as the produce of single sporangia. Sections of sporangia so cut as to traverse the annulus throughout its course show the wall as a single layer, but composed of more numerous cells than is commonly the case in Leptosporangiate Ferns (*D*). The same is seen by examining sporangia from without (*A*, *B*, *C*). Further, the short stalk is rather massive, and consists of a peripheral series of six or seven cells surrounding a central cell (*A*); this structure corresponds in its general features to that seen in the massive sporangia of *Gleichenia* and *Osmunda*.

The mature sporangium is rather variable in form, owing probably to pressures in the developing sorus. The annulus is incomplete and variable in position, consisting of 20 or more cells which form a sinuous group, but corresponding in the main with that in *Gleichenia*. The sporangia are often tilted right or left, as in Fig. 501, *B*, which represents two sporangia *in situ* as seen from the side facing the indusium. Sporangia in which the annulus is not tilted (*A*, *C*), show that it starts close to the stalk on one side, and pursues a sinuous course round the head, but stops short at some distance from the stalk at the side remote from the starting point. It is here that the dehiscence takes place, but there is no specialised stomium. The sporangium at the centre of Fig. 501 is ruptured in this way. Thus the sporangium of *Matonia*, while resembling that of *Gleichenia* in its general features, differs from it in the variability of its structure, its lateral dehiscence, and in the comparatively small output of its spores. While the general comparison is thus with the Gleicheniaceae, the correspondence of the details to those already described for *Platyzoma* is remarkable (p. 209). The abnormalities noted in that Fern serve to link the sporangial structure of *Matonia* with the Gleicheniaceous type: for they illustrate the interruption of the annulus at the insertion of the stalk, and the assumption of the lateral in place of the median dehiscence, though the stomium is ill-defined. The spore-output per sporangium was found in *Platyzoma* to be 16 or 32, in strong contrast to the much higher numbers in *Gleichenia*. But the output in *Matonia* has not sunk so low as this, being typically 64. These facts serve to connect the sporangial structure of the Matoniaceae with that of the Gleicheniaceae. The plan of the sorus is the same in both, the chief innovation appearing to be the presence of the indusium. But exuberant growth of the receptacle at the vacant centre of the sorus readily accounts for this biologically intelligible addition, while comparison may be made with the somewhat similar vegetative growth from the apex of the sporangiophores of *Helminthostachys* (Fig. 364, *G*).

Little is known of the prothallus of *Matonia*; a green prothallus of the usual type has been seen attached to young sporelings, but it has not been examined in detail. The embryology appears to be that usual for relatively advanced Ferns; the cotyledon is either simple or branched, and its venation is a scorpioid sympodium, as it is also in *Gleichenia* (Fig. 502).

Fig. 502. *Matonia pectinata*, prothallus and young plant, collected by Lang on Gunong Ladang, Malay Peninsula.. (×4.)

The genera *Laccopteris* and *Matonidium* are Mesozoic Ferns showing very close agreement with *Matonia*. The former had a scorpioid leaf like *Matonia*, with venation dichotomous and usually open. The sori were also of the same type, but the existence of an indusium is in doubt. *Laccopteris* was widely distributed in Rhaetic, Jurassic, and Lower Cretaceous Floras. Its pinnae show frequent anastomoses of the veins, while the sorus is also of the *Matonia*-type, but it contains sometimes 12–14 sporangia disposed in a ring, each with an oblique annulus (Fig. 503).

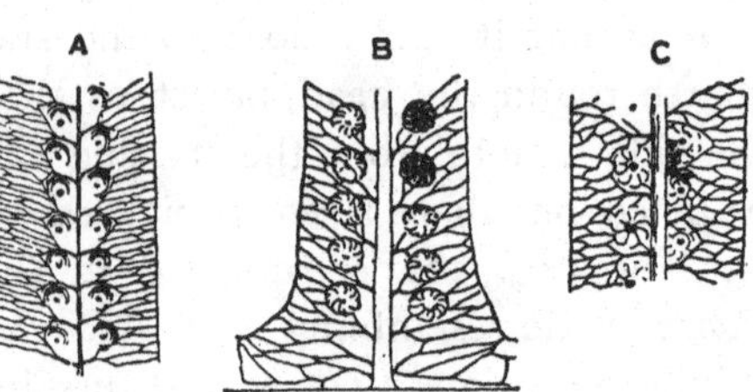

Fig. 503. *Laccopteris Woodwardi*, from inferior oolite of Yorkshire. *A*, a pinnule: the hemispherical bosses show the position of the sori (No. 257 Brit. Mus.). *B*, pinnule of *L. polypodioides*, with sori and soral impressions, from Gristhorpe Bay (No. 2522 Brit. Mus.). *C*, pinnule-fragment from the inferior oolite of Stamford (No. 52867 Brit. Mus.). (After Seward, from drawings by Miss G. M. Woodward.)

The conclusion which may be drawn from a comparison of the Matoniaceae with the Gleicheniaceae is that the former represent a type specialised early from a common stock with the latter. Both have a creeping rhizome, with stelar structure either protostelic or a solenostelic derivative from it: the difference is bridged over by *Platyzoma*, *Gleichenia pectinata*, and *Matonia sarmentosa*. The last of these leads to *M. pectinata* which is the most elaborate solenostelic Fern yet observed. The leaf architecture in both is along similar lines, with a strong scorpioid tendency in *Matonia* which appears also in certain species of *Gleichenia*, e.g. *Gl. Cunninghami*. In *M. sarmentosa* the arrested higher branchings in the leaf are strangely reminiscent of *Gleichenia*. The venation fluctuates between the open system of

Gleichenia, seen also in *M. sarmentosa*, and in *Matonidium Wiesneri* (Seward, *l.c.*, Fig. 265, *A*), and a coarse and irregular reticulum seen in variable degree in *M. pectinata*, and well represented in *Laccopteris*. The simple hairs of *Matonia* correspond to those of *Platyzoma*. The sorus is plainly Gleichenioid, but with modifications of the sporangia the most striking of which are already exhibited in the Gleicheniaceous *Platyzoma*. The distinctions between the two families are thus relatively slight: and a common parentage appears to be the natural explanation of the similarity that exists between them: but the divergence must have dated not later than early Mesozoic time. Seward ascribes the origin of the generic type of *Matonia* to the Northern Hemisphere, in the Triassic or early in the Jurassic Period. It reached its maximum development in the Mesozoic Era, and became restricted geographically towards the close of the Cretaceous Period. At the present day it survives only in the Malayan region.

BIBLIOGRAPHY FOR CHAPTER XXV

493. SEWARD. *Matonia pectinata*. Phil. Trans. cxci (Series B), p. 171. 1899.
494. BOWER. Studies IV. Phil. Trans. cxcii, p. 44. 1899.
495. ENGLER & PRANTL. Natürl. Pflanzenfam. i, 4, p. 343. 1902.
496. TANSLEY & LULHAM. *Matonia pectinata*. Ann. of Bot. xix, p. 475. 1905.
497. TANSLEY. Filicinean Vasc. Syst. New Phyt. Reprint. Cambridge. 1908.
498. COPELAND. Philipp. Journ. Sci. 3 (1908) Bot., p. 342, Tab. 2.
499. COMPTON. *Matonia sarmentosa*. New. Phyt. viii, p. 299. 1909.
500. SEWARD. Fossil Plants. Vol. ii, p. 155. 1910.
501. CAMPBELL. Mosses and Ferns. 3rd edn, p. 371. 1918.
502. SEWARD. Hooker Lecture, 1922. Linn. Journ. xlvi, p. 219.

CHAPTER XXVI

A GENERAL REVIEW OF THE SIMPLICES

In the foregoing Chapters those Ferns have been examined which fall under the designation of the Simplices. They have in common the feature that the sporangia of each sorus originate simultaneously, and come to maturity at approximately the same time. The sporangia are of relatively large size, and each produces numerous spores. Since the sori borne on any individual pinna, or even on a whole leaf, disclose only a slight acropetal succession in the time of their appearance, the physiological drain for their nutrition is practically simultaneous. This is regarded as a simple and primitive state as compared with the more elaborate arrangements seen in other Ferns, by which the demand for nutrition is spread over a long period. The Simplices include all the living Eusporangiate Ferns, together with some others which approach the Leptosporangiate state in the structure of their sporangia, though these are not always referable in origin to a single parent cell, which is the characteristic of that Sub-class. It is so with the Osmundaceae, Schizaeaceae, Gleicheniaceae, and Matoniaceae, Families which occupy an intermediate position between the Eusporangiate and the typical Leptosporangiate Ferns.

It is significant that all the Families included thus within the Simplices, with the exception of the Ophioglossaceae, have an early geological record. The Coenopteridaceae, known only as fossils, are the characteristic Ferns of the Palaeozoic Period. Ferns of the type of the Marattiaceae probably date back to the Upper Carboniferous and the Permian Rocks, and are represented in the Trias. The fossil Osmundaceae flourished from the Permian Period onwards, while the Gleicheniaceae and Schizaeaceae, though their typical features are strongly suggested by certain Palaeozoic fossils, undoubtedly formed with the Matoniaceae a prominent constituent in Mesozoic Floras. The only Family of the Simplices to which an early fossil history cannot be ascribed by direct observation is the Ophioglossaceae. But the characters which they bear leave little doubt that they are really related to other early Fern-types, and in particular to the Botryopterideae. We are now able to state on good palaeontological authority that "it is doubtful if the distinction between the Eusporangiate and Leptosporangiate Ferns existed in Palaeozoic times:—in other words, whether the development of the sporangium from a single cell had yet been arrived at" (Scott, *Studies*, 3rd Edn. p. 366). Kidston has also expressed himself in very similar terms. When this result is coupled with the positive existence of most of the types of the Simplices in the Palaeozoic Period, there can be little doubt that we are right in recognising these as the primitive representatives of the Class. This conclusion, thus fully substantiated from the Fossil Record, emerges equally as the result of comparative study. The broad ground of priority thus taken up is the natural starting point for phyletic seriation.

Making use of the criteria established in Volume I, it appears that the Coenopteridaceae, and especially the Botryopterideae, take their place as the most primitive of the Families of Ferns hitherto recognised as such. The dichotomous branching and solid protostele of the stem of *Botryopteris*, its monodesmic leaf-trace, oval in section and with a single protoxylem at its departure, the hairs consisting of a simple row of cells, the leaf with narrow dichotomous segments, and the Eusporangiate sporangia, marginal or distal, with large spore-output and complex annulus, are characters that define an extremely primitive type. When it is added that *B. antiqua* comes from the Petticur Beds (Calciferous Sandstone), it is clear that this primitive type was very ancient. It was also of small size, as were many other very early fossils. It has been suggested that *B. antiqua*, *cylindrica*, *racemosa*, and *forensis* form a series of progressive elaboration within the genus, as they do also in horizon, up to the Permo-Carboniferous of France.

The Zygopterideae are larger, and characteristically more elaborate than the Botryopterideae, though undoubtedly related to them. Stratigraphically they are recorded from the Upper Devonian, and they extended to the Permian, after which they appear to have died out. It is especially in the foliar development that they are remarkable, whether by vascular structure or by form. Indeed the Dineuroideae provide a unique type of leaf with four rows of pinnae, two rows on each side of the rachis which was probably vertical. This habit is also seen in *Stauropteris*, a common fossil of the Coal Measures, of which the axis, if indeed such a part existed, is still unknown. The stele of the stem of the Zygopterids is advanced both in form and in structure, in proportion to its size, beyond that of the Botryopterids. But though their whole vegetative system appears to have been more elaborate, the Eusporangiate sporangia were still distal or marginal, and the spore-output high, while mechanism of distribution was rudimentary. Thus notwithstanding the enhanced vegetative system, the Zygopterids are properly ranked with the Botryopterids as very primitive Filicales.

Comparison, unfortunately without the support of fossil history, justifies a position for the Ophioglossaceae in relation to the Coenopteridaceae. They share with them many characters, such as moderate size and simple form, with simple hairs as dermal appendages. The leaf is bifacial, but there is in the spike of *Helminthostachys* a structure which offers the nearest similarity to the Dineuroid or Stauropterid frond that is known among living plants. The stem has a simple stelar structure, but with more advanced medullation than in any Coenopterid; moreover in each of these families occasional cambial thickening may occur. The leaf-trace in *Helminthostachys* and *Botrychium* is always undivided, but the disintegrated stele in *Cheiroglossa*, and the divided leaf-trace in *Ophioderma* are features shared only with relatively advanced Ferns: moreover the reticulate venation of the whole genus *Ophioglossum* is itself a Mesozoic feature: it is not shared by *Helminthostachys* or *Botrychium*, nor by any of the Coenopteridaceae, all of whose fronds are of the open, Palaeozoic type of venation. Thus *Ophioglossum* shows definite signs of advance. The large sporangia in some Ophioglossaceae compare readily with those of certain Coenopterids, especially where, as in *Botrychium* and *Helminthostachys*, a vascular strand runs to the base of each, as it does also in *Stauropteris*. The opening mechanisms are simple. The spore-output is large: in *Ophioglossum* it is probably the largest in any Pteridophyte. But the mere numerical total of spores cannot be held to prove that genus to be the most primitive of all, as against the relatively advanced features of its vascular system. The sexual organs and the embryology in the Ophioglossaceae have been seen to be comparable to those of other living Eusporangiate Ferns. The sum of these characters justifies the general position here assigned to them, but with a special relation to the Coenopteridaceae.

The Marattiaceae bear typical Eusporangiate sporangia, disposed in sori of a primitive type, often synangial, and borne on leaves which are also primitive in many features, having as a rule open dichotomous venation and bearing dermal hairs: scales are present only in the synangial *Christensenia* and *Danaea*, which may be held as advanced on both of these grounds. Anatomically all the Marattiaceae are relatively advanced. Their sappy stems and leaf-stalks contain a vascular system in a high state of disintegration, which finds a feeble analogy in the sappy stem of *Ophioglossum palmatum*. But it is prefigured by the Psaronioid stems of the Permian Period, which have been held as related to the Marattiaceae. It seems uncertain whether these, and certain Fern-like fronds from the Upper Palaeozoic floras bearing sori of Marattiaceous type, should be included in the Family as at present defined. Seward states (Hooker Lecture, p. 235, 1922) that the oldest examples of fronds with fertile pinnae agreeing generally in habit with modern Marattiaceae are from the Upper Triassic beds, while the maximum development of the Family as we now know it seems to have been in pre-Cretaceous times. It is clearly an ancient stock still surviving. The characters of the gametophyte, sexual organs, and embryology accord with this conclusion.

The Osmundaceae which date from the Permian Period are, of all the early Ferns, to be held as most nearly representing a central stock from which many Leptosporangiate Ferns may have taken their origin. They have long been recognised as occupying a position intermediate between the Eusporangiate and the Leptosporangiate Ferns, and their early fossil history justifies this view. Not only are their soral and sporangial characters in accord with it, but their anatomy with its tentative advances towards medullation, solenostely, and in the largest types the final disintegration of the stele, the primitive trace widening into the horse-shoe, and the hairs as dermal appendages all support the position thus assigned. The gametophyte and sexual organs point rather to a Leptosporangiate affinity than to a Eusporangiate, though still with archaic features.

All the Families named in the foregoing paragraphs are regarded as specially primitive. None of them can be related as direct progenitors to any other living Family: so that it is in a general rather than a special sense that these Ferns may be held as prefiguring those later developments embodied in the Leptosporangiate Ferns. The Osmundaceae may, however, be regarded as synthetic forms in one feature which has stamped itself deeply on the later evolutionary progress. Of its two living genera *Osmunda* bears its sporangia normally in marginal tassels on the narrow fertile pinnae; while *Todea* bears them on the lower surface of its expanded pinnae. But where the fertile pinnae of *Osmunda* expand also, as they do sometimes abnormally, to a broad form, the sporangia are in like manner found on the lower surface. The two genera in their normal development prefigure on the one hand the marginal type of sorus, on the other the superficial type; and they supply a link between the two, while they suggest that an expanding leaf-surface encourages a superficial insertion. It is, however, a fact based upon wide observation that these two distinct soral positions are strictly maintained in many large groups of Ferns. For instance, in the Gleicheniaceae, Matoniaceae, Cyatheaceae (excl. Dicksoniaceae) the sori are all superficial, and remain so: in the Schizaeaceae, Hymenophyllaceae, Dicksoniaceae they are all marginal at least in their origin, and (excepting in the Schizaeaceae) they remain so up to maturity. Thus the distinction visualised within the Osmundaceae becomes one of the most stable characters for comparison of the early Leptosporangiates, and accordingly that ancient Family may be held to be synthetic in respect of it. It applies at once to the almost equally ancient Families of the Schizaeaceae and the Gleicheniaceae, the former being marginal, the latter superficial in origin of sori. Further, the relatively late and derivative Leptosporangiate Ferns constitute two main sequences related more or less directly to them. These sequences have been designated respectively the *Marginales*, retaining with greater or less persistence the marginal position as seen in the Schizaeaceae; this was probably the original position of the sporangia for all Ferns: and the *Superficiales*, which have their sori borne on the lower surface of the leaf, as in the Gleicheniaceae: this was probably in the first instance a derivative state consequent upon broadening of the foliar surface.

The distinction thus to be applied systematically depends for its value upon the high degree of constancy of the character used, and upon the relatively early establishment of it as shown by the fossil evidence. The leading types of the *Superficiales* have possessed superficial sori since Palaeozoic times. But the distinction is not absolute. As has been fully explained in Vol. I, Chap. XII, the passage of the sorus from the margin to the surface of the broadening sporophyll may be followed in the individual development among the Schizaeaceae of to-day, while comparison of living species and genera in the Dicksonioid-Davallioid series demonstrates a similar change of position as not unfrequent in them. The first impulse may naturally be then to rule out this distinction as systematically worthless. *But the point is that in certain Families the event happened as early as Palaeozoic time, and the result of it has remained constant ever since; in others the originally marginal position has been retained with high constancy, though in some instances it is even now in course of being departed from.* The former state is what we see in the *Superficiales* as exemplified by the Gleicheniaceae, and the same holds also for the Marattiaceae: the latter is what we see in the *Marginales* as exemplified by the Schizaeaceae, and the Ophioglossaceae retain persistently that primitive character. Ultimately both sequences are probably based upon a prevalence of narrow leaf-segments in the remote ancestry, with a distal sporangium borne upon the tip of each, as seen in *Stauropteris*. Webbing would result in a flattened leaf-surface with marginal sporangia: further widening, and slipping of the sporangia, earlier or later in descent, to the lower surface would give the state so frequently seen in modern Ferns. In the classification here adopted these Ferns will be ranked systematically according as they retained the primitive marginal position, or departed from it early or late in the history of descent.

Professor von Goebel holds that this distinction is an artificial one, in particular as applied to the Dicksoniaceae and Cyatheaceae (*Organographie*, ii, p. 1194, 1918). I hope that that opinion may be reconsidered by my esteemed colleague, when he takes into full account the palaeontological and anatomical facts here detailed, together with the high degree of constancy of the one soral position and of the other in the large series related on the one hand to the Schizaeaceae, and on the other to the Gleicheniaceae; also the very gradual steps of emergence of later terms of these series in relation to those very ancient stocks. It will then appear that the distinction is supported by a very great body of fact, and by wide comparisons. The stigma of artificiality cannot justly be placed upon the results of such wide comparative argument. The distinction according to soral position would rather appear to be a Natural Classification in the fullest sense of the word. It will be maintained and elaborated in this work, due attention being given to those exceptions upon which adverse criticism may be founded. Relatively recent transitions from a marginal to a superficial position may often be traced: but these do not prejudice the value of the distinction where it appeared in early geological time, and has been tenaciously held since.

In each of the two principal sequences or phyla of Leptosporangiate Ferns, prefigured respectively by the Schizaeaceae and the Gleicheniaceae, evolutionary advances may be traced involving many of the recognised criteria of comparison: especially those of anatomy, soral constitution individuality and protection, and of sporangial structure and spore-output: while the venation and dermal appendages provide accessory evidence. It will be found that the two main phyla, and certain other subsidiary sequences, run in parallel progression, culminating in types which are characteristic of the majority of Ferns living at the present day. The Schizaeaceae and Gleicheniaceae, like others of the *Simplices*, have sori inherently limited in their capacity for spore-production. In the Schizaeaceae the marginal sporangia are solitary, and each has been designated a "monangial sorus." The superficial sori of the Gleicheniaceae have at least a plurality of sporangia: but it has been seen how the construction of the sorus if advanced to its full capacity leads to a mechanical deadlock, as in *Gleichenia* (*Eu-Dicranopteris*) *pectinata*. Moreover in both that inherent weakness of the Simplices remains, viz. the simultaneous drain of all the sporangia for nutrition. It will be shown how the nutritional difficulty has been met by the introduction of the Gradate and the Mixed types of sorus in both phyla, together with other modifications. A succession of sporangia is thus provided, so that the drain shall be spread over a longer period. This important change is accompanied by advances in the vegetative system, which also provide material for comparative treatment. Such advances appear to have been initiated early in Mesozoic time. On the other hand, it may be stated broadly that the *Simplices* were typically Ferns of the Palaeozoic, though they overlapped into the Mesozoic Period, and some types have even survived to the present day.

CHAPTER XXVII

HYMENOPHYLLACEAE

THIS Family includes two large and comprehensive genera, *Hymenophyllum* and *Trichomanes*. The former is represented by 231 the latter by 228 species. Both have been variously sub-divided by different authors: but there is no need to enter here into these systematic details, for the morphological principles of their comparison are the present theme rather than their detailed application[1].

The Hymenophyllaceae are widely distributed, chiefly in moist and shaded spots, throughout the tropics: they extend as stragglers northwards, but more freely to the south, and there is a special centre of their prevalence in New Zealand. In point of time the records are doubtful, though Ferns with sori and sporangia of corresponding appearance have been traced from the Upper Carboniferous onwards. Various fossils have accordingly been described under the name *Hymenophyllites*: but Seward concludes from a critical examination of them that there is no evidence which can be adduced in favour of regarding the Hymenophyllaceae as Ferns of great antiquity[2]. Nevertheless they bear anatomical and other characters similar to certain Botryopterideae, while at least the position and nature of the sorus in *Hymenophyllites quadridactylites* (Gutbier), from the French Coal Measures, shows that the general type of their fructification existed in the Primary Rocks. An early record does not, however, seem essential to the view here entertained that the Hymenophyllaceae originated from protostelic Simplices, though it would readily accord with it.

The shoot of these Ferns is sometimes upright and radial, with spiral phyllotaxis, as in some species of *Trichomanes*: more commonly it is creeping and dorsiventral, with the leaves arranged distichously, and elongated internodes, as in many species of *Trichomanes* and all of *Hymenophyllum*. From the axis numerous roots arise in most species, but in some, and especially in the section *Hemiphlebium* of the genus *Trichomanes*, no adventitious roots are formed. The hairs borne on the shoot are filamentous, but sometimes stiffly branched as mechanical protections, as in *Trichomanes* (Fig. 504, *C*). In some species of *Hymenophyllum* of exposed habit the leaf is covered by a hairy felt, as in *H. sericeum*. Flattened ramenta are absent,

[1] The chief systematic works are cited in Engler and Prantl, *Natürl. Pflanzenfam*, i, 4, p. 91. Reference may also be made to Christensen's *Index Filicum*, pp. xiii to xvi, where the synonyms and sub-genera are given.

[2] *Fossil Plants*, vol. ii, p. 365. See also Kidston, *Fossil Plants of the Carboniferous Rocks*, Part IV, 1923, p. 279.

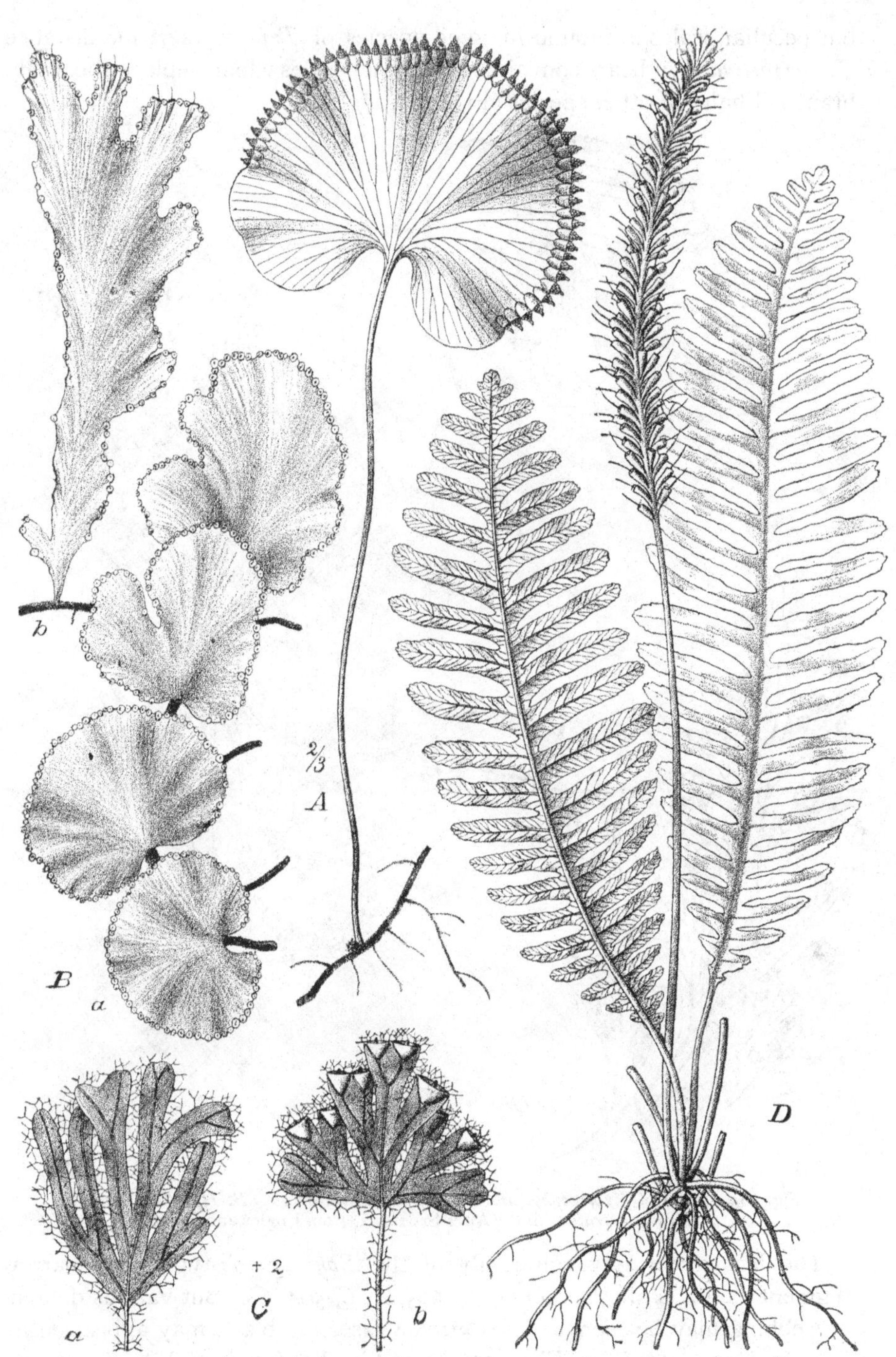

Fig. 504. Habit of *Trichomanes*. *A* = *T. reniforme* Forst. *B* = *T. membranaceum* L.; *a*, sterile, *b*, fertile. *C* = *T. Lyallii* Hook.; *a*, sterile, *b*, fertile. *D* = *T. spicatum* Hedw. (After Sadebeck, from Engler and Prantl.)

but peculiar scales are found in some species of *Trichomanes*: for instance *T. membranaceum* bears convoluted marginal scales which replace the stiffly branched hairs of other species (Fig. 504, *B*).

Fig. 505. Habit of *Hymenophyllum*. *A*, *Hym. cruentum* Cav. *B*, *Hym. dilatatum* Sw. *C*, *Hym. australe* Willd. (After Sadebeck, from Engler and Prantl.)

The leaves are very commonly of the *Sphenopteris*-type, each narrow segment having a single vein (Figs. 505, *B*, *C*, 504, *C*). But various degrees of webbing may be seen, and in extreme cases the blade may appear entire both in *Hymenophyllum* (Fig. 505, *A*), and in *Trichomanes* (Fig. 504, *A*, *B*).

The venation is open, without fusions. The only exception to this is *Trichomanes* (*Feea*) *elegans* Rudge, a Fern of Tropical America, which differs not only in the irregular anastomosis of the veins, but also in distinct sterile and fertile leaves: the former being pinnate and reticulate, but the latter entire, and with open venation. It may be held as an advanced type. It is an upright radial plant. The construction of the leaves is usually referable to dichopodial branching (see Vol. I, Chap. V). But the dichotomy may sometimes be remarkably equal, giving the form of a more or less regular webbed fan (Fig. 504, *A*, *B*). The similarity to the venation seen in *Schizaea* is marked, and the whole type of leaf-architecture suggests relatively primitive relations. In texture the leaves are "filmy": their translucent expanse is usually only one layer of cells thick: but in *T. reniforme* and *H. dilatatum* there may be three or four layers of cells, though without intercellular spaces or stomata.

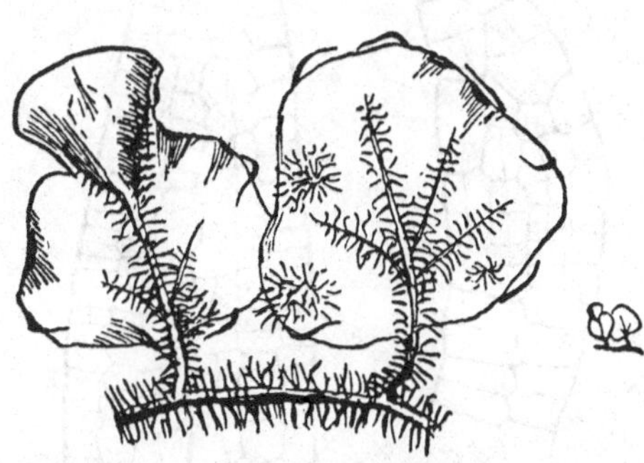

Fig. 506. *Trichomanes Goebelianum* Giesenhagen. To the right is part of a plant of normal size, to the left the same magnified. The plant is rootless, rhizoids serving for its attachment, which spring partly from the stem, and leaf-veins, partly even from the leaf-surface. Venezuela, 1890. (After von Goebel.)

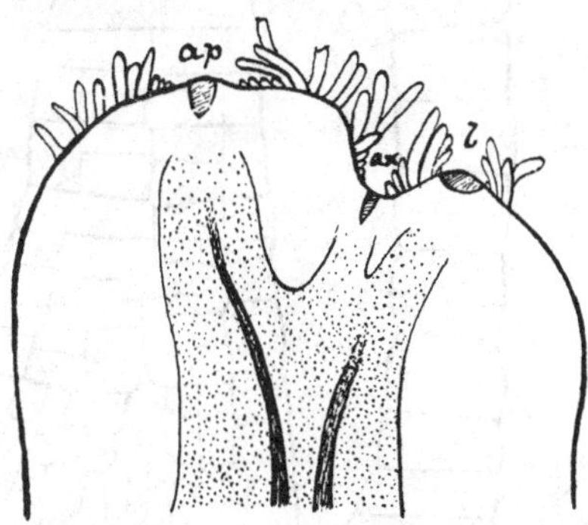

Fig. 507. *Trichomanes radicans*, longitudinal section of the apex (*ap*), with axillary bud (*ax*), and subtending leaf (*l*). (× 20.)

This filmy character is probably a secondarily acquired adaptation to the moist conditions under which these Ferns live (Vol. I, pp. 41, 115). It is accompanied by a simple construction of other parts: in particular, in certain leaves pseudo-veins are present, which can hardly be anything else than the vestigial remains of true veins no longer functional (Prantl, *Hymenophyllaceae*, p. 24). Cognate with this is the fact that the root-system may be reduced, or even absent in some species. Leafless branches of the rhizome covered with hairs resembling those normally found on the axis and leaf of rooted species, act as substitutes for true roots (Fig. 506). It may be expected that the vascular system of root and leaf will be relatively simple in these Ferns as compared with others, and examination shows that this surmise is correct.

Axillary branches occur very generally in the Hymenophyllaceae, but at many nodes the bud remains dormant. Its position and the relation of its vascular supply to that of the subtending leaf are shown in Fig. 507. The view may be held that it represents the weaker branch of a dichotomy, and that the subtending leaf may be the first produced by it. This view would also apply for the similar branching in *Zygopteris* and *Lyginopteris*, while the facts for *Botryopteris cylindrica* suggest that there is no obligatory relation with the subtending leaf. Here the special interest lies in the fact that the branching of the shoot in the Hymenophyllaceae closely resembles that in such primitive Ferns as the Botryopterideae and the Ophioglossaceae (see Vol. I, Chap. IV).

The apical segmentation of stem, leaf, and root is of the ordinary type for Leptosporangiate Ferns (Vol. I, Fig. 101). But the margin of the filmy

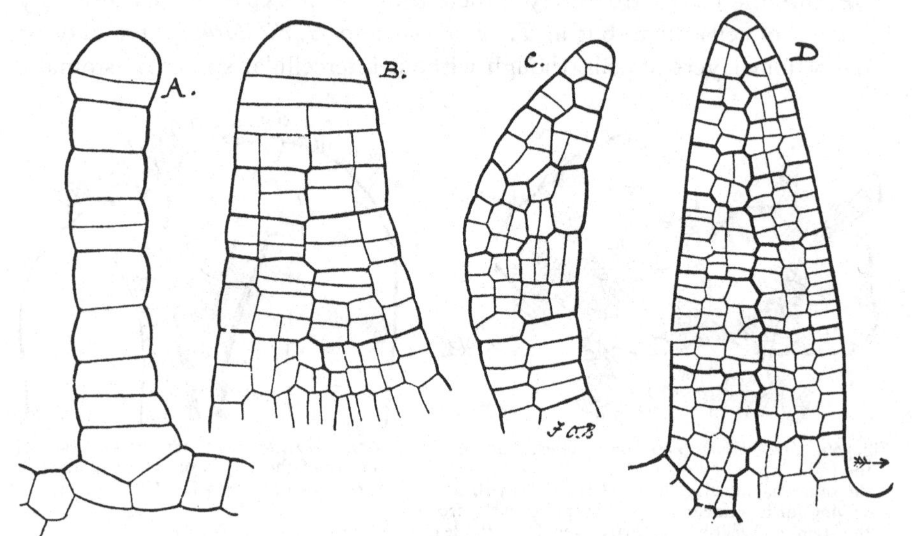

Fig. 508. Drawings of the segmentation at the margin of leaves of Leptosporangiate Ferns. *A* = *Trichomanes radicans*: *B* = *Trichomanes reniforme*: *C* = *Asplenium resectum*: *D* = *Phyllitis* (*Scolopendrium*) *vulgare*. The last is the most usual type: *A* and *B* are characteristic of Filmy Ferns.

leaf is occupied by a series of half-disc-shaped cells, with segmentation parallel to their inner face. In the absence of periclinal divisions this gives a filmy expanse of a single layer. In *T. reniforme*, where the expanse is four layers thick, the marginal cells have the same form and segmentation, but periclinal divisions follow (Fig. 508, *A*, *B*). Thus in both cases the segmentation differs from that in ordinary Leptosporangiate Ferns (*C*, *D*). But in the lower parts of the leaf both in *T. reniforme* and occasionally in *H. dilatatum* the marginal segmentation is by alternating oblique walls, as in *C*, *D*, a condition found also in *Todea superba*. These segmental facts further support the view that the filmy habit has been secondarily acquired: also they suggest that the thicker structure of *T. reniforme* has resulted from a tertiary return to a thicker structure.

The Hymenophyllaceae are all Ferns of relatively small size: some of them are very minute, for instance the rootless *T. Goebelianum*, with leaves only 3–4 mm. in length (Fig. 506). They are mostly inhabitants of wet forest, often occupying decaying stumps or rocks on the forest-floor, or spreading epiphytically up the stems and branches of trees, according to the hygrophytic conditions, to which they are very susceptible (see Forrest Shreve, *Bot. Gaz.* Vol. li, p. 184, 1911).

ANATOMY

The most obvious structural peculiarity of the Hymenophyllaceae is the "filmy" texture of their leaf-expanses. In accordance with this, and with the moist conditions under which they live, the vascular system, though not generally rudimentary, is nevertheless relatively simple. The stem is uniformly protostelic, and an undivided leaf-trace of simple outline is given off, naturally without any foliar gap, while the vascular supply to the axillary bud, if present, is united basally with the leaf-supply (Fig. 507). The stele varies considerably in structure in different species: in *Hymenophyllum* there is less variation than in *Trichomanes.* It is delimited by endodermis, which is lined by pericycle that extends sometimes to several layers, and surrounds a continuous phloem: the centre is occupied by a xylem-core, which is more or less parenchymatous towards the centre (Fig. 509). The protoxylems are sometimes peripheral, but usually mesarch, and are decurrent from the leaf-trace. There is rather wide variation according to size and habit. Fig. 509 is taken from *T. scandens*, a rather large species with wiry stem. In certain species of *Hymenophyllum* with large rhizomes, such as *H. scabrum* and *dilatatum*, the protoxylem lies centrally embedded in parenchyma, the metaxylem forming a ring round it (Fig. 510). The ring is often separated by parenchyma into two bands, their definition having relation to the origin of the root-supply. The lower of the two plates of xylem seen in *H. dilatatum* may be absent in smaller types. The phloem, that completely encircles the xylem in larger forms, may also disappear on the lower side in smaller types, and thus the stele may become collateral (*T. muscoides*). Finally, in *T. microphyllum* the xylem may be represented only by a single tracheid, while in *T. Motleyi* there is no xylem, and the stele consists of only a few cells of conjunctive parenchyma. These are plainly reduced types. The petiole in all cases

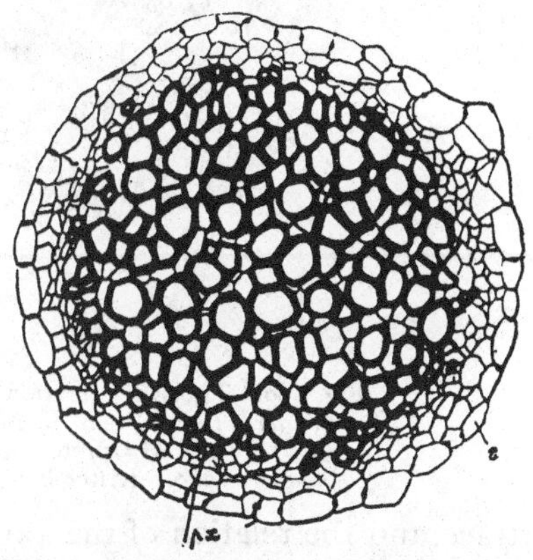

Fig. 509. Transverse section of the stele of *Trichomanes scandens*. *px* = protoxylem: *s* = endodermis. (After Boodle.)

receives only a single strand, which widens out upwards into a collateral structure, with the form of a more or less clearly curved horse-shoe.

Comparison of the anatomical structure of living species shows a close similarity among the larger species, such as *T. reniforme* and *H. scabrum* or *dilatatum* (Fig. 510). This structure finds its parallel in *Anachoropteris*, one of the Botryopterideae (see Vol. I, Fig. 117). If the small central tracheides of *Anachoropteris* are protoxylem, as seems highly probable, the agreement is very close, the differences being such as would be due to the two-fifths divergence and accommodation to the size-factor, as against a smaller size and distichous phyllotaxy. The protostelic origin of the leaf-

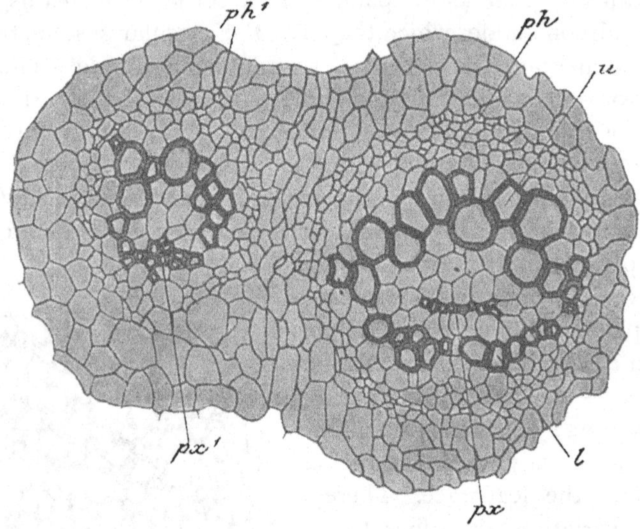

Fig. 510. Transverse section of a node of *Hymenophyllum dilatatum* v. *Forsterianum*. Stele of rhizome to the right, leaf-trace to the left. *ph* = phloem: *px* = protoxylem: *l* = lower xylem-bands: *u* = upper xylem band. (× 200.) (After Boodle.)

trace, and the relation of the axillary bud are also points of similarity. This resemblance to a very ancient Fern suggests the recognition of the living species named as probably primitive in the living family, while from that central point other species of *Hymenophyllum* and *Trichomanes* would have diverged along their several lines either of reduction or of other specialisation.

Sori and Sporangia

The receptacle of the sorus in the Hymenophyllaceae is always marginal in origin. It may be more or less elongated, and is traversed by the continuation of a vein of the lamina. In *Trichomanes* it persists after the sporangia are shed as a stiff bristle, hence the generic name. It is surrounded at the base by a cup-shaped, or two-lipped indusium. Upon this receptacle the numerous sporangia originate in basipetal succession, overlapping one another when mature like the shields in a Roman *testudo* (Fig. 511). The orientation of the sporangia is always with the peripheral or distal face outwards, and the central or basal face directed towards the receptacle. The annulus is oblique and the dehiscence obliquely lateral. The sporangia themselves are uniform in type for both genera: but they vary greatly in size in different species. In the larger species they are almost spherical, as in *H. dilatatum*: but in the smaller species they may be minute compressed

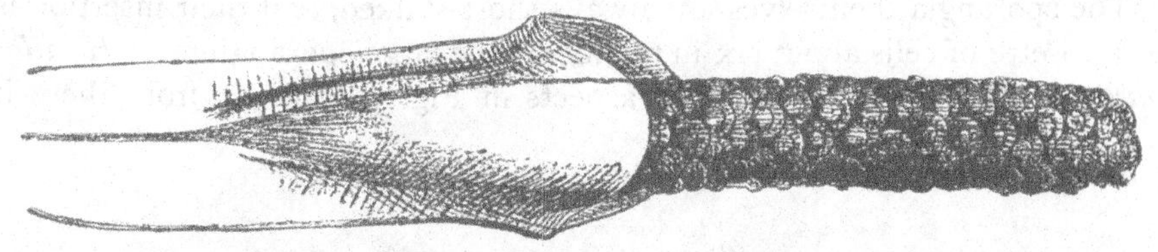

Fig. 511. *Trichomanes tenerum.* Sorus in surface-view: the placenta, bearing radially distributed sporangia, issues from the two-lobed beaker-like indusium. The annulus is visible in the several sporangia. (Magnified.) (After von Goebel.)

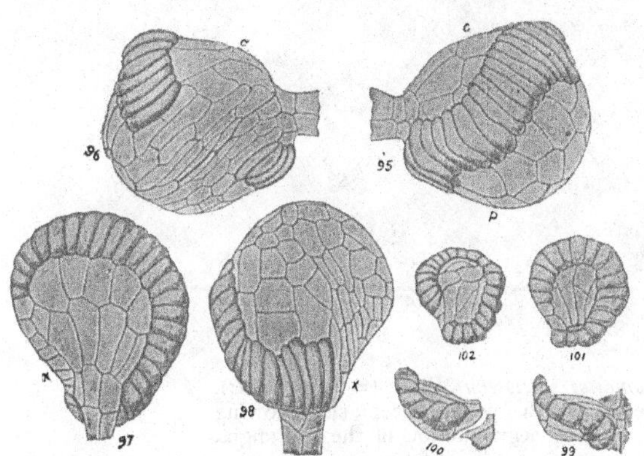

Fig. 512. Nos. 95, 96, 97, 98 = sporangia of *Hymenophyllum dilatatum* Swartz, seen respectively from their two sides (95, 96), and from the central (97) and peripheral (98) faces. Nos. 99, 100, 101, 102 are similar figures, to the same scale, of *Trichomanes radicans* Swartz, 99 and 100 show lateral views, 101 shows the central, and 102 the peripheral faces. (All × 50.)

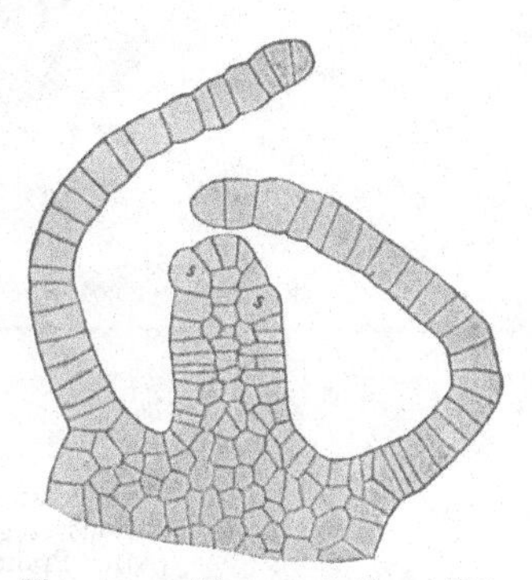

Fig. 513. *Hymenophyllum Wilsoni* Hk. Sorus in longitudinal section, showing the receptacle with divisions indicating intercalary growth, and the first sporangia (*s*) orginating near to the apex. (× 100.)

bodies, as in many species of *Trichomanes* (Fig. 512). The latter often show a low spore-output, while in the former the numbers may be large. These facts suggest that the family illustrates reduction, though the type of sorus and of sporangium is maintained. The development of the sorus shows that the sporangia are borne in strictly basipetal succession upon the more or less elongated receptacle. In a vertical section through a young sorus of *H. Wilsoni* the young sporangia (*s*) are already recognisable near its apex, while below there are clear indications of active intercalary growth (Fig. 513). The extent of this is usually greater in *Trichomanes* than in *Hymenophyllum*.

If the succession of the sporangia be strictly basipetal the result will be a sorus as in Fig. 511. Meanwhile the indusium grows obliquely upwards sheathing the base, and the youngest sporangia are thus protected in the angle between it and the receptacle. This is in fact an extreme example of the gradate sorus, the biological rationale of which has already been discussed (Vol. I, p. 212). It does not appear difficult to relate this with the marginal origin of the sporangia in the Schizaeaceae, while the indusium may find a correlative in the superficial protective flaps of *Lygodium*.

The sporangia themselves are always short-stalked, and their insertion is by a rosette of cells about six in number. The large sporangium of *H. dilatatum* is shown in four different aspects in Fig. 512, 95–98: from these it

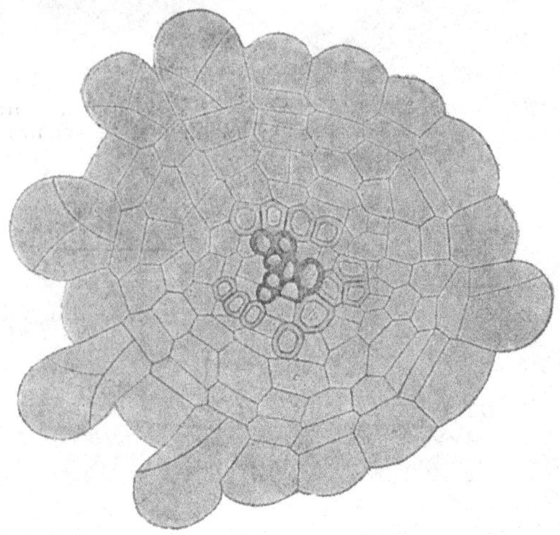

Fig. 514. *Trichomanes speciosum* Willd. (= *T. radicans*). Transverse section of the young receptacle, showing various stages of early segmentation of the sporangia. (After Prantl.)

may be gathered that there is an oblique annulus formed of a single row of cells, with lateral dehiscence by a long oblique slit: but the stomium is not a definite cell-group. The peripheral or distal face consists of numerous tabular cells, and is strongly convex. The central or basal face is also convex, but in less degree. The whole structure resembles the sporangium of some of the Simplices rather than that of advanced Leptosporangiate Ferns. In *T. radicans* (Fig. 512, 99–102) the sporangium though small is of the same type, but with fewer cells of the annulus, and a much simpler stomium.

The origin of the sporangium of the Hymenophyllaceae is from a single parent cell with a square base. It projects from the surface of the receptacle, and the first segment-wall strikes its basal wall (Fig. 514). This is a feature common in the sporangia of certain of the Simplices rather than in those of the more advanced Ferns. In particular it is seen in the Schizaeaceae.

Though apparently a minor detail this segmentation is actually an important point for comparison.

The sporangium projects horizontally from the surface of the receptacle, with the annulus directed obliquely towards the apex of it. The effect of this on the process of dehiscence is suggested by a diagram constructed from a section of a single sporangium of *Trichomanes speciosum* taken by Prantl (Fig. 515). Three such sporangia are represented one above another: the cells of the annulus traversed in the section are indicated by heavier lines. It is plain that on dehiscence taking place laterally, the distal side of the annulus has freedom to alter its form independently of the others, and it straightens itself out after the rupture. The oblique position of the annulus and its lateral dehiscence thus find their practical explanation, and this arrangement is even a necessity where all the stalks are short, and the sporangia overlap. Further, the dehiscence is aided by its occurring in the sporangia in basipetal sequence: for after the lateral slit is formed, the annulus is first reflexed, and then it recovers with a sudden jerk which often dislodges the whole sporangium. In *Trichomanes* the distal end of the receptacle may be thus stripped of its sporangia, while a succession of young sporangia may still be found around its base.

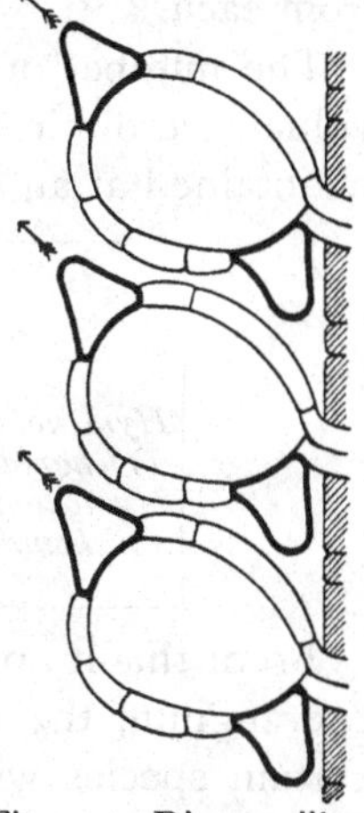

Fig. 515. Diagram illustrating the relative position of the sporangia on the receptacle in the Hymenophyllaceae. It was constructed from Prantl's section of a mature sporangium of *Trichomanes speciosum*.

The spore-output per sporangium is extremely variable in the family, as the following table shows:

Name	Result of counts	Typical number
Hymenophyllum tunbridgense	413, 416, 421	256–512
Trichomanes reniforme	247, 243	256
Hymenophyllum sericeum	216, 239	256
,, *dilatatum*	121, 127, 127, 127	128
,, *Wilsoni*	119, 121	128
Trichomanes crispum	51, 52, 59	64
,, *rigidum*	32, 48, 56	32–64
,, *radicans*	46, 58, 62	48–64
,, *javanicum*	38, 42, 48	32–48
,, *spicatum*	48	48
,, *pinnatum*	32, 48, 32	32–48

The limits of variation in number are here wider than it is found to be in any other family: ranging between such numbers as 421 and 32. The number of spores runs fairly parallel with the size of the sporangia. But on the other hand where the numbers are small there may be a correlative elongation of the receptacle, so that a larger number of sporangia may be formed in succession. On such facts the Hymenophyllaceae might be laid out in series, and it would be roughly a series of specialisation carrying with it a decrease in size of the sporangia, and in the number of spores produced from each.

The number of sporangia in the sorus tends to increase as their individual size diminishes: thus the actual output of spores per sorus may be maintained at an almost uniform figure, as the following table will show:

Name	Sporangia per sorus	Spores per sporangium	Output per sorus
Hymenophyllum tunbridgense	20	420	8,400
Trichomanes reniforme	40	256	10,240
Hymenophyllum dilatatum	90	128	11,500
Trichomanes radicans	140	64	8,960

From this it appears that, notwithstanding great variations of output per sporangium, the result per sorus may remain approximately uniform for certain species within this very natural family. The facts suggest a true biological progression: for the production of spores is a drain upon the resources of the plant: that drain is relieved by production of a succession of smaller sporangia within the sorus, the demand being thus spread over an extended period. This will be peculiarly important in Ferns growing as these do in deep forest shade, and with their leaves reduced to a filmy texture. But the principle of the Gradate Sorus is not dependent on such extreme conditions. It is found to be illustrated in several distinct families of Ferns of normal habitat. It may be held as a biological amendment upon the primitive state of the Simplices and as marking, where it occurs, an evolutionary advance upon those Ferns which Palaeophytology and comparison both indicate as the earliest.

THE GAMETOPHYTE

The simple structure of the sporophyte of the Hymenophyllaceae finds its correlative in the simplicity of the gametophyte. But here it may be a question whether this is due to reduction or to a primitive state. In *Trichomanes* the prothallus consists of coarse branched filaments, in habit like a green Alga (Fig. 516). Sometimes, as in *T. alatum*, the filament may widen out into flattened expanses, one layer of cells in thickness: but these are of very irregular outline. They may have a definite apical segmentation at first, but they do not take the cordate form usual in Leptosporangiate Ferns. They may be regarded as amplifications of the simple filament (Fig. 517, *A*, *B*, *C*). In *Hymenophyllum* this form appears to be dominant, and the prothallus consists of strap-shaped expanses one layer of cells in thickness: but there may be a thickening of the margins where the gametangia are produced. Like many epiphytic Liverworts and some other Fern-prothalli there may be a vegetative propagation by gemmae. These are habitually produced in *T. alatum*, *venosum*, and *pyxidiferum*. They are borne distally, each appearing as a swelling on the tip of a flask-shaped sterigma. Extending laterally, and undergoing segmentation, this forms a spindle-shaped body easily detached, and thereafter it may germinate into a new filamentous gametophyte (Fig. 517).

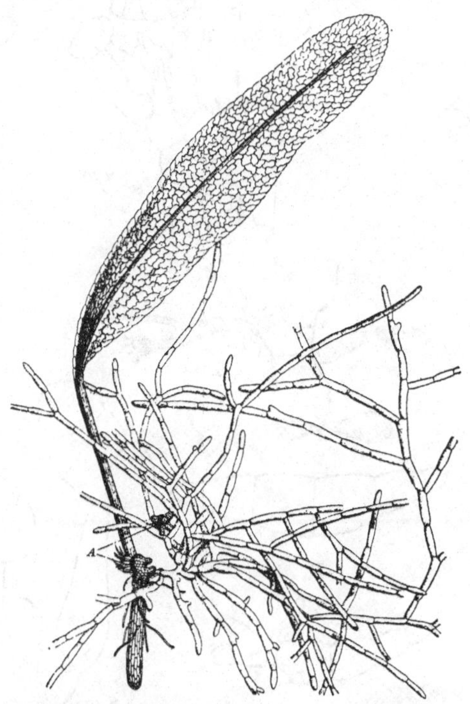

Fig. 516. *Trichomanes rigidum* Sw. Habit of a prothallus of which only a small part is represented, with archegoniophores (*A*), on one of which (the lower) an embryo plant is seated. (× about 50.) (After von Goebel from Engler and Prantl.)

The gametangia of *Trichomanes* appear on the ends of dwarf branches (Fig. 518). The antheridia undergo a segmentation similar to that in Leptosporangiate Ferns: by position they closely resemble those of *Schizaea*. The number of spermatocytes in section (19 in *T. pyxidiferum*), and the two divisions of the cap-cell (Vol. I, Fig. 283, *f*) accord with a position between primitive and highly specialised Ferns. The archegoniophore of *Trichomanes* appears as a short massive lateral branch, bearing a plurality of archegonia of an ordinary type. In *Hymenophyllum* the antheridia and archegonia may be borne together upon a lateral lobe of the strap-shaped thallus, which may probably correspond to a flattened archegoniophore of *Trichomanes* (Fig. 519). These facts are chiefly interesting as points for comparison with *Schizaea*, where the gametophyte is also filamentous; but in some respects, such as the archegoniophore, its structure there is even simpler. It must, however, be remembered that the prothalli of *Anemia*, *Mohria*, and *Lygodium* are of the ordinary flattened type. A circumstance which may have its bearing in

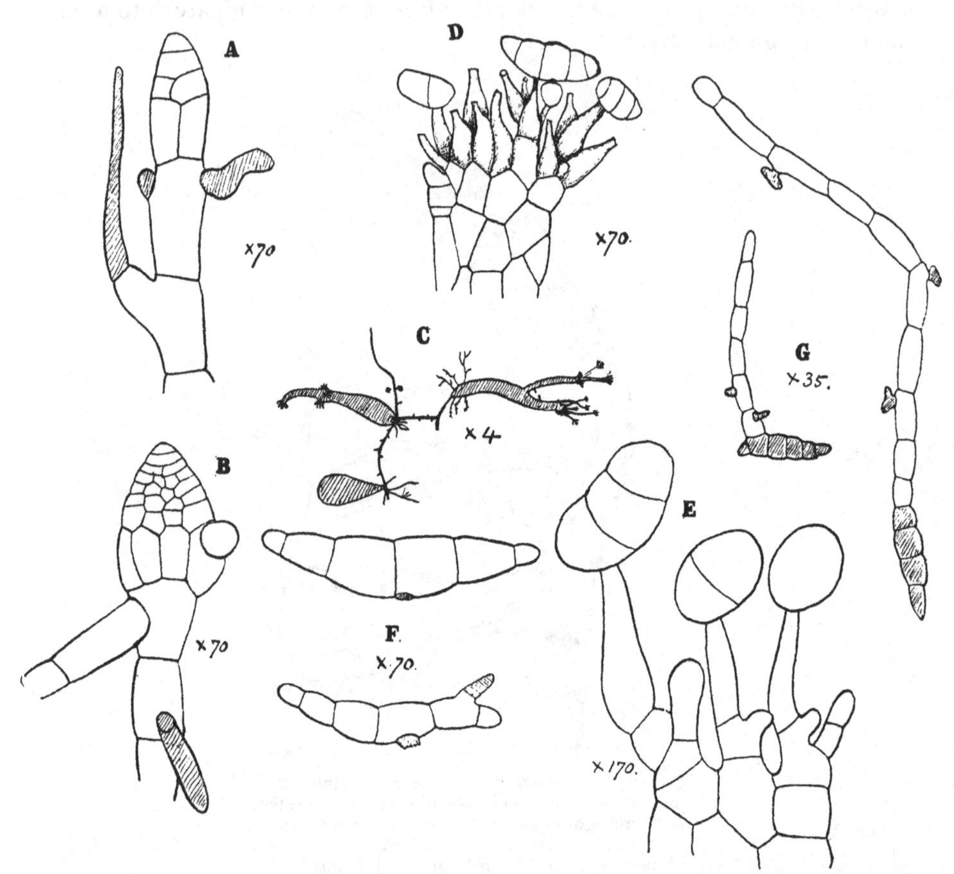

Fig. 517. *A—F*, *Trichomanes alatum*. *A*, *B*, show the formation of a flattened prothallus from a filament. *C*, relation of filaments to the flattened expansions which bear sterigmata distally. *D*, sterigmata bearing gemmae. *E*, very young gemmae. *F*, mature gemmae detached, the lower beginning to germinate. *G*, result of germination of gemmae of *Trichomanes Kaulfussii*.

various directions is that both in *Schizaea* and in a less degree in *Trichomanes*, a fungus is present, and in the former there is a definite mycorhizic relation. The embryo of *Trichomanes* has been observed in a relatively advanced state, showing the usual relation of parts, and with a simple spathula-shaped, single-nerved cotyledon (Fig. 516). But the embryogeny has not yet been worked out.

Lastly, it should be stated that both apospory and apogamy have been observed. The latter is definitely seen in *T. alatum*, but the embryo in Fig. 516, of *T. rigidum*, can hardly have been produced otherwise than by normal syngamy.

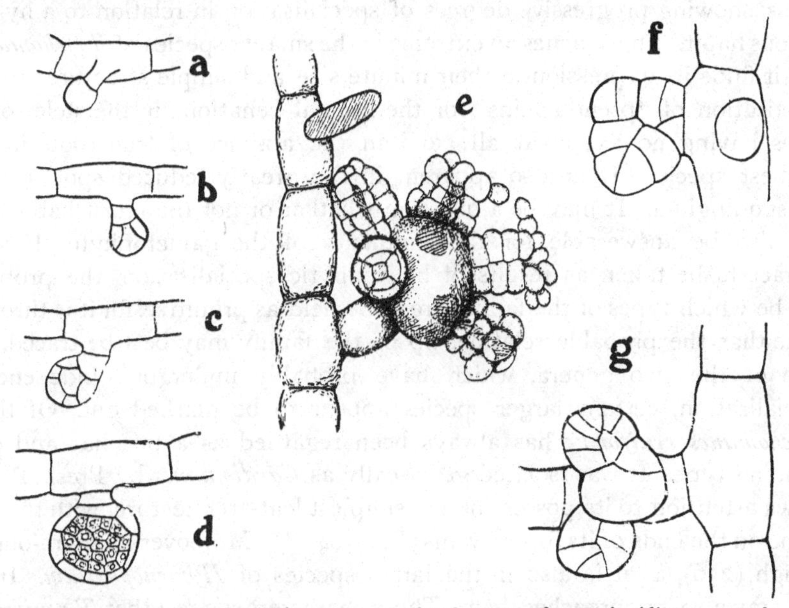

Fig. 518. *Trichomanes pyxidiferum*. *a*—*d* = development of an antheridium: *e* = archegoniophore with five archegonia: *f*, *g* = development of archegoniophore. (× 150.)

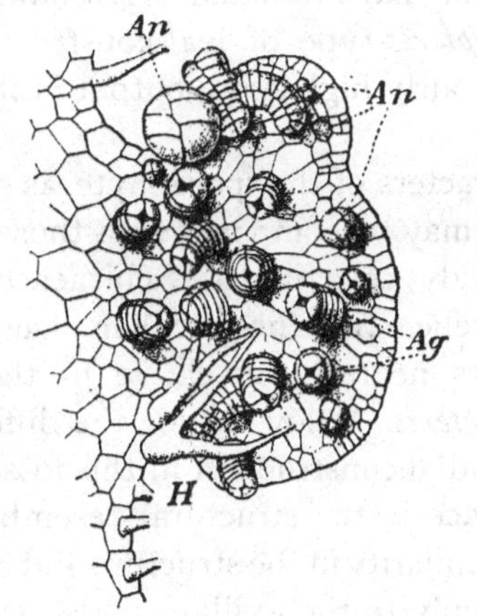

Fig. 519. *Hymenophyllum dilatatum* Sw. Lobe of a prothallus with a group of archegonia (*Ag*). *An* = antheridia: *H* = hairs. (After von Goebel from Engler and Prantl.)

Comparison

From what has been stated above there is no reason to regard the Hymenophyllaceae as anything more than a relatively primitive family of Ferns, showing progressive degrees of specialisation in relation to a hygrophilous habit. This reaches an extreme in the smaller species of *Trichomanes*, and it finds its expression in their minute size, and simple structure: in the substitution of "pseudo-veins" for the normal venation: in the stele sometimes having no xylem at all: and in the absence of true roots in the smallest species. It is also apparent in the greatly reduced spore-output per sporangium. It may be a question whether or not the moist habit may not also be answerable for the simplicity of the gametophyte. If such characters be taken as results of hygrophytic specialisation, the problem will be which types of the family are to be held as primitive, for it is through these that the probable relationship of the family may best be traced. In each of the two genera, which have probably undergone independent specialisation, certain larger species appear to be marked out. Of these *Trichomanes reniforme* has always been regarded as a peculiar and outstanding type. It was isolated generically as *Cardiomanes* by Presl. Prantl draws attention to its possessing the simplest leaf-architecture, with the sori borne on the ends of its forked veins (Fig. 504, *A*). Moreover its spore-output is high (256), as it is also in the larger species of *Hymenophyllum*. In its anatomy also it approaches them. These characters suggest that *T. reniforme*, together with some large species of *Hymenophyllum*, such as *H. dilatatum*, may be held as taking a central position in the Family, from which divergent lines of specialisation may have radiated. *H. dilatatum* is confirmed in that position by its *Sphenopteris*-type of leaf-construction, its well developed vascular structure, and fairly high spore-output (128): but *T. reniforme* has its leaf fully webbed.

Taking then the characters of the sporophyte, as seen respectively in these two central types, they may be compared with those of other Filicales. The creeping habit is already seen in the Schizaeaceae and Gleicheniaceae, which are also protostelic. But the peculiar structure of the stele in the species named finds its nearest correlative in the Botryopterideae, and especially in *Anachoropteris*. Allowing for the differences which follow on an upright habit and radial construction in this fossil and the creeping habit of most Hymenophyllaceae, the structural resemblance is very close: and with it goes a striking similarity in the structure and insertion of the leaf-trace, and in the mode of supply to the axillary buds. *H. dilatatum* and *T. reniforme* are both species with a lamina several layers of cells in thickness, and occasionally showing alternate segmentation of the marginal cells, as in Leptosporangiate Ferns. It may be a question what interpretation these facts will bear, but at least they serve to connect the more definitely hygrophilous filmy Ferns with the latter in this structural detail.

The marginal position of the sorus is shared with the Schizaeaceae, Botryopterideae, and *Osmunda*. But the Hymenophyllaceae differ in the basipetal sequence of the sporangia, the elongated receptacle, and the cup-like indusium. The basipetal sequence may be held as a secondary condition, advantageous in spreading the drain of spore-production over a longer period. The intercalary elongation of the receptacle is an almost necessary condition of its adoption. The basal indusium is foreshadowed in principle though not in detail by the Schizaeaceae, and it serves an obvious biological purpose. It is thus possible to conceive the origin of the Hymenophyllaceous sorus from some Fern-type with marginal sporangia, or perhaps tassels of sporangia, by initiation of a basipetal sequence and the establishment of a protective indusium. The type from which they might have originated would probably be found among protostelic types with large sporangia marginally produced, of which the Botryopterideae, Osmundaceae, and Schizaeaceae are the known representatives.

A comparison of the sporangia themselves is naturally directed to the Simplices rather than to Leptosporangiate Ferns. Among the Schizaeaceae it is *Anemia* that corresponds most closely, for there the distal face is of considerable size though flat (see Fig. 451, *F*, *G*). If this were enlarged and convex, the basal face reduced, the annulus oblique instead of transverse, and the line of dehiscence swung obliquely to one side but maintaining the indefinite stomium, the sporangium of *Hymenophyllum* would be the result, while the typical spore-number (128) is the same. It is not suggested that *Anemia* was an actual progenitor of *Hymenophyllum*: what is suggested is that the two sporangia have essential points of similarity. But in spore-number certain Hymenophyllaceae approach still more nearly to the earlier Simplices. *H. tunbridgense* with 256–512, *T. reniforme* and *sericeum* with 256, show figures which find no correlative among ordinary Leptosporangiates, a fact which strongly supports the suggested relation with the Simplices. On the other hand the low spore-output of certain species of *Trichomanes* is held to be evidence of their advanced hygrophytic adaptation.

The comparison of the gametophyte of the family with that of *Schizaea* is striking, though some may still doubt whether it has not some homoplastic origin. But Heim's comparison of the antheridia of the Hymenophyllaceae with those of the Gleicheniaceae (*Flora*, p. 363, 1896) again directs attention in a matter of detail to the Simplices.

The result of a general comparison of the Hymenophyllaceae with other Ferns is to recognise that they approach most nearly to certain Simplices, with which they agree in many characters both of the sporophyte and of the gametophyte. Those of the sporophyte are the more distinctive:

they show similarity to several protostelic families of the Simplices, but more particularly to the Schizaeaceae. The Hymenophyllaceae are to be looked upon as probably of early origin, but ending as a blind line of descent, characterised by specialisation of both generations to a hygrophilous habit, which takes the form of simplification. The more robust species with large spore-output per sporangium, such as *Trichomanes reniforme* and *Hymenophyllum dilatatum*, appear to occupy a central position phyletically: *Trichomanes* shows the greater specialisation, and is on that account to be regarded as the farthest removed from the original source.

The Hymenophyllaceae comprise two substantive genera:

I. *Trichomanes* Linné, 1753. Species 228.

The prothallus is usually filamentous: the indusium tubular or flask-shaped.

II. *Hymenophyllum* Smith, 1793. Species 231.

The prothallus is a flat expanse or ribbon: the indusium two-lipped.

These genera have been variously sub-divided, and the numerous species grouped: but with few exceptions the sub-genera or groups appear to be of use rather for the recognition and designation of species than indicative of phyletic relations. Therefore they need not be detailed here, but reference may be made to systematic works. Certain groups or species that bear characters of phyletic significance may however be noted. For instance in *Trichomanes* the section *Cardiomanes*, containing the isolated, but probably central species *T. reniforme*, with its primitively equal dichotomy of the veins, but fully webbed lamina, and many layered structure: its robust texture and vascular system, and high spore-output. Also the section *Feea* Bory, comprising four heterophyllous species, all of the New World, of which the most remarkable are *T. spicatum* Hedw. (Fig. 504, *D*), and *T. elegans* Rudge, the latter with erect, tufted shoot, sterile leaves pinnate and webbed, with vein-fusions, and narrow fertile leaves with open venation. This is morphologically the most advanced type of all. On the other hand thalloid leaves are found in *T. membranaceum* L. and *Lyallii* Hk. (Fig. 504, *B*), while many are minute epiphytes, culminating in *T. Goebelianum* (Fig. 506). Many of the species are characterised by very low spore-output.

Hymenophyllum shows less variability, and a prevalent creeping habit, with "Sphenophylloid" leaves. *H. cruentum* Cav. is exceptional in having a fully webbed lamina (Fig. 505, *A*). The spore-output is usually higher than in *Trichomanes*, i.e. more primitive: that of *H. Tunbridgense* is unusually high (256–512), showing that prolific production does not necessarily follow the size of the plant.

BIBLIOGRAPHY FOR CHAPTER XXVII

503. ENGLER & PRANTL. Natürl. Pflanzenfam. i, 4, p. 91, where the earlier literature is very fully cited.

504. CAMPBELL. Mosses and Ferns. 3rd Edn. p. 372 etc. 1918, where there is a full citation of literature.

505. PRANTL. Die Hymenophyllaceen. Leipzig. 1875.

506. CHRIST. Die Farnkraüter der Erde. Jena. 1897.

507. BOODLE. Ann. of Bot. xiv, p. 455. 1900.
508. BOWER. Origin of a Land Flora. p. 575. 1908.
509. TANSLEY. Lectures on the Filicinean Vasc. Syst. New Phyt. Reprint, p. 27. 1908.
510. FORREST SHREVE. Bot. Gaz. Vol. li, p. 184. 1911.
511. SEWARD. Fossil Plants. Vol. ii, p. 365.
512. VON GOEBEL. Organographie. II Teil. Heft 2. 1918.
513. SCOTT. Studies in Fossil Botany. 3rd Edn. Part i. 1920.
514. CHRISTENSEN. Index Filicum. pp. xiii–xvi. 1906.
515. KIDSTON. Fossil Plants of the Carboniferous Rocks. Part iv, p. 279. 1923.

CHAPTER XXVIII

LOXSOMACEAE

THIS Family includes two genera, *Loxsoma* and *Loxsomopsis*, of the affinity of which there can be no doubt. The former is represented by the single species *L. Cunninghami* Br., endemic in New Zealand: the latter by three

Fig. 520. Leaf of *Loxsomopsis notabilis* Slosson. From a photograph of the type specimen, kindly supplied from the United States National Herbarium. (Reduced.)

species discovered since 1900, widely spread through Central America, viz. *L. costaricensis* Christ, from Costa Rica, *L. Lehmanni* Hier, from Equador, and *L. notabilis* Slosson, from Bolivia. The isolation of the habitats of the two genera, their paucity of species, and their peculiar characteristics reveal the family as one of the most interesting problems in Fern-morphology, while in themselves they suggest antiquity and survival. The attempt to place *Loxsoma* in relation to other Ferns has led to very divergent results, which are summarised elsewhere (*Phil. Trans.* Vol. 192, p. 47, 1899). But its distinctiveness of character from other Ferns clearly indicates the wisdom of Presl (*Hymenophyllaceae*, p. 98), and of Bommer (*Bull. Soc. Bot. de France*, Vol. 20, p. 35) in forming for it a separate Family of the Loxsomaceae, in which the more recently discovered *Loxsomopsis* finds a natural place. These Ferns take a position about the limit between the Simplices and the Gradatae, and appear to be related to the Hymenophyllaceae and

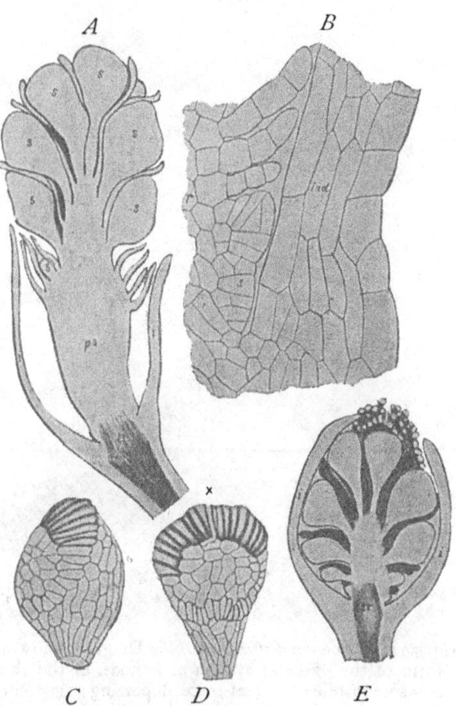

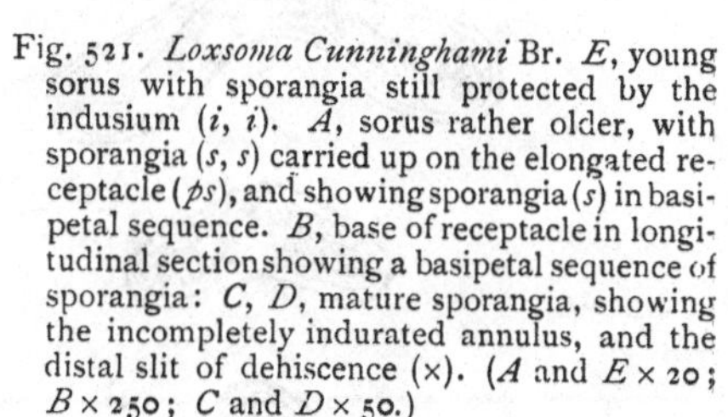

Fig. 521. *Loxsoma Cunninghami* Br. *E*, young sorus with sporangia still protected by the indusium (*i*, *i*). *A*, sorus rather older, with sporangia (*s*, *s*) carried up on the elongated receptacle (*ps*), and showing sporangia (*s*) in basipetal sequence. *B*, base of receptacle in longitudinal section showing a basipetal sequence of sporangia: *C*, *D*, mature sporangia, showing the incompletely indurated annulus, and the distal slit of dehiscence (×). (*A* and *E* × 20; *B* × 250; *C* and *D* × 50.)

Dicksonieae. *Loxsoma* was at one time definitely included in the Hymenophyllaceae (*Synopsis Filicum*, 1874), but the new facts for *Loxsomopsis* suggest that a separate position is preferable.

Loxsoma is an elegant Fern with a stout rhizome bearing irregular roots, and firm coriaceous, long-stalked leaves at intervals of about an inch. They are 1–2 feet high, glabrous, 2 or 3 pinnate, and glaucous beneath, with open venation. *Loxsomopsis* is a more stately Fern of Bracken-like habit, its slim stiff fronds rising to a height of eight feet (Fig. 520): but its rhizome is thinner than that of *Loxsoma.* Both bear their sori terminal on the veins, marginal in origin, but deflected downwards in the course of development. The sorus is of the type of *Trichomanes*, with a basipetal sequence of the sporangia borne on an elongated receptacle: a basal cup-like indusium protects the youngest sporangia. But at maturity the receptacle elongates, so that the sporangia are raised above the lip of the indusium, and there dehisce (Fig. 521). But the sporangia of the two genera differ in detail, notwithstanding the similarity of their sori as a whole. In both genera the dermal appendages are hairs. They appear of two types: soft uni-seriate hairs borne upon the leaf-blade, and stiff bristle-like hairs which widen downwards into a multicellular conical base: these are found on the leaf-base, and densely cover the rhizome. There are no flattened scales or ramenta.

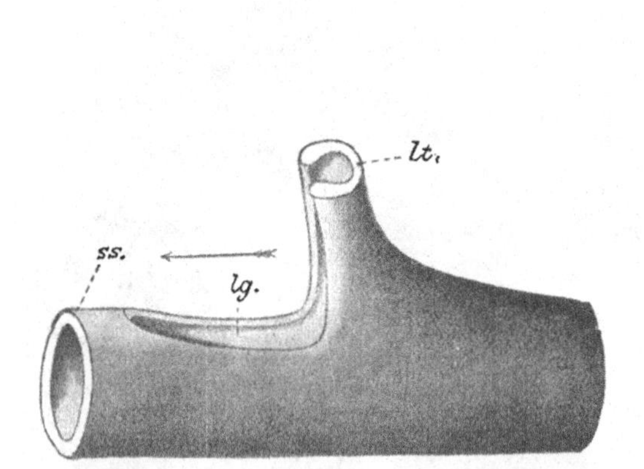

Fig. 522. *Loxsoma Cunninghami*. Diagram showing the form of the vascular system at a node of the rhizome. *ss*=solenostele: *lt*=leaf-trace departing: *lg*=leaf-gap: the arrow points towards the apex of the rhizome. (After Gwynne-Vaughan.)

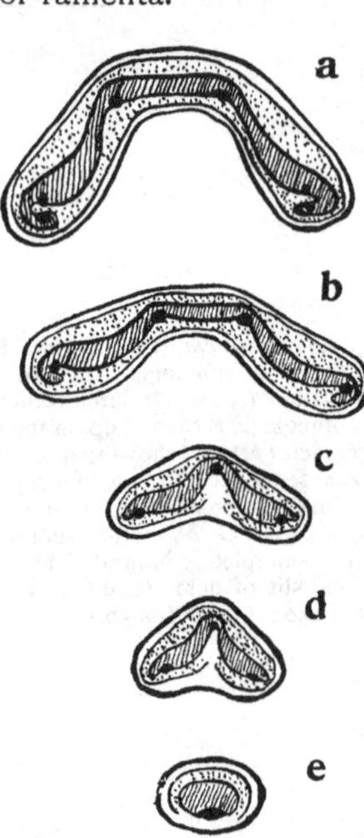

Fig. 523. An acropetal series of sections of the meristele of *Loxsoma* showing its modifications towards the apex of the leaf. (After Gwynne-Vaughan.)

ANATOMY

The vascular anatomy of *Loxsoma* has been fully worked out by Gwynne-Vaughan (*Ann. of Bot.* xv. p. 71). The rhizome contains a typical solenostele completely surrounding a central pith in the internodes, but opening at each node by a leaf-gap to give off an undivided leaf-trace, hollowed on the acroscopic side (Fig. 522). An unusual feature is the occurrence of islets of parenchyma in the sclerenchymatous masses of the stem, a peculiarity shared with certain species of *Dicksonia*. The leaf-trace flattens upwards, taking a rather open curve with the usual adaxial hooks at the margins. It diminishes in size as the pinna-traces pass off from its margins: higher up the hooks disappear, and distally it contracts to an oval form with a single protoxylem (Fig. 523). *Loxsomopsis* has essentially the same stelar structure on a smaller scale: these facts strengthen the comparison with *Loxsoma*, and indicate anatomically a primitive position for both.

The ontogenetic origin of this solenostele was traced, at my suggestion, by Dr McLean Thompson (*Trans. Roy. Soc. Edin.* Vol. lii, p. 728, 1920). It was found to correspond in essentials though not in detail to that of *Gleichenia pectinata* as regards the primary medullation and the formation of an internal phloem (see Vol. I, Fig. 134). The further steps to complete solenostely are suggested by the diagram, Vol. I, Fig. 135, p. 145. The facts have already been discussed there in relation to the theory of the stele.

SORUS AND SPORANGIA

The chief interest of these Ferns centres in the sorus. A vertical section through that of *Loxsoma* when young is seen in Fig. 521, *E*. The cup-like indusium encloses the short receptacle, the sporangia, and the hairs that accompany them: there is an obvious basipetal sequence of the sporangia (Fig. 521, *B*). The orientation of the sporangia is constant. They are pear-shaped, and rise obliquely upwards, each having a complete annulus (*C*, *D*), but the induration is very unequal; the cells of the distal half are larger and fully thickened, the lower half consists of smaller thin-walled cells, hardly differing from those of the rest of the wall except in form and arrangement: it has evidently ceased to be functional. When mature a longitudinal slit appears in a median plane, dividing the indurated group into two equal halves (*D*). There is no differentiation of a stomium. The annulus straightens on either side of the slit, and may even become reflexed, gaping like the covers of an open book (see Vol. I, Fig. 206, also Hooker, *Genera Filicum*, Pl. 15). The shedding of the spores is on the principle of the inverted pepper-pot.

This anomalous sporangium of *Loxsoma* suggested earlier a comparison with *Gleichenia* on the ground of the median dehiscence. But the facts for *Loxsomopsis* lead to a different interpretation. Here the plan of the sorus is the same. The sporangia are pear-shaped as before, and each has a complete oblique annulus. But it is indurated all round, except for a lateral stomium, well differentiated. A noteworthy feature is that the stomium may be either on the right or left side of the sporangium (compare Fig. 524, *A*, *B*, *C*). This instability of position of the stomium in *Loxsomopsis*, right or left, suggests that the annulus is not strictly standardised, i.e. that various parts of the ring may be thin-walled, or may be indurated. This has its bearing on the question of the swing of the stomium from a median position as in the Simplices, to a lateral position as in most Gradatae. The stalk in both genera is short, being composed of about six rows of cells. The dehiscence of *Loxsoma* being distinctly anomalous, while that of *Loxsomopsis* is closely comparable with that of *Dicksonia* and other Gradate Ferns, it seems right to seek an interpretation of the sporangia of the older genus in terms of the normal type. A simple explanation seems to be that *Loxsoma* originally had a sporangium like that of *Loxsomopsis*, with a lateral stomium: but that as the cup-like indusium fitted close round the receptacle in relation to its protective function, the lateral dehiscence became mechanically ineffective. The lower hoop of the annulus, including the stomium, then became vestigial, as we actually see it to be (Fig. 521, *C*, *D*). Rupture was, however, provided by a new slit in the middle of the indurated distal curve of the ring, a position

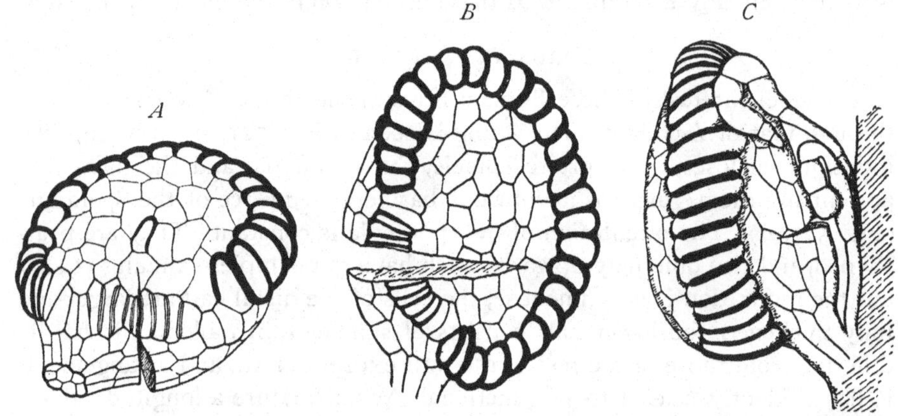

Fig. 524. Sporangia of *Loxsomopsis notabilis*, from three different aspects. *C* shows the sporangium in natural position on the receptacle (shaded), with attendant hairs, as in *Loxsoma*. The distal face is to the left. *A* has its distal face upwards, and shows the stalk as in *Loxsoma*, but the stomium is lateral and well formed. *B* has the distal face to the right and dehiscent stomium to the left. Note that the dehiscence *B* and *C* is to the left of the sporangium, and that in *A* to the right. Thus it is not constant. (× 250.)

that proved more convenient. If this were so then the median dehiscence of *Loxsoma* is not a survival of that of the Simplices, but a new feature. Certainly the slit occurs at the point most convenient for the shedding of the spores (compare Vol. I, Fig. 206). If this view be correct, one of the most interesting points in *Loxsoma* for comparison with the Simplices falls away, and the Loxsomaceae appear as ordinary Gradate Ferns with basipetal sequence, and originally lateral dehiscence, right or left, as in *Loxsomopsis*. But in *Loxsoma* a mechanical difficulty has been met by adopting secondarily a distal rupture.

The number of spores in *Loxsoma* is 64, and they are of a large size.

The Gametophyte

In a family so interesting as the Loxsomaceae it is important to know the characters of the gametophyte, and von Goebel has supplied an excellent drawing for *Loxsoma* (Fig. 525). It has the usual cordate form, but it is notable that when old it may bear on its lower surface bristles with distended base, similar to those on the rhizome. Von Goebel compares these with those of the Cyatheaceae (incl. Dicksonieae). Since in *Loxsoma* the appendages are of the nature of bristles and not actually flattened ramenta, their presence does not appear to be decisive as to phyletic relationship. Moreover, it is not uncommon to find the same type of appendages on both generations in Ferns (Vol. I, p. 198, Fig. 185). The

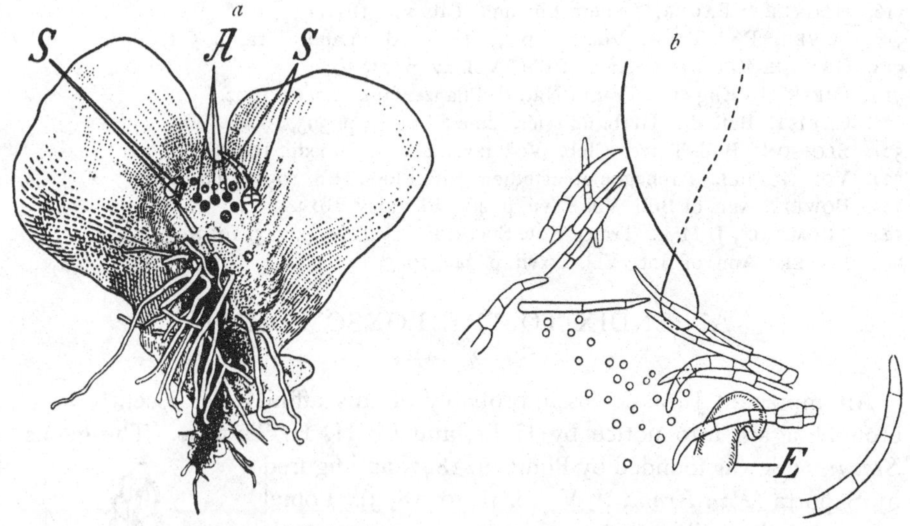

Fig. 525. *a*, prothallus of *Loxsoma* seen from below. (Enlarged.) *S*=appendages described by von Goebel as "schuppenhaare": *A*=archegonia. *b*, apical groove of a prothallus of *Loxsoma*, more highly enlarged than (*a*), and showing the bristle-like hairs, *E* is the leaf of a sporeling. (After von Goebel.)

structure of the antheridium of *Loxsoma* does not differ essentially from that of the Polypodiaceae. Thus the gametophyte does not greatly help in establishing phyletic relations (see below, p. 267).

Comparison

The Loxsomaceae appear as a generalised type of Ferns. In habit and anatomy they resemble the *Dicksonia-Dennstaedtia* series, with which also the marginal position of their sorus and their dermal appendages agree, while the sporangium of *Loxsomopsis* is very like that of *Thyrsopteris*, or of a short-stalked type of *Dicksonia*. On the other hand there is a general similarity to the sorus and sporangia of the Hymenophyllaceae, though the spore-output is lower than that of their more robust forms. The indusial protections of the Schizaeaceae show a general but not a close analogy, as also do the sporangia of *Anemia*. It seems reasonable to contemplate an origin from some Schizaeoid source. But the sorus of the Loxsomaceae is a radiate type: for this character it would be necessary to look back among marginal types to the tasselled sorus of *Zygopteris*, or of *Corynepteris* or of *Osmunda*. But these are without indusial protection. It is the absence of any direct or distinctive indication of affinity downwards that points the family out as generalised. But the nearest affinity upwards is clearly with the Dicksonioid series, and in particular with *Thyrsopteris*.

BIBLIOGRAPHY FOR CHAPTER XXVIII

516. HOOKER & BAUER. Genera Filicum. Tab. xv. 1842.
517. BOWER. Phil. Trans. Vol. 192, p. 47, 1899. Also Land Flora. p. 571, 1908.
518. GWYNNE-VAUGHAN. Ann. of Bot. Vol. xv, p. 71, 1901.
519. DIELS. In Engler u. Prantl, Natürl. Pflanzenfam. p. 112, 1902.
520. CHRIST. Bull. de l'Herb. Boissier. 2me Série, iv, p. 393, 1904.
521. SLOSSON. Bull. Torrey Club. Vol. xxxix, p. 285, Pl. xxiii, 1904.
522. VON GOEBEL. Archegoniatenstudien. xiv, Flora. Bd. 105, p. 33, 1912.
523. BOWER. Ann. of Bot. Vol. xxvii, p. 463, Pl. xxxiv, 1913.
524. THOMPSON, J. McL. Trans. Roy. Soc. Edin. Vol. lii, p. 715, 1920.
525. BOWER. Ann. of Bot. Vol. xxxvii, p. 349, 1923.

APPENDIX TO THE LOXSOMACEAE

Stachypteris

An important Jurassic fossil, probably of this affinity, has recently been brought again into notice by Halle, and by H. H. Thomas. The genus *Stachypteris* was founded by Pomel in 1847, and figured by Saporta (*Pal. Franc.* II, Vol. i, p. 49, 1873). Pomel compared the spikes that are borne on the margins of its Sphenopterid type of leaf with those of *Lygodium*. Additional material was found by Halle in 1910 at Whitby, and again by H. H. Thomas in 1912 near Saltburn. The leaf is 2 to 4 pinnate, and the pinnae of the second order bear pinnules sometimes terminated by a fertile spike. There are indications that the spike was of the type of *Lygodium*, but the existence of an indusium for each sporangium is uncertain. The sporangia were arranged in 2 or 3 rows, each with a slightly oblique or vertical annulus, which is certainly not of a Schizaeaceous type. The spores were rounded or tetrahedral (Fig. 526). The result of these observations, though not decisive, opens interesting suggestions. Thomas concludes that *Stachypteris* does not possess close affinities with any modern group of Ferns, but suggests a position intermediate between the Schizaeaceae and the "Cyatheaceae"—doubtless using the latter, in the sense of the

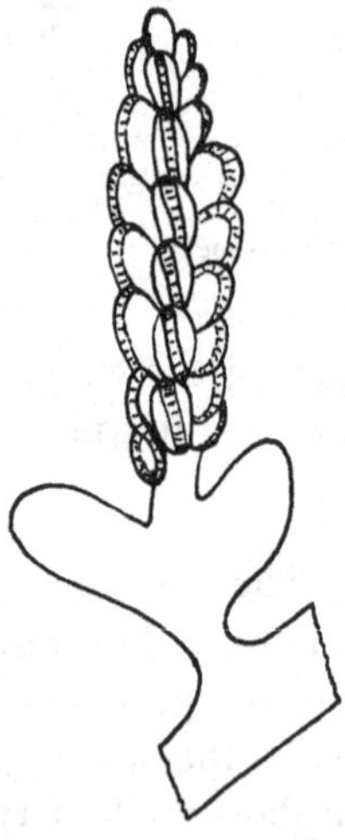

Fig. 526. A somewhat diagrammatic drawing of the marginal sorus of *Stachypteris Hallei*. (After H. H. Thomas. Enlarged.)

Synopsis Filicum, to include the Dicksonieae. The Fern obviously has nothing to do with the Cyatheae. It appears to find its nearest relation with *Loxsomopsis*, for it shares an elongated marginal sorus, and oblique or almost vertical annulus, and almost certainly a lateral dehiscence: but there is no cup-like indusium. On the other hand, a near relation with *Lygodium* appears improbable, since the character of the annulus is strongly against it. No fixed opinion is possible, but the discovery of *Loxsomopsis* has certainly made a nearer comparison possible with living Ferns of archaic type. The existence of such a Fern as *Stachypteris* in Jurassic time is a fact of great interest, as establishing the early occurrence of Ferns with marginal and almost certainly gradate sori.

BIBLIOGRAPHY

526. SAPORTA. Pal. Franc. II. Vol. i, p. 49, 1873.
527. THOMAS. *Stachypteris Hallei.* Proc. Camb. Phil. Soc. Vol. xvi, Part vii, Plate iv, p. 610, 1912.

CHAPTER XXIX

DICKSONIACEAE

I. Thyrsopterideae

THIS sub-family is now represented only by the genus *Thyrsopteris* including the single species *T. elegans* Kze, endemic in Juan Fernandez: but many authors have shown evidence that Ferns of this type existed in the Jurassic Period (Seward, *Jurassic Flora*, Vol. I, p. 98, etc.). The isolation of this single living species suggests that it is an ancient survival. It is a Fern with an upright axis, sometimes as thick as a man's thigh, and three to five feet high. It spreads by runners, from which shoots come up at some distance from the parent plant. The leaves have thick stalks, and are 3–4 times pinnate: the upper pinnae are sterile and leathery, the lowest pairs are slender and fertile (Fig. 527). Each pinnule of these terminates in a sorus, the whole giving the appearance of a complicated thyrsus. The dermal appendages are mucilaginous hairs, apparently unbranched: on the rhizome there are also stiff brown bristles, but there are no scales. The thin runners are solenostelic: but where larger, a single strand or sometimes more may be found in the pith. In a larger stem this has been found to expand, as in *Dennstaedtia*, into a medullary system with a compensation-strand filling each leaf-gap (Fig. 528). It would be interesting to see the structure of one of the largest stems, for it might well show a structure not unlike some "Psaronieae." The leaf-trace passes off as a corrugated horse-shoe: in larger leaves it becomes more corrugated higher up, and divides opposite the lateral pneumatodes into three straps, still preserving the horse-shoe form (see Vol. I, Fig. 161, 5). These characters accord with what is seen in some Dennstaedtiinae.

The sori are marginal in origin, and they retain that position. Each has a cup-like indusium surrounding the receptacle, which bears a basipetal sequence of sporangia (Fig. 527, *D*, *E*). The indusium which originates as a massive superficial outgrowth on each leaf-surface, below the marginal receptacle, is at first slightly two-lipped, but this is not obvious in later stages. The formation of sporangia soon follows, the first appearing on the extreme margin (Fig. 529): but the receptacle is really a flattened lobe, and a series of sporangia occupies its margin as in *Schizaea*: these are followed by others in gradate sequence.

The sporangia are large with a short massive stalk, which shows about eight cells in transverse section. The head bears an annulus with 50 or more cells forming a curiously twisted hoop (Fig. 529, *E*, *F*). Rather more than half of the cells are indurated, while the rest form an obliquely lateral stomium, the structure of which is complex, and not strictly defined. Rupture, the exact position of which may vary, is lateral. The typical number of spores is 48 to 64. The parent-cell of the sporangium is not uniform in shape: it commonly has a square base, and the first segment-wall passes obliquely to the basal wall as in Schizaeaceae, the second being inserted

Fig 527. *Thyrsopteris elegans* Kze. *A* = part of a leaf, above a sterile, and below a fertile pinna. *B* = part of a sterile pinnule. *C* = part of a fertile pinnule. *D* = sorus with indusium. *E* = sorus with the indusium cut through. *F* = sporangium. (After Kunze, from Engler and Prantl.)

obliquely on the first (*A*, *C*). The result is that the stalk is short and thick. Sometimes the parent-cell is wedge-shaped (*B*). Thus *Thyrsopteris* wavers between a more primitive and a more advanced type of segmentation of the sporangium. The orientation of the sporangia is usually as in the Hymeno-

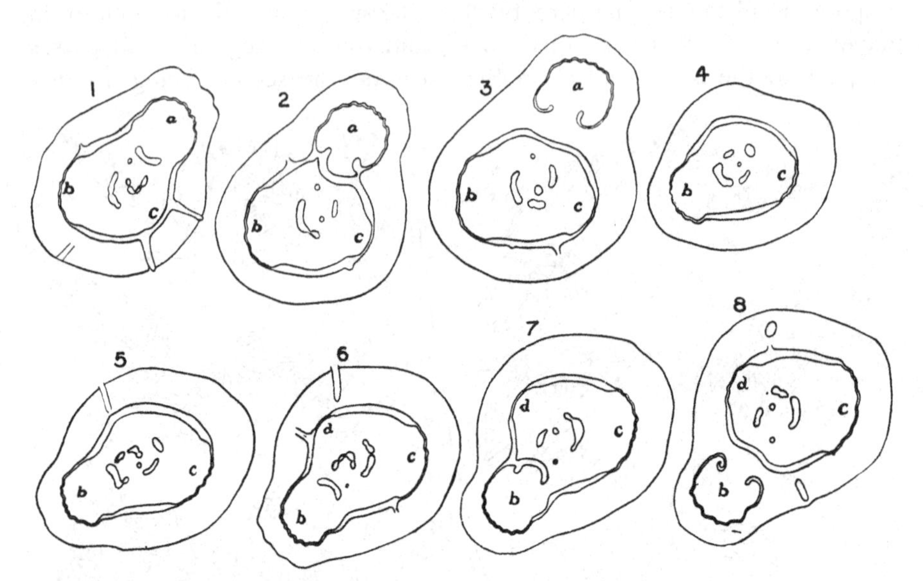

Fig. 528. Series of transverse sections. 1—8, arranged in ascending sequence, from the rhizome of *Thyrsopteris elegans*, showing the relatively primitive, but polycyclic vascular system. The leaf traces *a*, *b* depart without opening the solenostele, owing to the action of compensation strands that fill up the gaps.

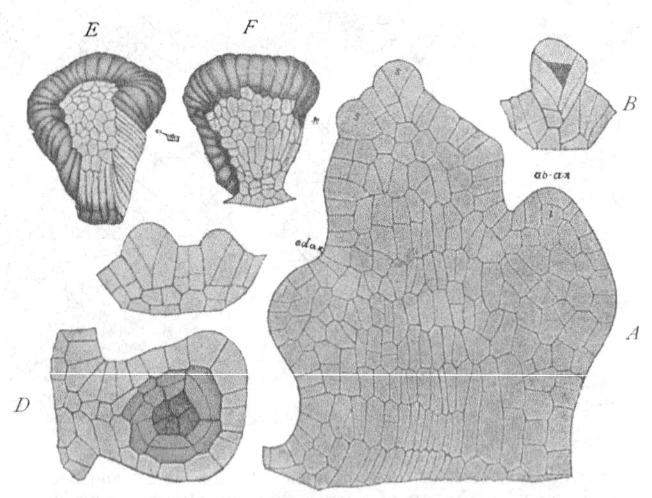

Fig. 529. *Thyrsopteris elegans* Kze. *A* = longitudinal section through the young sorus, showing the two-lipped indusium *i*, *i*, and sporangia *s*, *s*, seated on the receptacle, the oldest being at the distal limit of it. *C* = two young sporangia. *B* = one rather more advanced. *D* = a sporangium with the tapetum and sporogenous groups shaded. *E*, *F* = mature sporangia. (*A*—*D* × 200; *E*, *F* × 50.)

phyllaceae, the lateral dehiscence is thus effective as in other Gradatae. But the succession of the sporangia is not long continued.

As regards its sorus and sporangium *Thyrsopteris* appears to be one of the least specialised of Leptosporangiate Ferns, and to show analogies with *Loxsomopsis* and with *Hymenophyllum*. The distal curve of the annulus runs almost in a horizontal plane, and thus is reminiscent of *Anemia*, while the low organisation of the stomium also indicates a primitive state. On the other hand there are distinct analogies with *Dicksonia*, with which its dendroid habit and dermal hairs relate it. There appears little reason to link *Thyrsopteris* with the Cyatheae, from which the absence of scales, and the strictly marginal sorus definitely remove it. It finds its best place as the sole survivor of a separate and archaic family, related to the Dicksonieae, and it is probably akin to *Coniopteris*, a Fern of the Jurassic Period.

Prothalli of *Thyrsopteris* have been raised by Boodle at Kew, and by Miss Stokey at Holyoke College, Mass. The latter has not found them to bear either hairs or scales: but naturally rhizoids are present. A detailed description will be published shortly.

II. Dicksonieae

This sub-family is here treated in the same limited sense as in the *Origin of a Land Flora*, p. 592: that is to say, excluding §*Patania* (see *Synopsis Filicum*, p. 52), but retaining *Balantium* Kaulfuss, *Dicksonia* L'Héritier, and *Cibotium* Kaulfuss, as they are recognised by Diels (*Natürl. Pflanzenfam.* I, 4, p. 119). The Ferns thus grouped vary in habit from low creeping stocks to lofty tree-ferns. Their leaves are repeatedly pinnate, and bear numerous marginal sori, but without any clear differentiation of sterile from fertile pinnae. The sori are protected by a two-lipped indusium, but the lips are more or less unequal, and their characters have been used in generic distinction. The receptacle is itself marginal, the lips of the indusium originating as outgrowths of the upper and lower surfaces of the pinnule as in *Thyrsopteris*, the Hymenophyllaceae, and ultimately the Schizaeaceae also (Fig. 530). There is thus a strict local correspondence between the sori of all of these Ferns. The differences are mainly in habit and size. All of them are alike in having hairs as their dermal covering, while scales (ramenta) are consistently absent.

Creeping stems such as those of *Cibotium Barometz*, and *Balantium Culcita* are massive, and if cut transversely at an internode show a very large solenostele surrounding a distended pith (Fig. 531). From this an undivided meristele passes off to each leaf. The leaf-gaps may overlap, so that more than one may be traversed, and the structure may be technically dictyostelic. The leaf-trace is a broad ribbon with infolded margins, and very shortly after its departure from the stele it breaks up into a number of isolated strands arranged horse-shoe fashion, as seen in transverse section

(Fig. 532). The point of disintegration of the meristele varies: it is clearly an advance upon what is seen in *Thyrsopteris*. There are no accessory strands in the pith or leaf-stalk. Where the stem is upright or dendroid as it is in *Dicksonia antarctica*, the stelar structure is similar, but more complicated by the dense insertion of the leaves. It becomes thus a highly

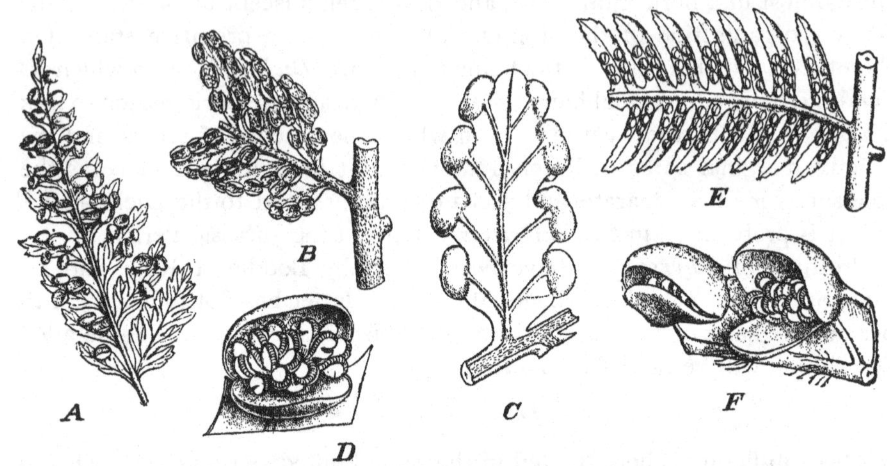

Fig. 530. Illustrations of the sori of Dicksoniaceae. A = *Balantium Culcita* (L'Hér.) Kaulf.: part of a pinnule of third order. B–D = *Dicksonia arborescens* L'Hér.: B = lower part of a pinna of the first order; C = pinna of the second order; D = sorus with indusium. E, F = *Cibotium Barometz* Link: E = lower part of a pinna of first order; F = part of a segment with two sori. (From Engler and Prantl. B–F after Hooker.)

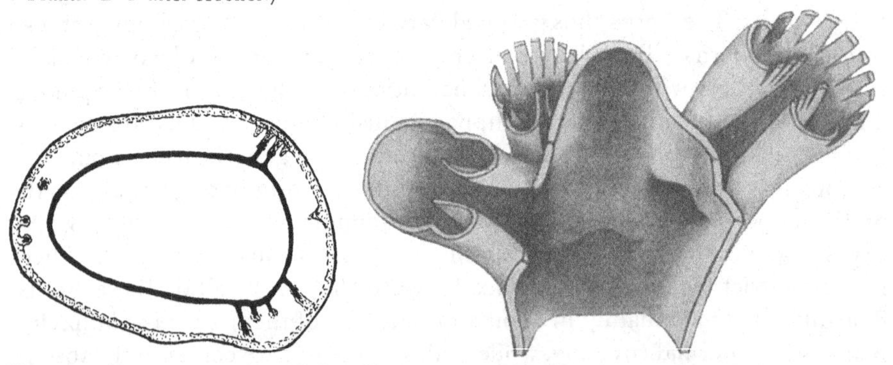

Fig. 531. Transverse section of the solenostelic rhizome of *Cibotium Barometz* (L.) J. Sm. (Natural size.)

Fig. 532. *Cibotium Barometz*. Portion of the vascular system of the stem, seen from within, and showing the departure of three leaf-traces. (After Gwynne-Vaughan.)

elaborated dictyostele, but with very narrow and short leaf-gaps (Fig. 533). It is further complicated by the deeply sinuous corrugations of the stele, and of the sclerotic sheaths that invest it internally and externally (compare Vol. I, p. 196, Fig, 184). Doubtless this is a source of additional mechanical strength, but it also gives greatly increased surface-area to the vascular system, and particularly to the xylem. The pith is of unusual size, three inches or more in diameter in a large stem; but still there is no medullary vascular system: this is one of the largest examples of a purely parenchymatous pith.

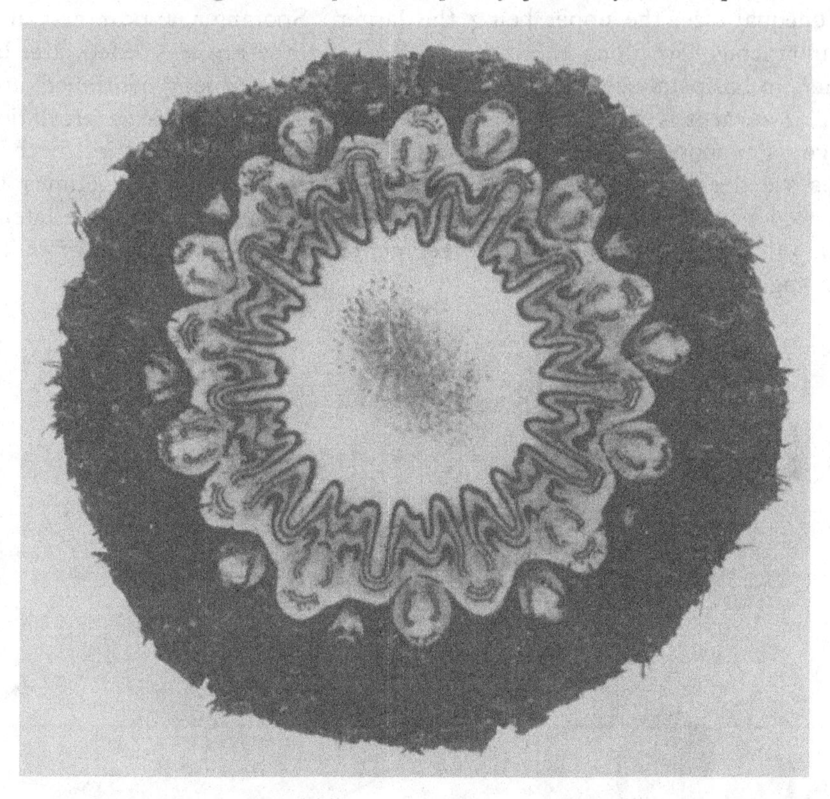

Fig. 533. *Dicksonia antarctica* R. Br. Transverse section of stem, showing highly corrugated stele, and large pith without any medullary vascular system. (½ natural size.)

SORUS AND SPORANGIA

The proof that the receptacle of the Dicksonieae is really marginal in origin is of considerable importance in their comparison with other Ferns. It is shown to be so by the segmentation of the very young fertile pinnule (Fig. 534). It is then seen that the marginal cell (*R*) projects between the upper and lower indusia (*U*, *L*), which originate by upgrowths of its segments. The correspondence with *Hymenophyllum* (Fig. 513), and *Thyrsopteris* (Fig. 529) is very close. A later stage is shown in Fig. 535, *A*, where already the marginal receptacle (*m*) is covered in by the massive indusial lips. A section parallel to the leaf-surface (*z*, *z*) shows that the receptacle is flattened and elongated (*B*), while a transverse section at the

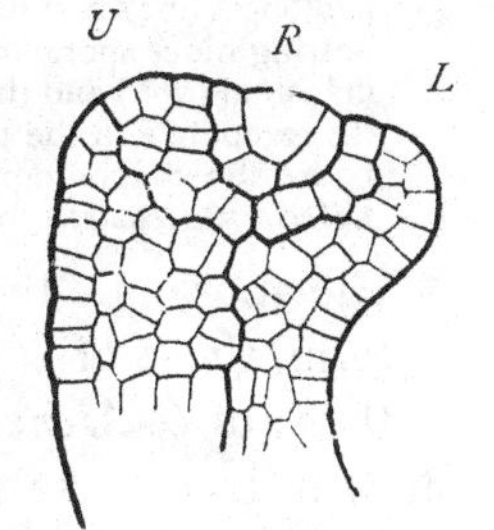

Fig. 534. Vertical section through a young sorus of *Cibotium Scheidei*. (×200.) *R* = receptacle. *U* = Upper indusium. *L* = Lower indusium.

same age demonstrates its relation to the indusium, and that the receptacle is structurally like a normal leaf-margin (*D*). Already the indusial lips are of unequal size, the upper being the larger. Sporangia appear first in a simultaneous row along the margin (*B*, *C*): these are succeeded later by others in basipetal sequence, but the succession is not long continued, and in *B. Culcita* it is not clearly marked. There is, however, no intercalation of younger sporangia between those first formed. The sporangial mother-cells are deeply sunk, and the first segmental wall may sometimes be inserted upon the basal wall (3, Fig. 535, *C*), in others upon the lateral wall (4). Speaking generally the segmentation is characteristic of bulky sporangia.

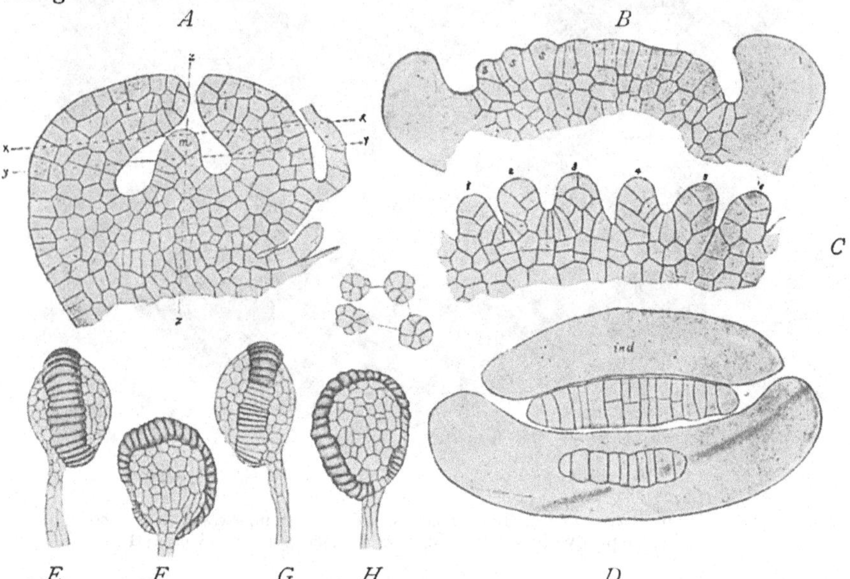

Fig. 535. *Cibotium Scheidei* Baker. *A* = section through a young sorus perpendicular to the leaf-surface: *i*, *i* = indusium; *m* = cell of marginal series. *B* = section of sorus parallel to the leaf-surface as along a line *i*, *i* in Fig. *A*: it shows the receptacle bearing sporangia *s*, *s*. *C* = a similar section bearing older sporangia. *D* = transverse section of a young sorus showing the two lips of the indusium (*ind.*) and the receptacle between them, as along a plane *y*, *y* in Fig. *A*. A section of the receptacle in the plane *x*, *x*, in Fig. *A* is superposed on the lower indusial lip. The central figure shows sporangial stalks cut transversely. (*A*–*D* × 200.) *E*, *F*, *G*, *H* adult sporangia of *Cibotium Menziesii* from four different aspects. (× 50.)

The sporangia themselves are large and relatively long-stalked: they vary in form, those of *B. Culcita* being pear-shaped with very oblique ring, while in those of *C. Menziesii* it is almost vertical. The stalks are elongated, and show in transverse section six or seven cells (Fig. 535). The sporangia of *C. Menziesii* are represented in four aspects in Fig. 535, *E*, *F*, *G*, *H*, from which it appears that except for the long stalk, and the almost vertical annulus, they are substantially like those of *Thyrsopteris* or of *Hymenophyllum*. There is, however, a more definite stomium, associated on both sides with connective cells of the ring that are not indurated. But the number of the cells of the annulus and of its parts does not appear to be constant. The orientation of the sporangia is not uniform, especially at the margins of the flattened receptacle: but the majority have the distal face turned outwards from the receptacle. The output of spores is not large: the typical number is 64.

Hitherto the prothalli of the Dicksonieae appear to be imperfectly known. But recently prothalli of *Cibotium Barometz* and of *Dicksonia antarctica* have been raised in pure culture by Prof. A. G. Stokey of Holyoke College, Mass., U.S.A. She will shortly publish her results: but she permits me to state that the former species, after two years' culture, shows no bristles ("borsten"): in the latter no hairs or scales (excepting rhizoids) appear even on old and branched prothalli. Heim (*Flora*, 1896, p. 362) states, however, that the prothalli and antheridia of the Dicksonieae resemble those of the Cyatheaceae, and bear on margin and surfaces glandular hairs transitional to those of Polypodiaceae. Von Goebel describes bristle-like cell-plates ("borstenförmige Zellflächen") on the margin of prothalli of *Balantium antarcticum* (*Organographie*, II Aufl., ii, p. 950). These statements do not accord with the results of Prof. Stokey, and the matter awaits further detailed demonstration.

Both structurally and in the characters of the sorus and sporangium the Dicksonieae, in the restricted sense, occupy a position related closely to *Thyrsopteris*, and less directly to the Loxsomaceae and Hymenophyllaceae. But turning from these relatively primitive types they show also a relationship to the Dennstaedtiinae, so close indeed that Hooker included *Dennstaedtia* as a sub-genus of *Dicksonia*, under the name of §§§*Patania* Presl. The signs of advance which these Ferns show as compared with the *Dicksonieae* fully justifies their separation, with *Microlepia* and some others, as a natural sub-tribe. It will be seen that certain definite characters of advance in their sori uphold this treatment.

It seems probable that some of the Jurassic fossils grouped under the name *Coniopteris* may have really been early Dicksonioid types.

BIBLIOGRAPHY FOR CHAPTER XXIX

528. HOOKER & BAUER. Genera Filicum. Tab. 44, *A*, 1842.
529. HOOKER. Species Filicum. i, p. 64, 1846.
530. KUNZE. Farnkräuter. Leipzig. 1847.
531. ENGLER & PRANTL. Natürl. Pflanzenfam. i, 4, p. 122.
532. BOWER. Studies in Spore-producing Members. IV. Phil. Trans. Vol. 192, p. 67.
533. BOWER. Origin of a Land Flora. 1908, p. 588.
534. BOWER. Studies in Phylogeny. III. Ann. of Bot. xxvii, p. 454, 1913.
535. SEWARD. Fossil Plants. ii, p. 367, 1910.

CHAPTER XXX

DICKSONIACEAE

III. Dennstaedtiinae

UNDER the name Dennstaedtieae, Kuhn grouped together the genera *Hypolepis*, *Microlepia*, *Leptolepia*, and *Dennstaedtia*, basing the relationship chiefly upon the simple hairs which form their dermal covering (Kuhn, *Die Gruppe der Chaetopterides*, Berlin, 1882). With the addition of *Saccoloma*, and on a broader basis of comparison, Prantl adopted this grouping under the name Dennstaedtiinae (*Abhandl. d. Königl. Bot. Gart. Breslau*, I, i). But the old genus *Dennstaedtia* of Bernhardi had previously been merged by Sir W. Hooker into *Dicksonia*, as §§§*Patania* Presl, a fact which sufficiently indicates the similarity it bears to that genus, and particularly to the species with a creeping habit: for the Dennstaedtiinae are themselves chiefly prostrate. However, the detailed study of the Dennstaedtiinae as now constituted, while it fully bears out the close affinity with *Dicksonia*, shows features of advance which justify the maintenance of Prantl's distinct grouping, while they also point in the direction of the two series characterised by *Davallia* and by *Pteris*. Though *Dennstaedtia* was thus included by Hooker under *Dicksonia*, *Microlepia* had been placed as a section of *Davallia* (*Syn. Filic.* p. 97). The numerous synonyms which they have borne from time to time in the writings of other systematists show how closely the two genera are allied, and justify Prantl's grouping of them together. They occupy a peculiarly interesting position as connecting links between the basipetal sorus seen in *Dicksonia*, and the mixed sorus of *Davallia*. In other respects *Saccoloma* is also a leading outlier pointing especially toward *Pteris*. For these reasons the Dennstaedtiinae will demand careful attention as an important transitional series.

These Ferns have for the most part a creeping habit, with solitary, long-stalked, repeatedly pinnate leaves. The dermal appendages are hairs, excepting in *Saccoloma*, which has broad scales. The sori are marginal, as a rule distinct from one another, and slightly turned downwards, with the upper protective flap larger than the lower. It has long been known that the axes of *Dennstaedtia* and of *Microlepia* show the relatively primitive solenostelic state. The vascular relations of leaf and axis in different species of *Dennstaedtia* are indicated by Fig. 536, *A*, *B*, *C*. In all of them the leaf-trace is an undivided ribbon-like strand. The stelar tube opens at its insertion forming the leaf-gap, which is here short and soon closes (*A*). This simple structure holds for most species, though the vascular supply for a lateral axis may be inserted close to the leaf-base. But in *D. adiantoides* a local thickening appears internally at the margins of the leaf-gaps, from which a projecting ridge extends into the cylinder, and is continuous from

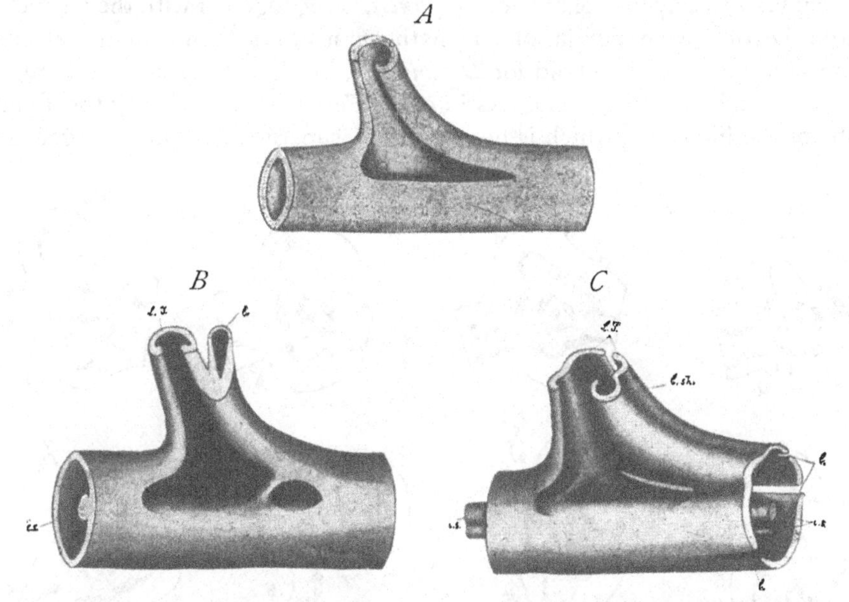

Fig. 536. *A*, *B*, *C*, Diagrams of stelar structure in the Dennstaedtiinae. *A*, *Dennstaedtia punctilobula*. The stele including a node, and the base of a leaf-trace. The upper surface of the rhizome would face the observer. It is a simple solenostele with undivided leaf-trace. *B*, a similar diagram of *Dennstaedtia adiantoides*: *l.sh.* = lateral shoot arising from the basiscopic margin of the leaf-trace *l.t.* *i.s.* = ridge upon the internal surface of the solenostele. *C*, a similar diagram of *Dennstaedtia rubiginosa*: *l.sh.* and *i.s.* as in *B*. *l*, *l* = lacunae (or perforations) in the solenostele not related to the departure of the leaf-trace. See text. (All after Gwynne-Vaughan.)

one leaf-base to another (*B*). In *D. rubiginosa* this ridge is represented by a separate strand, which maintains its connexion with each leaf-gap, but may divide into several distinct rods that traverse the cylinder (*C*). In a very large rhizome of *D. dissecta* (Sw.) Moore, such rods are replaced by a second complete solenostele, within the first (Fig. 537). *Saccoloma elegans*, which has an upright stock with crowded leaves, goes further than this in stelar elaboration. It initiates a third system within the outer two, while it is also dictyostelic, and the leaf-trace is already disintegrated before it is detached (Fig. 538). These stelar elaborations have their special interest for comparison with the essentially similar structure seen in the Pterideae. One further point is that perforations, not to be confused with leaf-gaps, are found in *D. rubiginosa* (Fig. 536, *B*). It will be

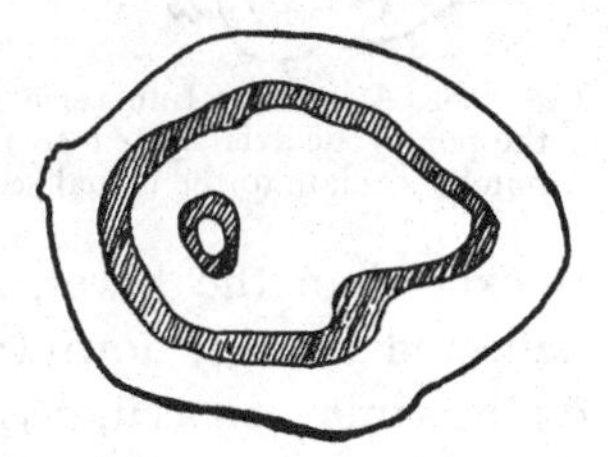

Fig. 537. Transverse section of rhizome of *Dennstaedtia dissecta* (Sw.) Moore, showing two complete solenosteles. Nat. size.

seen presently that there is a progression in soral characters in the Dennstaedtiinae which runs parallel with that in their vascular system.

The origin of the sorus in all of these Ferns is marginal. This has been shown very clearly by Conard for *D. punctilobula*, together with the basipetal sequence of the sporangia, of which the first arises from a marginal cell. The same was found to hold for *D. apiifolia*, in which the sequence is regularly sustained (Fig. 539, *A*). As in other Ferns of this affinity the upper lobe of the indusium, which is here again of superficial origin, is larger and

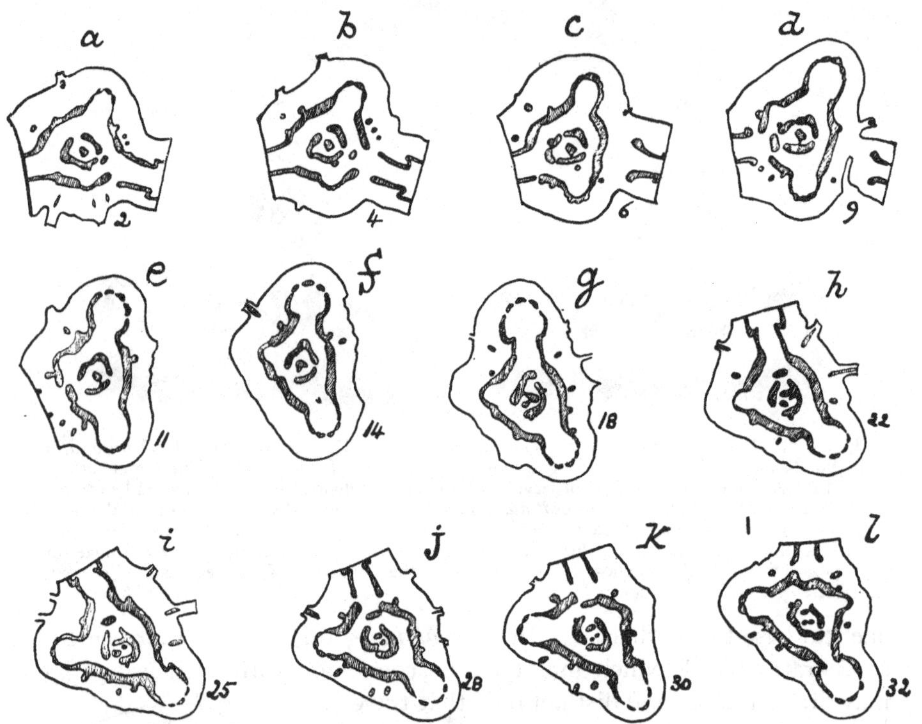

Fig. 538. A series of transverse sections arranged in ascending sequence, showing the relations of the polycyclic axial system to the leaf-traces in the upright stock of *Saccoloma elegans* Klf. The numbers relate to the actual sections cut. (Natural size.)

thicker than the lower, and as development proceeds the whole sorus is deflected strongly downwards. In fact, in the species named it is constructed on the same plan as in *Hymenophyllum*, *Loxsoma*, *Thyrsopteris*, and *Dicksonia*, though smaller and more strongly deflected than in the latter. But the sporangia themselves are smaller: the stalk is elongated as in *Dicksonia*, but it consists of only three rows of cells, while the head is flattened on either side of the almost vertical annulus (Fig. 539, *C*). The ring of cells runs round the margin of the sporangium almost in a vertical plane, but at the stalk it is slightly diverted to one side: usually the sequence of its cells is not wholly interrupted by its insertion; in such cases it is actually continuous at the base, as it is in the more primitive sporangia with their conspicuously oblique ring. But not uncommonly the ring is actually interrupted at the stalk. In fact these sporangia show only faint traces of obliquity of the annulus, and these inconstantly. The annulus is almost vertical, and its dehiscence transverse. Sections of the young sporangium show that the number of spore-mother-cells in each sporangium is variable: eight, twelve, or sixteen have been observed. Countings of the mature spores have shown that though sometimes the full number of 64 may be produced, the tale is frequently less. The orientation of the sporangia is not constant, but perhaps this may be due in part to twisting of the long stalk.

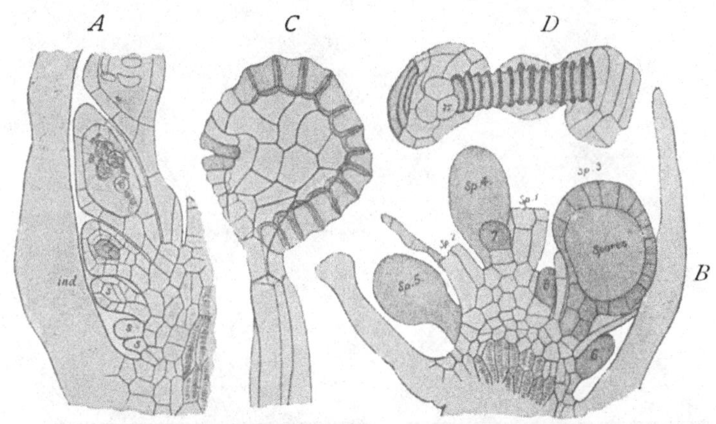

Fig. 539. *A*, *Dennstaedtia apiifolia* Hook. Sorus showing basipetal succession of sporangia: *ind* = indusium. *C*, dehiscent sporangium of the same, showing very slightly oblique annulus. *B*, *Dennstaedtia rubiginosa* Kaulf.: sorus in vertical section showing that it has been at first basipetal, but a mixed character has supervened. *D*, dehiscent sporangium of the same, seen from the base, showing that the annulus stops short on either side of the insertion of the stalk (*st*). (All × 100.)

All these features indicate for *Dennstaedtia* an advance upon the marginal types so far described. But there is another still more important for comparison. Deviations from the basipetal succession of the sporangia have been observed in *D. davallioides* (Br.) Moore, and in *D. rubiginosa* (Kaulf.) Moore. Even in *D. apiifolia* isolated cases occur of a sporangium near to the apex of the receptacle in a less developed state than those below. In *D. davallioides* such cases are fairly common; but in *D. rubiginosa* the sorus bears upon the shortened receptacle sporangia without definite order of succession (Fig. 539, *B*). The sorus has, it is true, some signs of a basipetal succession at first, but later it takes clearly the "mixed" character, with younger sporangia interspersed irregularly between those that are older: and correlated with this we find the receptacle short, but wide. Moreover, the annulus is here vertical, and definitely interrupted at the stalk (Fig.

539, *D*). Similar conditions are occasionally seen also in *Microlepia hirta*, though in that genus the basipetal succession appears to be the rule (Fig. 540). The importance of these deviations will become apparent as our comparisons proceed. Meanwhile they should be carefully noted, with the general remark that they point in the direction of that large body of Ferns, of which *Davallia* is a central type, in which the marginal sorus is of the "mixed" type.

Saccoloma which was associated by Prantl with the other Dennstaedtiinae, stands out from the family as advanced in its vascular system, and in its flattened dermal scales. It is also rather distinct in habit, the stock being

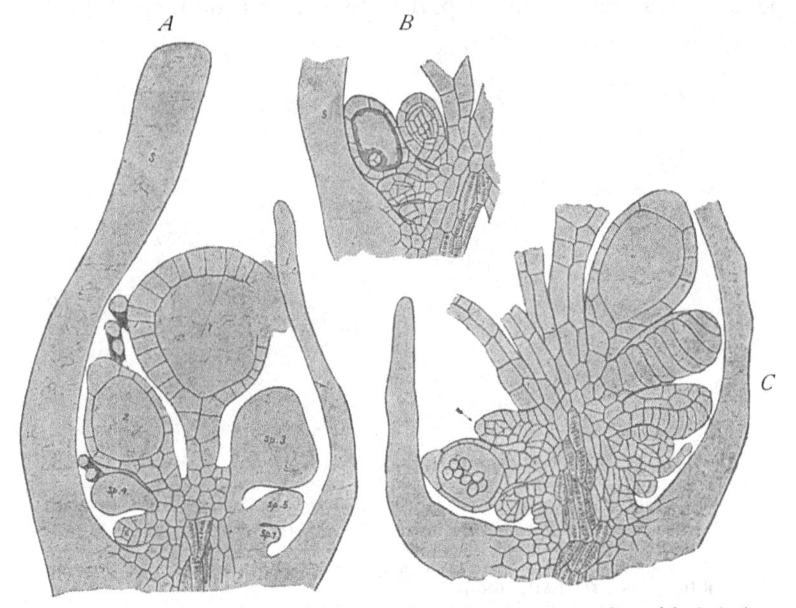

Fig. 540. *A*, *Microlepia speluncae* (L.) Moore. Sorus showing unequal lips of the indusium, and basipetal succession of the sporangia. (× 100.) *B*, *C*, *Microlepia hirta* Kaulf. Similar sections to *A* but showing departures from the strict basipetal succession. (× 100.)

upright and radial, while the leaves are simply pinnate. The pinnae are as much as a foot long, traversed by parallel, occasionally forked veins, upon the ends of which in fertile pinnae the sori are borne in an apparently intra-marginal row. The relations of the sori have been elucidated by von Goebel (*Organographie*, II Aufl., p. 1143). What appears as a continuous leaf-margin is really composed of the upper indusial flaps of a row of sori fused laterally. The upper flaps are larger than the lower, and the latter appear intra-marginal, covering the still distinct receptacles. Seam-like swellings of the upper flap indicate the limits of the partially fused sori (Fig. 541, *A*). Comparing this structure with that seen in *Microlepia platyphylla*, and these again with *Nephrolepis* or with *Lindsaya*, we see in these Dennstaedtiinae ferns which are advanced in various of their structural details, and also in their sori: and that those advances towards lateral linkage of sori, and of transit to the lower surface respectively, are such as point in the direction of those large genera, *Nephrolepis*, *Lindsaya*, and *Pteris*. Here no more can be done than to indicate this as a special interest attaching to the Dennstaedtiinae. They may be held as advanced types from the Dicksonioid affinity, serving as synthetic links with large bodies of Leptosporangiate Ferns.

But still even in *Saccoloma* the receptacle is itself truly marginal. It has been found on examination of very young fertile pinnae that the marginal cells give rise, as in all the early Marginales, to the receptacle; while upgrowths from the upper and lower surfaces develope into the indusial

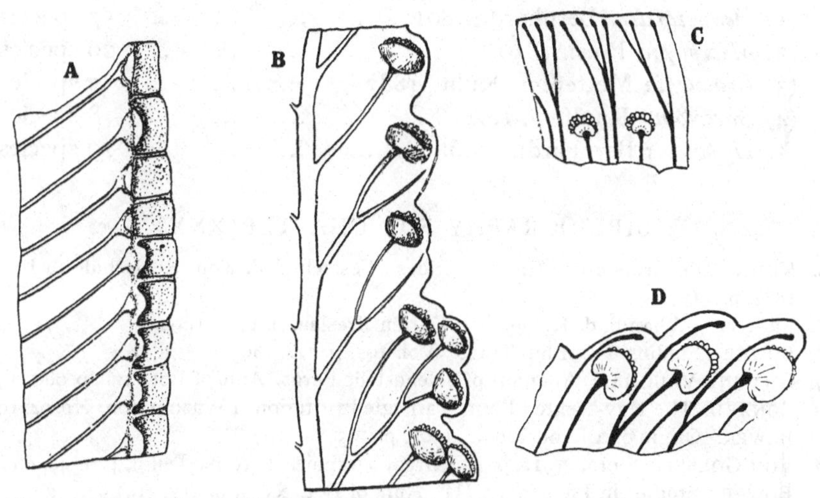

Fig. 541. Group of sori, which have become apparently intra-marginal by the upper indusial lip of each fusing with its neighbour, and simulating the rest of the leaf-surface. *A* = *Saccoloma elegans*. (After von Goebel.) *B* = *Microlepia platyphylla*. (After von Goebel.) *C* = *Nephrolepis acuta*. (After Christ.) *D* = *Nephrolepis cordifolia*. (After Christ.)

flaps, the sporangia arising in a one-sided gradate sequence ("Studies in Phylogeny, III," *Ann. of Bot.* p. 457, Pl. xxxiii, Fig. 16, 1913). It will be shown when the Pterid and Davallioid Ferns are taken up, that these are typically Ferns with a mixed sorus, a condition already faintly foreshadowed by *Dennstaedtia* and *Microlepia*. Accordingly we shall recognise not by one character only but by several distinct characters that the Gradate Marginales occupy an intermediate place between the marginal Simplices, and certain large families of Ferns which have advanced to a mixed condition of their sori.

The fifth genus of Prantl's family, viz. *Hypolepis*, shows still more definite advance not only to a mixed sorus, but to one without any obvious indusial protections: and thus it would rank under the old comprehensive title of

Polypodium. The discussion of this will be left over for the present. But meanwhile the existence of such features still further enhances the interest in these Ferns as synthetic types.

TABLE OF GENERA OF DICKSONIACEAE

I. THYRSOPTERIDEAE.
  (1) *Thyrsopteris* Kunze, 1884. ... ... 1 species
II. DICKSONIEAE.
  (1) *Balantium* Kaulfuss, 1834. ... ... 3 species
  (2) *Dicksonia* L'Héritier, 1788. ... ... 17 species
  (3) *Cibotium* Kaulfuss, 1820, 1824. ... ... 9 species
III. DENNSTAEDTIINAE Prantl.
  (1) *Dennstaedtia* Bernhardi, 1801. ... ... 57 species
  (2) *Microlepia* Presl, 1836. ... ... 29 species
  (3) *Leptolepia* Mettenius, Kuhn, 1882. ... ... 2 species
  (4) *Saccoloma* Kaulfuss, 1820. ... ... 8 species
  (5) *Hypolepis* Bernhardi, 1806. ... ... 29 species

BIBLIOGRAPHY FOR CHAPTER XXX

536. KUHN. Die Gruppe der Chaetopterides. Festschrift d. Kön. Realschule zu Berlin. 1882, p. 321.
537. PRANTL. Abhandl. d. König. Bot. Garten Breslau. i, p. i. 1892.
538. BOWER. Studies. IV. Phil. Trans. Vol. 192, p. 71. 1899.
539. GWYNNE-VAUGHAN. Anatomy of Solenostelic Ferns. Ann. of Bot. xvii, p. 689. 1903.
540. CONARD. The Hay-scented Fern. Carnegie Institution of Washington. No. 94, 1908.
541. BOWER. Origin of a Land Flora. 1908, p. 595.
542. VON GOEBEL. Flora. 1912, p. 47. Organographie. II Aufl., Teil ii, p. 1140, etc.
543. BOWER. Studies in Phylogeny. III. Ann. of Bot. xxvii, p. 457. 1913.

CHAPTER XXXI

PLAGIOGYRIACEAE

THE single genus included in this family has by reason of its curiously mixed characters been greatly misunderstood. The outstanding feature is the presence of an oblique annulus in a Fern having the habit of a *Lomaria*. So long as those external features which constitute "habit" dominated classification such misunderstandings were natural. But when so pronounced a feature as an oblique annulus is combined with a relatively primitive anatomy and with dermal hairs, it is high time to reconsider the value of habit as a guide.

The genus *Plagiogyria* was first distinguished by Kunze (*Bot. Zeit.* p. 867, 1849), who regarded it as a section of *Lomaria*. Later it was the subject of special examination by Mettenius (*Farngattungen*, ii, Frankfurt, 1858), who on the ground of the oblique annulus placed it with the Cyatheaceae. But Sir W. Hooker, following Kunze, still retained it with *Lomaria*, where it appears in the *Synopsis Filicum*, 1874, p. 182. Christ (*Farnkräuter*, pp. 6, 175), ranks it as a substantive genus with the Pterideae, placing it between *Pteris* and *Blechnum*. Diels ranks it in that rather incongruous group, the Pterideae-Cheilanthinae (*Natürl. Pflanzenfam.* i, 4, p. 281). In Christensen's *Index* it appears as the last genus of the Cheilanthinae, in close relation with *Cryptogramme* and *Llavea* (p. xliii).

There is no doubt that the sporangium with its oblique annulus must be given full weight, but the general characters of this remarkable genus must also be taken into account before any final position can be assigned to it. The fact that *Plagiogyria* alone of described Ferns combines the oblique annulus of some Simplices and of the Gradatae, with certain general characters of the Mixtae, at once commands attention and suggests for it a position as a synthetic type. As at present described there are eleven species: two are found in Central America, and the rest are Eastern, widely spread in the Malayan region.

These Ferns have an erect, radial woody stem, covered by the crowded leaf-bases, and giving the effect of a stunted tree-fern: in fact they have the habit seen in the Osmundaceae. Occasionally the axis bifurcates, particularly in *P. semicordata* Baker: but this has not been seen in *P. pycnophylla* Kunze, a species which, however, bears stolons. These vary in length, taking at first a horizontal course as thin runners bearing scale-leaves: but sooner or later they expand at the apex into a massive upright trunk with dense foliage. There is reason to regard the dichotomy and the formation of stolons as complementary modes of branching: the latter may even be a modification of the former. The facts for *Plagiogyria* are distinctly suggestive in relation to the branching of Ferns (see Vol. I, Chap. IV). The leaves are simply pinnate, forming a dense rosette: the broader sterile leaves are distinct from the narrow fertile leaves, which are often longer (Fig. 542). The venation is always open and forked (Bower, "Studies, I," *Ann. of Bot.*

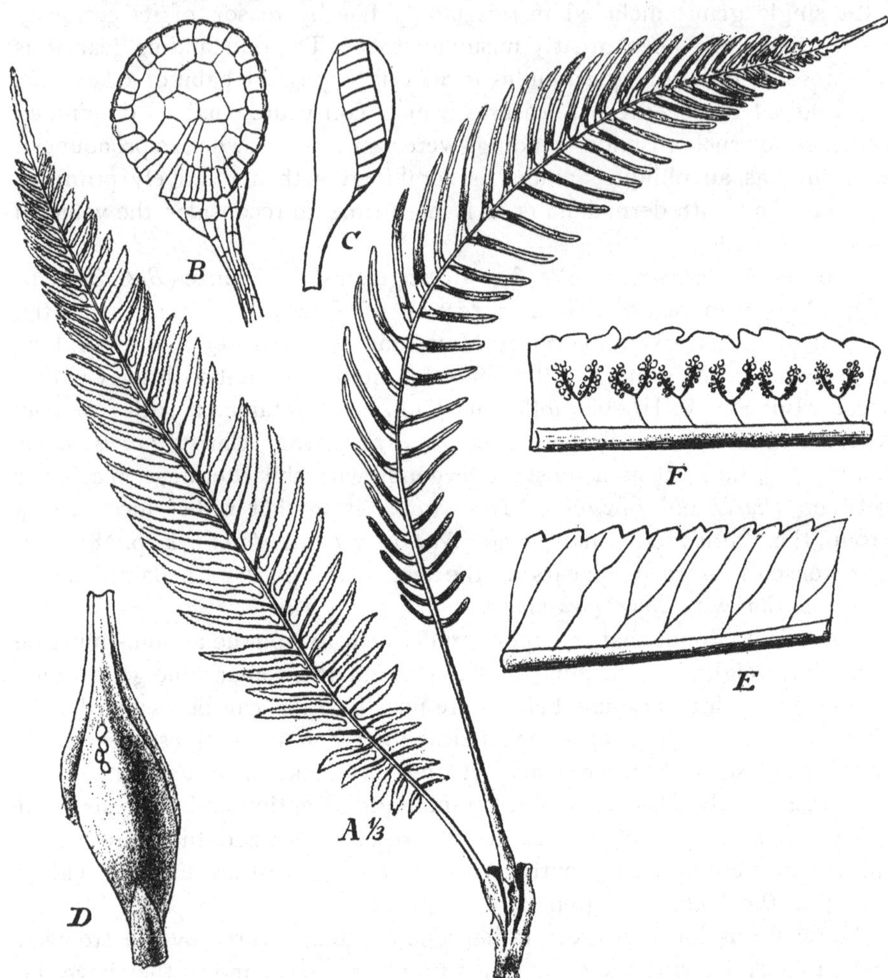

Fig. 542. *Plagiogyria* Kze. *A–F*=*P. semicordata* (Presl) Christ. *A*=habit: *B*, *C*=sporangia in face, and lateral view: *D*=base of the leaf: *E*=portion of a sterile leaf with venation: *F*=part of a fertile leaf with venation and sori. (After Mettenius, from Engler and Prantl.)

1909, p. 428). The leaf-stalks are enlarged at the base as in the Osmundaceae: they bear on the abaxial face two rows of pneumatophores (Fig. 543). Similar organs are also found on the upper leaf, and are specially prominent alternating with the pinnae in the young circinate state in *P. pycnophylla*. The young leaves are densely covered with mucilaginous hairs, as in the Osmundaceae. Scales are absent. The pneumatophores are probably important for aeration through this impervious covering. The sori appear superficially upon the veins of the fertile pinnae, the receptacle being elongated and slightly enlarged. There is no indusial protection, but the margins of the fertile pinnae are strongly curved inwards. The sporangia are rather large and long-stalked, with oblique annulus, and lateral dehiscence. They are almost simultaneous in time of maturing, but in origin the sorus shows a "mixed" character (see Vol. I, Fig. 250, *c*, where the sporangium of *P. pycnophylla* is placed in relation with those of other Ferns).

With such features as these the anatomy calls for very careful consideration, and it shows relatively primitive characters. A transverse section of a large stock of *P. pycnophylla* discloses a stelar structure not far removed from solenostely, with **V**-shaped leaf-traces departing from narrow leaf-gaps, each with an attendant strand of sclerenchyma on the adaxial side. The

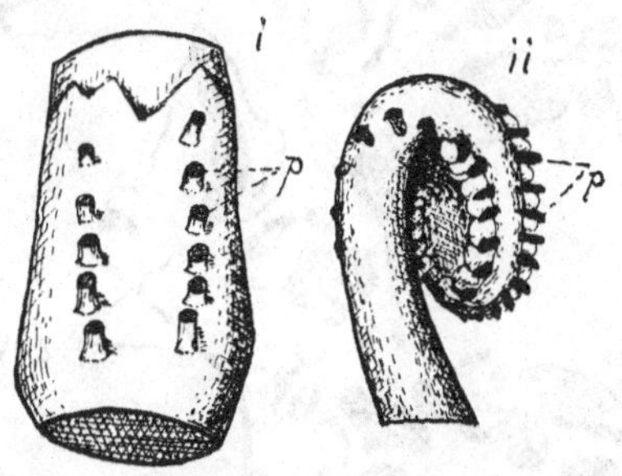

Fig. 543. Pneumatophores of *Plagiogyria*. i, shows the persistent base of the leaf of *P. glauca* the abaxial face bearing two rows of them (*p*). ii, the circinate apex of the leaf of *P. pycnophylla* with the pneumatophores (*p*) alternating with the pinnae, and projecting through the covering of mucilaginous hairs.

centre of the pith is occupied by a large sclerotic core with which they appear to connect, while a shell of sclerenchyma covers the whole outer surface of the stem (Fig. 544, *A*). The stelar ring is delimited externally and internally by a brown endodermis, which is continuous through the foliar gaps, and completely surrounds all the vascular tracts. The protoxylem is mesarch, and there are usually four groups of it in each separated leaf-trace. As the trace passes obliquely out through the cortex it enlarges, taking in the leaf-base a wider form. The relation of the pneumatophores (*p*) to the trace is seen in Fig. 544, *E*. Meanwhile the axillary sclerotic strands also widen greatly, taking gradually a semilunar form, while in the middle of each strand a large air-space appears, formed by actual involution of the outer surface of the stem at the depressed leaf-axil (Fig. 544, *A*, *inv*). This resembles what is seen in *Anemia*, and some other Ferns, and has been discussed in

Vol. I, p. 149. Roots arise with some degree of regularity right and left of the departing leaf-trace, each with a sclerotic sheath.

The anatomical similarity of these features to what is seen in the more advanced types of the Osmundaceae cannot be missed. If allowance be made for the larger number of more closely disposed leaves, together with a different distribution of the sclerotic tissue, the primitive dictyostelic ring of *Plagiogyria* closely resembles the still more primitive dictyoxylic ring of *Osmunda cinnamomea*, where there is also an inner endodermis and phloem,

Fig. 544. *A*, transverse section of the stem of *Plagiogyria pycnophylla*, showing sclerenchyma black, vascular tissue shaded. (× 2.) *B*, tangential section of leaf-bases, one of them (*st*) converted into a stolon. (Natural size.) *C*, a dichotomy in *P. semicordata*, showing steles, leaf-traces, and sclerenchyma. (× 2½.) *D*, transverse section of a solenostelic stolon of *P. pycnophylla*. (× 4.) *E*, base of a petiole of *P. pycnophylla*, showing pneumatophores (*p*). (Enlarged.)

and the protoxylem is mesarch as in *Plagiogyria* (compare Fig. 126, Vol. I). One point of difference is that in *O. cinnamomea* the inner endodermis does not connect with the outer at the xylic gaps, though it does at the "ramular gaps": in *Plagiogyria* connection is made at each foliar gap. This, however, is found to occur, though irregularly, in the large *Osmundites Carnieri*, if as is believed the line of delimitation of the stele in that Fern is really endodermis (Vol. I, Fig. 127). The similarity extends in structure, but not in outline, to the leaf-traces, though those of *Plagiogyria* show at their first origin a more advanced state than in *Osmunda*. There are thus differences of detail, but the underlying similarity is patent, extending as it does to the behaviour on dichotomy (Fig. 544, *C*).

In *P. pycnophylla* the sections show that occasionally leaves may appear to be substituted by stolons (*st*, in Fig. 544, *A*, *B*), and these may at first have a protostelic vascular supply. But this expands later into a solenostele (*st*, Fig. 544, *A*, and Fig. 544, *D*): later as the stolon bears leaves the structure becomes dictyostelic (see Vol. I, Fig. 138). Sometimes a leaf seems to be merged into the stolon, but various relations occur which may be held as depending upon a balance between the leaf-primordium and the stolon-primordium, the resultant structure taking the character of the predominant partner (compare Goebel, *Organography*, Engl. Edn., Vol. ii, p. 240). This condition resembles what has been described by Lang for *Osmunda* (*Mem. Manch. Lit. and Phil.* Vol. 68, Part I, p. 53), and such facts have their bearing on the question of the primal relations of axis and leaf (Fig. 545).

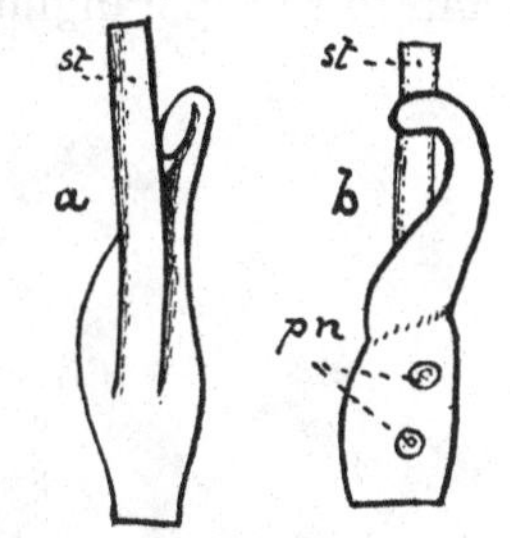

Fig. 545. A leaf of *P. pycnophylla*, itself arrested in its growth, and bearing on its adaxial face a stolon (*st*). *pn* = pneumatophores. *a*, seen in frontal view; *b*, from the side.

SORUS AND SPORANGIUM

The drawing by Mettenius (Fig. 542, *F*) gives a fair idea of the sorus as a whole, with the sporangia borne superficially on the forked veins, a position the same as that seen in *Todea barbara*. But whereas in the latter the sporangia are formed simultaneously, in *Plagiogyria* there is a succession of them: not in any gradate sequence, but with different ages intermixed, younger sporangia being interpolated between the older. The interpolation is not long continued, and as they mature the sporangia of a sorus appear to be almost of the same age. They are protected by the incurved margins, not by any specialised indusium. The sori are thus superficial, but "mixed."

The sporangium originates from a single cell, with variable segmentation, like that of *Alsophila* or *Schizaea* (Vol. I, Fig. 238, *c*, *d*). This provides for a stalk with 5 or 6 rows of cells. The further segmentation is after the Leptosporangiate type, resulting in *P. pycnophylla* in 12 spore-mother-cells, with a typical output of 48 tetrahedral spores. This small output goes, however, with a structure of the sporangial head not highly specialised, while the slit may open either right or left-handed, as in *Loxsomopsis* (Fig. 546). The stalk is unusually long for a sporangium with oblique annulus, though it is matched by *Dicksonia*. The head is pear-shaped, as in *Loxsoma*, and the oblique and rather variable ring consists of about 30 cells, surrounding the distal face. Dehiscence is by a lateral slit with a stomium fairly well organised, but not always exact in position, or in the number or size of the cells which determine it. A comparison of the sporangium shows a definite resemblance to that of *Dicksonia* and some other Gradatae, while a further reference may be made to the more distant *Anemia*. The resemblance in detail to the sporangium of the Osmundaceae is not close.

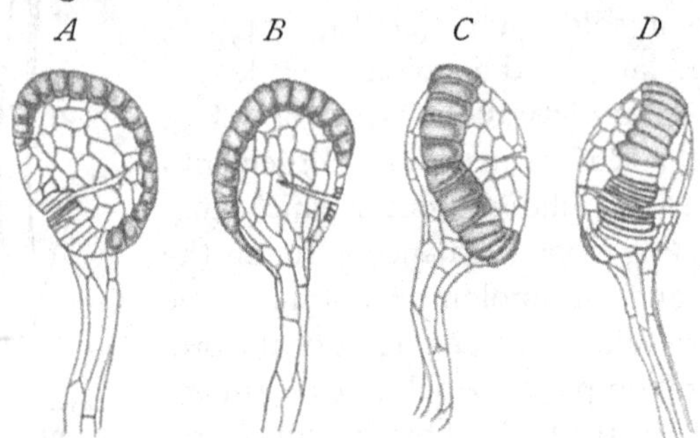

Fig. 546. Mature sporangia of *Plagiogyria euphlebia* Kze. *A* = presenting the distal face: *B* = the basal: *C* = the indurated side of the annulus: *D* = the stomial side. (× 50.) *A* dehisces to the left, but *C* and *D* dehisce by a right-handed slit, as viewed from the distal face. Compare Fig. 524 of *Loxsomopsis*.

The prothalli are of the ordinary Leptosporangiate type, symmetrical in form, and without distinctive hairs or glands. The antheridia are of the type of the Polypodiaceae, with which *Anemia* and *Mohria* were grouped by Heim as regards the detail of rupture (*Flora*, 1896, p. 349).

COMPARISON

The similarity of the stele of the large stock and of the leaf-trace to that of *Osmunda* has already been noted, and this together with the enlarged leaf base, and the dense covering of the youngest parts by mucilaginous hairs forms a ground of resemblance that cannot be missed. On the other hand, the Schizaeaceae include protostelic, solenostelic and dictyostelic types, and all of these states may be seen in the single runner of *Plagiogyria*. Moreover the deep axillary involutions are a rare peculiarity which is shared with *Anemia*. These indications are very suggestive. The open venation so usual in primitive Ferns is shared by both Osmundaceae and most Schizaeaceae, and by the Dicksonieae. But the superficial sorus, without any indusium, points definitely to *Todea*, rather than to any Schizaeoid or Dicksonioid Ferns, where its origin is marginal. The mixed character of the sorus is a feature of advance seen in none of the Ferns named. In its structure the sporangium compares most nearly with that of *Dicksonia*, and both of these may be traced in essentials to a Schizaeoid source, such as that of *Anemia*. The low output of spores (48) is a sign of advance far removed from

Osmundaceae or Schizaeaceae: but it is not very materially smaller than that of *Dicksonia* (64). Finally, the cordate prothallus, and the dehiscence of the antheridia would accord with various relatively primitive Ferns, and does not appear distinctive. The sum of all of these considerations leads to the conclusion that *Plagiogyria* is a relatively primitive type, but not very closely allied downwards to any one of the known primitive Ferns. The relation to the Schizaeaceae suggested by the axillary involutions in the dictyostelic stem, and in some degree by the structure of the sporangia, is vitiated by their superficial position in *Plagiogyria*: that to the Osmundaceae is indicated by the basal swellings of the leaves, the stelar structure and dichotomy of the upright stock, the dermal hairs, and particularly by the superficial sorus, as in *Todea*: but it is negatived by the difference in sporangial structure, and the small spore-output. The relation to other Superficiales such as the Cyatheae does not appear cogent. On reconsidering the question as a whole it seems necessary to constitute *Plagiogyria* as the only living genus of an independent Family of Plagiogyriaceae, its position being somewhat isolated, but with relations downwards to such Families of the Simplices as the Osmundaceae and Schizaeaceae.

The relations upwards have already been recognised by various writers. They are with such genera as *Llavea* and *Cryptogramme*: while these lead on to a large number of Superficiales of the type of *Gymnogramme*, which constitute a phylum to be discussed later under the collective name of the Gymnogrammoid Ferns. It will, however, be seen as the study of their sori is developed, that there is no near phyletic relation either to the Pteroideae or to the Blechnoideae, notwithstanding the superficial similarity which the fertile pinnae appear to possess in all these three sequences.

Incidentally the existence of a "mixed" character of the sorus, indicated rather than fully developed, suggests that this state may have been derived directly from a "simple" sorus, without the intervention of a gradate condition. It will be seen later that evidence of a like direct progression is afforded within the genus *Dipteris*.

Family. PLAGIOGYRIACEAE Bower, 1924.

1. Genus. *Plagiogyria* Kunze, 1849 ... ... 11 species.

BIBLIOGRAPHY FOR CHAPTER XXXI

544. KUNZE. Bot. Zeit. 1849, p. 865.

545. METTENIUS. Ueber einige Farngattungen. II. *Plagiogyria*. 1858.

546. HOOKER. Species Filicum. iii, p. 2.

547. DIELS. Natürl. Pflanzenfam. i, 4, p. 281.

548. BOWER. Studies in Phylogeny. I. Ann. of Bot. xxiv, p. 423. 1910.

549. LANG. Mem. Manch. Lit. and Phil. Vol. 68, Part I, p. 53. 1924.

CHAPTER XXXII

PROTOCYATHEACEAE

IN Chapter XXVI a general review of the Simplices has led naturally to the distinction of two main sequences, or ascending series of Ferns. The first includes those Ferns which retain the originally marginal position of their sori (Marginales): of these the Schizaeaceae are a primitive example and the preceding chapters have shown advances while retaining for the most part a strictly marginal position of the sorus. The second comprises those which bear their sori superficially (Superficiales): of these the Gleicheniaceae are a primitive example. It was further recognised that the distinction by soral position is not absolute, for it is based upon an adaptive change that appears to have followed, earlier or later in descent and probably polyphyletically, as a consequence of the progressive broadening of leaf-surfaces. It was further indicated that the relatively late and derivative Leptosporangiate Ferns may be laid off as continuations of those two main sequences, being related for the most part more or less directly either to the Schizaeaceae or to the Gleicheniaceae. In Chapters XXVII to XXXI the earlier examples of the derivative Marginales have been described, and reasons have been given for ranking *Plagiogyria* with them, notwithstanding its superficial sori. All of the Ferns treated in those Chapters may be held as related to such primitive stocks as the Schizaeaceae and Osmundaceae, the latter showing a peculiarly interesting divergence of its two genera in respect of the soral position: for *Osmunda* retains the marginal position, while *Todea* foreshadows the Superficiales.

Turning now to the Superficiales themselves we shall examine those earlier derivative Ferns which appear consistently to have superficial sori. They may be held as included in or related to the primitive stocks of the Gleicheniaceae and Matoniaceae, Ferns in which we may believe that the sori were established relatively early in a superficial position. The reality of such sequences will depend upon the demonstration of the closeness of the ndividual links. In the present instance a careful comparison will first have to be made between the most advanced Gleicheniaceae, and any outstanding and apparently primitive forms usually classed with the Cyatheaceae. Later a similar comparison will have to be made between the Matoniaceae and certain more advanced types which suggest relation with them, and also share the superficial sorus: viz. the Dipterids.

It will be remembered that two species of *Gleichenia*, separated from the rest as the sub-genus *Eu-Dicranopteris*, are characterised primarily by bearing hairs and no scales: one of them (*G. pectinata*) has advanced beyond the rest in several distinct features: for instance it has attained structurally to solenostely: further its sorus bears more numerous sporangia than the rest of the family, but the spore-count from each is relatively low. The sporangia having only short stalks and median dehiscence are of a type that requires "elbow-room" for the shedding of the spores, which the crowding of the sorus mechanically prevents. In fact the sorus of *G. pectinata* suffers from a condition of dead-lock. An escape from this difficulty would, however, be possible by one or more modifications such as these: (i) by replacing median by lateral dehiscence, (ii) by increasing the length of the sporangial stalk, (iii) by extending the area of the sorus, (iv) by elongating the receptacle. None of these have been adopted by the Gleicheniaceae, and *G. pectinata* shows in consequence a soral state advanced to the point of inefficiency. But the Cyatheaceae all have a lateral dehiscence: in this and in other ways they have resolved the difficulty of a dead-lock. We may be prepared to find among them, as still relatively primitive Superficiales, some outlying species which share certain characteristics of the more advanced Gleicheniaceae, but have escaped from their disabilities. Such links exist in *Lophosoria* and *Metaxya*, which have been variously treated by systematists; they are related directly with *Alsophila*, a genus which in soral character may be held as the most primitive of the Cyatheaceae. Such outlying Ferns thus appear as synthetic types, connecting the Gleicheniaceae with the Cyatheaceae. On these grounds they call for careful examination.

Lophosoria quadripinnata Gmel.

The Fern so named has always been regarded as an outstanding species: but, as its synonymy shows, it has been variously classified ("Studies in Phylogeny, II," *Ann. of Bot.* xxvi, p. 279). Though it was constituted the type of a substantive genus by Presl (1848), it was placed by Sir W. Hooker in *Alsophila* (*Species Filicum*, i, p. 47), as *A. pruinata* Kaulf. Its characters are, however, so distinctive that it will be best to uphold Presl's genus *Lophosoria*, of which it is the sole species. The best description of this Fern is given by Jenman in his *Synoptical list of Jamaican Ferns*. It is widely spread in the Western Tropics as a low-growing Tree-Fern of forest shade, with short upright stem, only about 3 feet high, and about 3–4 inches in diameter, bearing spirally arranged leaves 6–10 feet long, and repeatedly pinnate, but with fine segments, and open venation. The large leaf is of the Cyatheaceous type, which with the upright stem gives the plant the appearance of a low-growing *Alsophila*. But the dense vestiture is of soft hairs, no scales being present even on the leaf-bases. Associated with each leaf-base is usually an abaxial bud, which may remain dormant in the upper leaves, but at the base of the plant the buds develope into runners, which after growing at first horizontally with arrested leaves, turn upwards as new radial shoots like the parent (Fig. 547).

The rachis bears a dense felt of hairs even when mature, but it has neither basal pinnae nor armature of thorns, though in large leaves some low

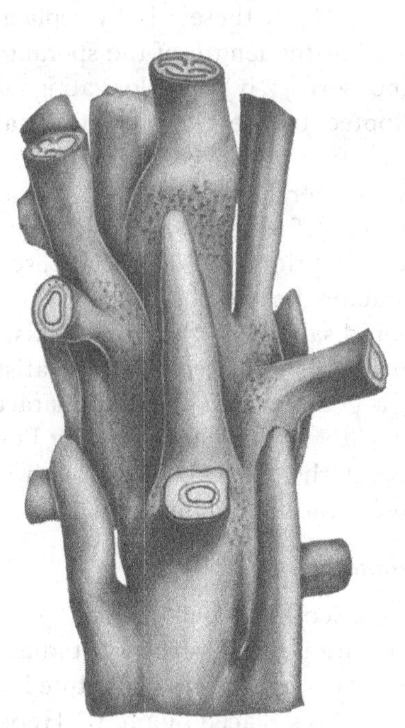

Fig. 547. Upright stock of *Lophosoria*, after removal of the hairy covering. The erect appendages are all leaves, of which those inserted lower down are abortive, while those higher up are developed. From the base of several of them, but not of all, arise horizontal solenostelic runners. Note the emergences on the bases of the fully developed leaves. (Half natural size.) (From a drawing by Dr J. M. Thompson.)

Fig. 548. Pinnule of *Lophosoria* showing the venation and sori. (Enlarged.) (Drawn by Mr Maxwell.)

emergences may appear at the extreme base. Hairs persist on the veins of the under side of the pinnules, but the general surface is glaucous. The pinnules themselves are narrow, with crenate margins turned downwards: the venation is forked but sympodially developed, without fusions, while the circular sori are inserted with regularity upon the lowest anadromic veinlet of the ultimate pinnule (Fig. 548). All this is just as it is seen in the pinnules of *G. linearis* or *pectinata*. The number of the sporangia in each sorus is small (7–10), arranged in two tiers, and orientated as in *Gleichenia*, the upper tier being less regular than the lower (Fig. 549). But as the stomium is lateral the shedding of the spores presents no mechanical difficulty. Simple hairs are associated with the sporangia. From this description it appears that *Lophosoria* holds a position between the Gleicheni-

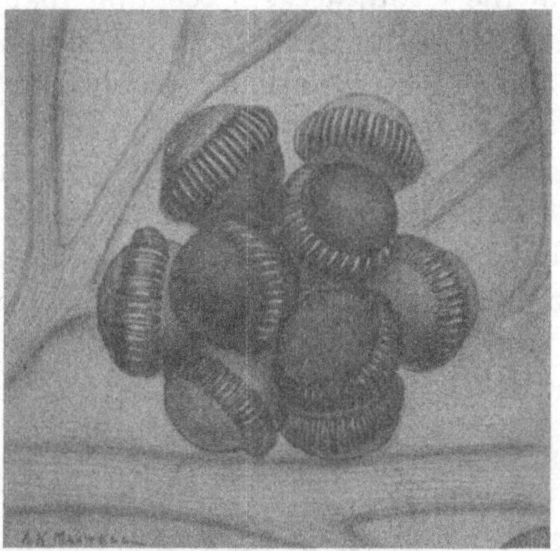

Fig. 549. A single sorus of *Lophosoria*, showing the small number of sporangia, with their regular orientation. Each sporangium has its annulus apparently complete: the stomium is lateral, and therefore out of sight. (Enlarged.) (Drawn by Mr Maxwell.)

aceous and Cyatheaceous types, in respect of the external characters: but with special similarity to *G. pectinata*. There is, however, an entire absence of those consequences of apical arrest which give to the leaves of *Gleichenia* the appearance of a false dichotomy.

ANATOMY

The axis and leaf-stalk are strengthened externally by a dark-coloured band of sclerenchyma, without lenticels. Sclerotic tissue also forms a ring within the solenostele, and straps of it accompany the leaf-traces outwards, lining their concave face. In certain sections there is a perfect solenostele, as may frequently be seen in the runners if cut between the leaf-insertions. The ring opens to give off a leaf-trace in the usual way, and closes almost at once (Vol. I, Fig. 156, p. 163). But in the upright stock where the axis is shorter, there may be an approach to dictyostely, which however is seldom fully realised. A transverse section of a large stock will commonly show a single leaf-gap, but already the stele may be thinning out at another point preparatory to giving off a second leaf-trace, or even a third (Fig. 550). The leaf-trace departs as a single wide gutter-shaped strand, not contracted as in *Gleichenia*, and it soon assumes the usual horse-shoe with deep involutions opposite the lateral pneumatodes, which are here present as in most large petioles. In large leaves it may divide into three portions, but these unite again upwards, showing as in many large petioles a corrugation corresponding to the constituent "divergents": but it does not disintegrate again, showing in this its primitive state. The pinna-traces are extra-marginal in origin, as in the Cyatheaceae and the Gleicheniaceae (Vol. I, Fig. 170, p. 174). The leaf-structure is as in the Cyatheaceae, excepting that vein-fusions never occur in *Lophosoria*. The vascular supply to the stolons arises externally from the leaf-bases, and as they are usually solenostelic from the first, each appears as a diverticulum from the axial stele, just below the leaf. In all these features *Lophosoria* is like a primitive Cyatheoid, which had not departed far from typical solenostely. But comparison with the solenostelic *G. pectinata* shows similarity in all essential points, allowance being made in the latter for the structural effect of the contracted leaf-base. These two Ferns form in fact a structural bridge between the protostelic Gleicheniaceae and the dictyostelic Cyatheaceae.

Fig. 550. Transverse section of stem of *Lophosoria*, from which a leaf-trace has just departed, and preparation is being made by thinning of the solenostele for a second, and even a third. (Natural size.) The sclerenchyma is dotted, and vascular tissue black. The leaf-trace is divided into three parts.

SPORANGIA

The sporangia not only of a sorus but of a pinnule of *Lophosoria* are formed simultaneously, a state characteristic of the Gleicheniaceae and other primitive Ferns rather than of the Cyatheaceae. Each sporangium is almost spherical, and is seated on a short thick stalk, which shows six or more cells in transverse section: this also is comparable with Gleicheniaceae. The annulus is oblique, and appears as a complete ring defining the almost equal distal and basal faces. It is usually a single row of cells, but it may occasionally be doubled near to the stomium, as happens occasionally also in *Gleichenia* (Fig. 551). The stomium is lateral, and not highly differentiated, its cells being variable in number and in position. The development of both sorus and sporangium is very similar to that of *Gleichenia*: but here only 16 spore-mother-cells are formed, and the tale of spores is typically 64: a much smaller number than in any *Gleichenia.* Nevertheless in *Platyzoma* still smaller numbers have been found pointing to a typical number of 32, or sometimes of only 16 spores in each sporangium. Thus in development and in soral relation the sporangia of *Lophosoria* resemble those of

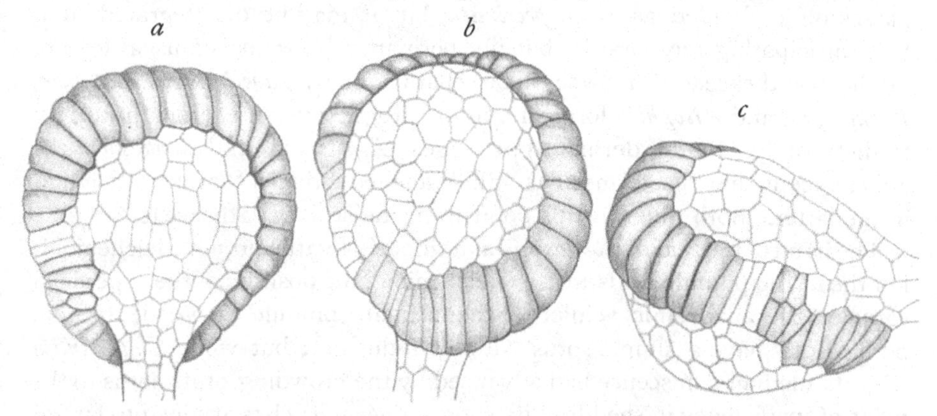

Fig. 551. Sporangia of *Lophosoria*, from three different aspects. *a*, presenting the distal or peripheral face. *b*, presenting the proximal or basal face. *c*, seen from the side, and showing the irregular stomium, with partial duplication. (× 100.)

Gleichenia, but they differ in their lateral dehiscence, while in their spore-output they point rather to the Cyatheaceae.

The spores are large and tetrahedral, and on germination they produce a prothallus at first filamentous, which widens into a spathulate form with unequal lobes, as is seen in *Alsophila.* The antheridia are of the usual type, and open by the extrusion of a single cell.

Comparison

It must be understood that in this whole discussion the Dicksonieae are excluded, their probable relation being with the Marginal Schizaeaceae. Wherever the Cyatheaceae are named, the word is used in the restricted sense, i.e. exclusive of the Dicksonieae.

It is seen from the above description that *Lophosoria*,—which systematists have usually ranked with gradate Cyatheaceous *Alsophila*, though it has really a simple sorus,—takes an intermediate place between the Gleicheniaceae and the Cyatheaceae in respect of its various characters. While *G. pectinata* may be held as the most advanced species of the creeping genus *Gleichenia*, *Lophosoria* is the most primitive of the upright Cyatheoids; but in many features these Ferns approach one another. It has been argued at length elsewhere ("Studies on Phylogeny, II," *Ann. of Bot.* xxvi, p. 292) that the primitive Gleichenias foreshadowed the probably prone ancestry of the upright Cyatheoid series, and that of living Ferns they most nearly represent the forerunners of that great dendroid Family. *Lophosoria* shows the creeping habit in its runners: but in their upturned tips they demonstrate a direct passage to the dendroid state. The vascular system supports this, for while the runners are solenostelic, the upright axis shows steps towards the dictyostely seen in *Alsophila* and *Cyathea*. The leaf-trace of *Lophosoria* passes off undivided as in *G. pectinata*, but it may be disintegrated later, thus anticipating temporarily, but not permanently, what is a usual feature of the Cyatheaceae. In these stelar characters *G. flabellata*, *G. pectinata*, *Lophosoria*, and *Alsophila* form a naturally progressive series from protostely to dictyostely. In the dermal appendages *G. pectinata* and *Lophosoria* are alike, except that the former has stiff branched bristles that are not found in the latter. Both differ in this point from the scaly Cyatheaceae.

The habit of the leaf of *Lophosoria* as a whole is Cyatheoid, not Gleichenioid: but the fertile pinnule is distinctly Gleichenioid, the position of the superficial sorus on the anadromic veinlet of the ultimate pinnule being identical in both. Both have a simple sorus without indusium: but while *G. pectinata* with its median dehiscence had advanced by the crowding of its sorus to the point of inefficiency in shedding its spores, *Lophosoria* has attained to lateral dehiscence, so that there is no risk of a mechanical dead-lock. In both the sporangia remain short-stalked. In point of spore-number while *Gleichenia* stands high, *Lophosoria* has an output reduced to the lower number seen in many Cyatheoids. Lastly the prothallus of *Lophosoria* corresponds to that of *Alsophila*. The whole effect of these comparisons is to show that *Lophosoria*, though clearly one of the Cyatheoids, stands aloof from them, and corresponds rather with *G. pectinata*. It may be held as one of the advanced Simplices, of similar ancestry to the Gleicheniaceae, which while retaining many of the same features as they possess, has adopted two distinctive innovations, viz. an upright habit, and a lateral dehiscence of its sporangia. But these are prominent characteristics of the Cyatheaceae, and notably of *Alsophila*, with which genus it has habitually been classed.

Metaxya rostrata Presl

Metaxya rostrata Presl is a handsome South American Fern, with leaves a metre long, borne on its creeping stem. Buds are frequently borne on the abaxial side of the leaf-base, which may grow out into branch-rhizomes. The leaves are simply pinnate, with broadly lanceolate pinnae 6–12 inches long, and serrate near to the tip, suggesting an origin by webbing from a doubly pinnate type. The veins are parallel and occasionally forked, but never fused. The sori form an irregular series on either side of the midrib: on some veins the sorus lies nearer to, on others further from the margin, and not uncommonly two or more may be borne upon a single vein. Their position is consequently irregular (Fig. 552). Their form is flat, and copious long hairs are mixed with the very numerous sporangia. Apart from the sori the mature leaf is destitute of hairs, but the rhizome is permanently covered by a dense brown felt. The hairs are unbranched, and bear no glands. Scales are absent, nor are there any massive emergences or branched hairs, such are seen in *Gleichenia pectinata*.

Metaxya has been variously ascribed generically to *Polypodium*, *Aspidium*, *Amphidesmium*, and *Alsophila*. Though Presl (*Tentamen*, 1836) constituted it a substantive genus, Sir W. Hooker merged it into *Alsophila*, as *A. blechnoides* Hk. But this tends to disguise its synthetic nature, and it is best to retain it under Presl's name, as the sole representative of a substantive genus. It will be seen that it takes an interesting place as intermediate

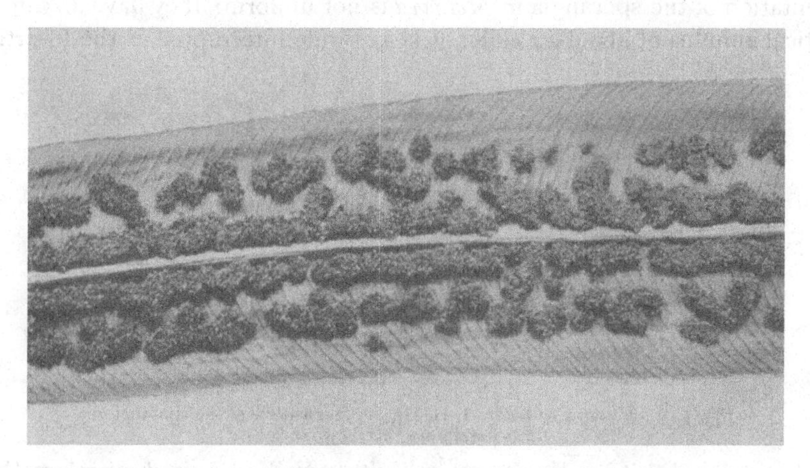

Fig. 552. Part of a pinna of *Metaxya rostrata* Pr. (= *Alsophila blechnoides* (Rich.) Hk.) showing the relation of the sori to the veins. More than one sorus may be borne on a single vein. (× 2.)

between the Gleicheniaceae and Cyatheaceae, with suggestions also towards the Matonineae.

The habit of *Metaxya* is always creeping, as in *G. pectinata.* Sections of the rhizome disclose a distended solenostele, ½ an inch in diameter, with voluminous pith, no medullary vascular strands, and no sclerotic bands (see Vol. I, Fig. 149, I). The leaf-trace is undivided, and soon takes the form of a crinkled horse-shoe, while the pinna-traces are of extra-marginal origin. The supply to the branch-rhizomes appears as a diverticulum from the base of the leaf-trace, and it is solenostelic from the first, so that the pith of the branch is continuous with that of the rhizome, as in *Lophosoria*: this fact suggests unequal dichotomy. The whole vascular system of the leaf including the venation is of a relatively primitive type. The petiolar meristele in *G. pectinata* is of a contracted form in accordance with the narrow leaf-stalk, but that of *Metaxya* appears as a wide gutter. It has been seen that the stele of *Lophosoria* has advanced towards dictyostely in the shortened upright stock, with a tentative disintegration of the meristele near to the leaf-base. But notwithstanding these differences of detail in essentials the system is comparable in all of these ferns. They are all advanced from protostely to solenostely, but have not arrived definitely at dictyostely or disintegration.

The sori of *Metaxya* are oval in outline, with a flat receptacle, upon which 50 to 100 sporangia are closely grouped, together with long occasionally branched hairs. The number of the sporangia is in strong contrast to the 16, or often much less, in each sorus of *G. pectinata*, *Lophosoria*, and the other Gleichenias. The development of them all is simultaneous, as in the Simplices, to which, including *Metaxya*, they all technically belong. The orientation of the sporangia in *Metaxya* is not uniform: they have an almost vertical annulus of about 27 cells: it is as a rule interrupted at the insertion

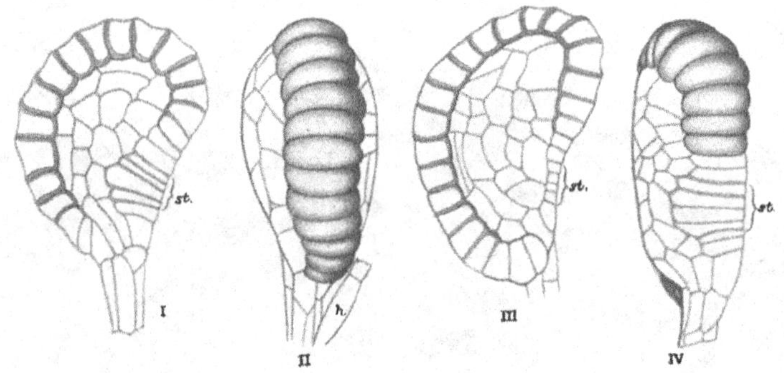

Fig. 553. *Metaxya rostrata*. I, II, III, IV, four aspects of the sporangium. For details see text. (× 100.)

of the stalk, and it is almost vertical with a well organised stomium (Fig. 553). The slit of dehiscence is in a transverse plane, which is convenient in a flat sorus where the sporangia originate simultaneously. In this it differs from the Gradate Cyatheaceae, where the slit is oblique. The spore-output is 64. An interesting feature is the four-rowed stalk, giving a characteristic transverse section (Fig. 554). This may be held as an indication of the mode of segmentation of the sporangium. It would naturally follow from a two-rowed segmentation, in place of the three-rowed which is usual in advanced Ferns. This proves to be the fact in *Metaxya*, a point of some comparative importance (Fig. 555). Nevertheless the annulus is formed just as in sporangia with the three-rowed segmentation, and it provides a pregnant instance of the independence of segmentation from morphological initiation (see Vol. I, p. 246). The gametophyte of *Metaxya* is unfortunately unknown

These features provide interesting material for comparison with a view to phyletic seriation. In respect of form *Metaxya* retains the creeping habit of the Gleichenias, and has not like *Lophosoria* assumed an upright pose: but the underlying scheme of them all is the same. The vascular anatomy accordingly remains primitive also, as well as the dermal appendages. The fertile pinnae and sori, however, are advanced. The former have resulted from a condensation of more elaborate branching into broad simple pinnae, though relics of a more elaborate structure are still

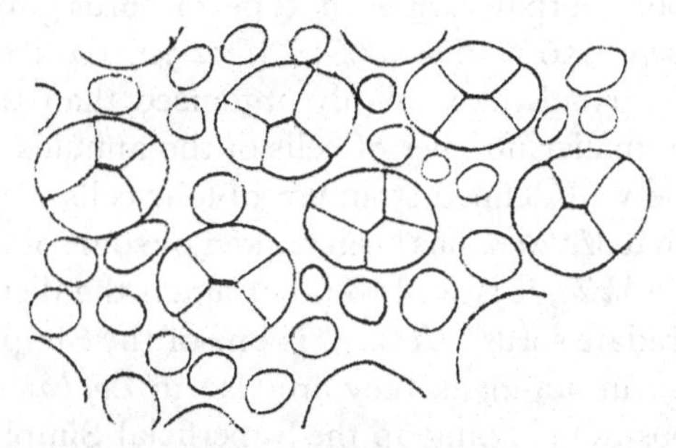

Fig. 554. Tangential section through part of a young sorus of *Metaxya rostrata*, showing the hairs and sporangial stalks cut transversely. The stalks consist each of four cell-rows and are orientated with regularity. (× 250.)

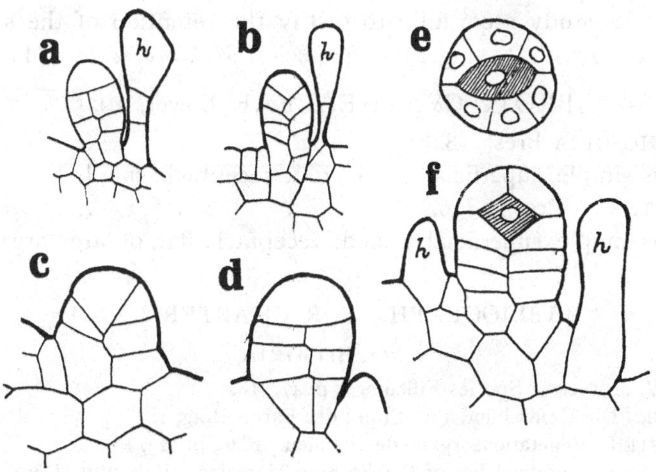

Fig. 555. Illustrations of the two-rowed segmentation of Fern sporangia. *a*, *b*, young sporangia of *Cheiropleuria*, showing aspects at right-angles to one another. (× 165.) *c*, *d*, similar drawings of *Metaxya*. (× 200.) *e*, is a rather older sporangium of *Metaxya* seen from above: *f*, seen from the side. *h*, *h* = hairs. (× 200.)

seen in the marginal serrations, indicating the pinnules which are webbed to form the wider expanse. But it is in the sori and sporangia that *Metaxya* shows its most advanced features. The mechanical dead-lock of *G. pectinata* has been resolved along three lines of advance. The receptacle is not elongated, as in the Gradatae, but remains about the same height as in

Gleichenia: its area is, however, extended so as to accommodate 50–100 sporangia, as against 16 or less. These sporangia have lateral as against median dehiscence, thus removing the need for elbow-room: the stalk also is slightly elongated. All these features, together with the protecting hairs, make an extended sorus with numerous sporangia practically efficient, without loss of spore-output; for each type of sorus produces about 5000 spores (*G. pectinata*, 256 × 16 = 5096; *Metaxya*, 64 × 100 = 6400). The sporangium of *Metaxya* is more highly organised than that of *Lophosoria*. This is seen in the smaller number of cells of the annulus, its almost vertical position, and in the well-defined stomium of four cells.

Remembering that *Metaxya* has been ranked by some of the best authorities as a species of *Alsophila*, it is well to point again the distinction that while *Alsophila* has a Gradate sorus, *Metaxya* is one of the Simplices, its sporangia being simultaneous in origin, as they are also in *Lophosoria*. These genera take a synthetic position, linking up the Superficial Simplices and Gradatae, and in particular the Gleicheniaceae, with the Cyatheaceae: they correspond in many features of form, structure, and soral character. But *Metaxya* also points clearly in the direction of the Matoniaceae and Dipteridaceae, which may be held as probably related to a primitive Gleicheniaceous stock. These aspects of its study seem fully to justify the retention of the substantive genus of Presl with its single living species, *Metaxya rostrata* Presl.

PROTO-CYATHEACEAE Bower, 1925

I. LOPHOSORIA Presl, 1848. ... ... 1 species
Sorus simple, superficial and naked: receptacle small.

II. METAXYA Presl, 1836. ... ... 1 species
Sorus simple, superficial, naked: receptacle flat, of large area.

BIBLIOGRAPHY FOR CHAPTER XXXII

LOPHOSORIA

550. Sir W. HOOKER. Species Filicum. i, p. 47. 1846.
551. PRESL. Die Gefässbündel im Stipes der Farn. Prag, 1847, p. 36.
552. KARSTEN. Vegetationsorgane der Palmen. 1846, p. 123, Pl. ix.
553. JENMAN. Synoptical list of the Ferns of Jamaica. Bull. Bot. Dept. of Jamaica, 1890–1898.
554. BOWER. Origin of a Land Flora. 1908, p. 604, Fig. 336.
555. BOWER. Studies in Phylogeny. II. Ann. of Bot. xxvi, p. 279. 1912.

METAXYA

556. PRESL. Tentamen. 1836, p. 59, Tab. i, Fig. 5.
557. Sir W. HOOKER. Genera Filicum. 1913, Tab. xlii, B.
558. BOWER. Studies in Phylogeny. III. Ann. of Bot. xxvii, p. 443. 1913.

CHAPTER XXXIII

CYATHEACEAE (EXCL. DICKSONIEAE)

IN the strict sense only three genera should be included under this title, viz. *Alsophila*, *Hemitelia*, and *Cyathea*, which are mostly Ferns of dendroid habit, with chaffy scales, and they all possess gradate sori. But as already suggested *Lophosoria* and *Metaxya*, which are technically Simplices, are so closely allied to *Alsophila* as to have been included by most systematists in that genus: nevertheless, following Presl, they are best placed as distinct genera, and even ranked as an intermediate family of Proto-cyatheaceae. On the other hand, *Woodsia*, *Diacalpe*, and *Peranema* have always been held as related to the Cyatheaceae: it will be seen later that they show gradual steps of character which can only indicate a further phyletic progression towards a mixed sorus. The family of the Cyatheaceae forms in fact a middle term of transition from the superficial Simplices such as the Gleicheniaceae, through a gradate phase which they represent, to the mixed type of sorus. Parallel with this progression in soral characters go also other features, both vegetative and propagative, which confirm the comparisons between these naturally related Ferns. A strong phyletic interest thus attaches to the Cyatheaceae (excl. Dicksonieae).

The three genera include species of varying size. Though such Tree-Ferns as *Alsophila excelsa* and *Cyathea medullaris* rise with a single trunk to a height of 60 or even 80 feet, many are of low stature. Some have numerous lateral branches, which are related to the main axis and the leaf-bases exactly as are the runners in *Lophosoria* (see Fig. 547). This has been worked out in *Alsophila aculeata* by Stenzel (*Nova Acta d. D. Akad.* 1861), who demonstrated the vascular connections and showed in the branches a transition through solenostely to dictyostely, as it is seen in the runners of *Lophosoria*. The suggestion thus comes clearly, from the comparison of *Gleichenia*, *Metaxya*, and *Lophosoria* with *Alsophila*, that the upright habit is derivative from the creeping rhizome. The result of such branching as is seen in *A. aculeata* and *Cyathea mexicana* is a more or less shrubby habit. But most Cyatheoids have the single columnar trunk, though it may occasionally bifurcate (compare Frontispiece to Vol. I). The continuity of the massive pith upwards into the two equal limbs of such a fork indicates that they originated by dichotomy (Fig. 556). A transverse section of a large stem low down also demonstrates how great a proportion of the apparent bulk of the columnar stem is due to the massing of adventitious roots, which gives added mechanical strength to a stem itself incapable of growth in thickness. In these Ferns, as also in the Dicksonioids, the best has been made of an unpractical scheme of construction, viz. an unlimited upward growth with a constantly increasing head of leaves, but without secondary thickening. Its origin from a creeping habit is clearly seen here, through comparison with *Lophosoria*. But the dendroid habit has also been adopted by the Osmundaceae and the Blechnineae. It is therefore concluded that it may arise polyphyletically, and is in itself no index of affinity between Dicksonioids and Cyatheoids.

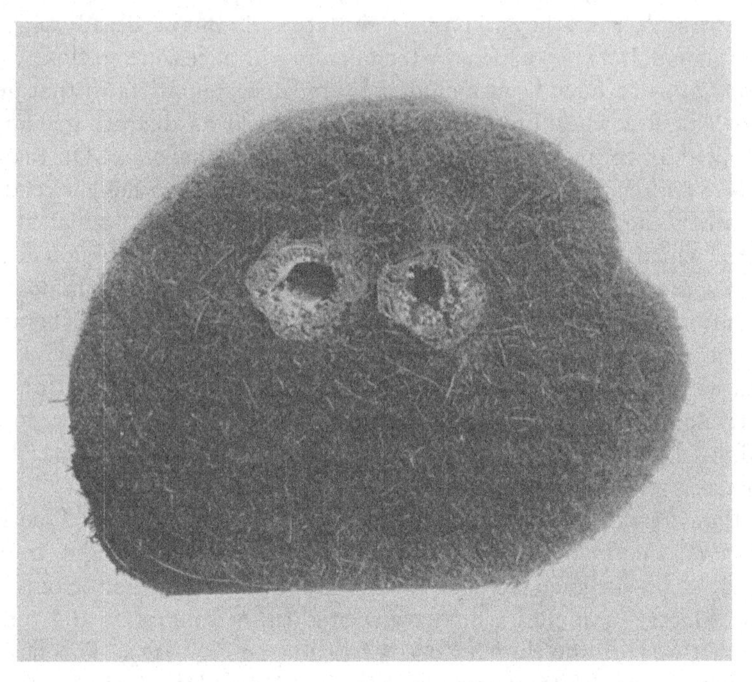

Fig. 556. Transverse section of a large bifurcated trunk of *Cyathea medullaris* showing the small size of the twin stems and the large bulk of the adventitious roots that embed them. (Much reduced.)

The leaves of the Cyatheaceae are as a rule large and repeatedly pinnate. Many have narrow segments, often of "Pecopterid" type as in *Gleichenia*, and the venation is usually open. But various degrees of condensation of the leaf-structure may be seen, and some steps towards vein-fusion. For instance, while many Cyatheoids have leaves 3 or 4 times pinnate, there is only a double pinnation in *A. Taenitis*. A single pinnation appears in *A. phegopteroides*, or in *Hemitelia grandifolia*, though in the latter the pinnae are themselves pinnatifid. Lastly, in *Cyathea sinuata* there is a simple leaf, the origin of which by condensation is suggested by the open pinnate venation (Fig. 557). These may all be held as secondary results of simplification of a pinnate branching, combined with webbing of the segments.

Such vein-fusions as are found in some Hemitelias and Alsophilas are obviously consequences of that condensation: they commonly appear as loops parallel to the midrib, while the rest of the venation is open. But occasionally fusions exist also nearer the margin. Nevertheless the venation is always readily referable to a primitive Pecopterid type (Fig. 558).

Fig. 557. *Cyathea sinuata* Hook. *A*, shows the habit, with simple leaves, unique in the genus. *B*, part of a fertile leaf. *C*, part of a fertile leaf more highly magnified, showing the venation and sori. (After Hooker, from Engler and Prantl.)

The surfaces of the stem and of the leaf-bases are covered by a hard sclerotic coating that would be impervious for gas-interchange, were it not for pneumatodes which appear on stems and leaf-bases as oval areas filled with spongy tissue, as in *Alsophila crinita* (Vol. I, Fig. 193). On the leaf-stalks they appear as the usual pale lateral lines. But the most characteristic surface-features are the broad chaffy scales which form a dense covering over leaf and stem while young. They are present in all the three genera, and form a distinctive feature of the Family. But *A. pubescens* Baker, is described as "not scaly," and it is possible that some others may, like *Lophosoria* and *Metaxya*, bear only hairs. This feature contrasts strongly with the hairs of the Dicksonieae, in which Family scales are absent. In many of the Cyatheaceae the scales are borne upon peg-like outgrowths, comparable to those bearing the stiff branched hairs of *Gleichenia pectinata* (compare Vol. I, Figs. 191, 192). It is probable that the scales are the result of a webbed and flattened development of such hairs. The pegs persist after the scales fall away, and constitute the "armature" of hard woody spines present in certain species belonging to all three genera, but most frequently in *Hemitelia* and *Alsophila* (see Fig. 567, also Vol. I, Fig. 191).

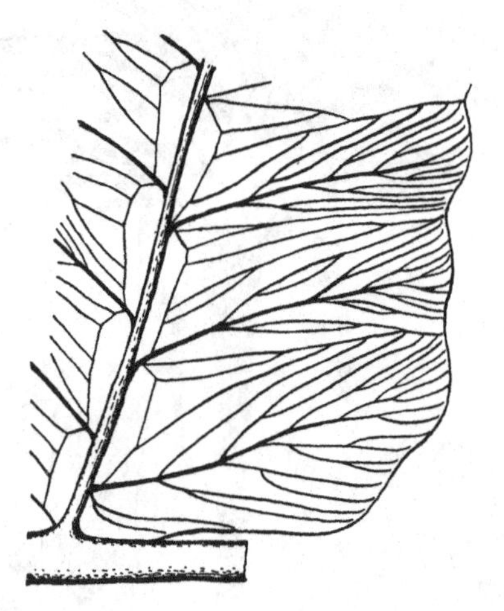

Fig. 558. Part of a pinna of *Hemitelia Karsteniana* Kl. showing many vein-fusions. (After Mettenius.)

ANATOMY

The stems of the Cyatheaceae have a highly complex structure. But it is readily intelligible in terms of the simpler states seen in *Gleichenia* and *Lophosoria*: for the former genus gives the steps from protostely to a fully developed solenostele, while the latter leads on to the dictyostelic state. A well-developed dictyostele is present in the adult stem of all the three genera, giving off highly disintegrated leaf-traces and numerous root-traces. But there is in addition a medullary, and sometimes also a cortical accessory system. Thus a high complexity is attained. It will be unnecessary to repeat here the detailed descriptions already given elsewhere (compare De Bary, *Comp. Anat.*, Engl. Edn. pp. 291–294. *Origin of a Land Flora*, p. 604. Also *Ferns*, Vol. I, p. 156, Figs. 150, 151). But the inconstancy of the accessory strands may be noted. A medullary system is present in the adult stem in all three genera. But while a cortical system also is present in *Cyathea Imrayana* (Vol. I, Fig. 150), it is absent from *Hemitelia setosa* (Vol. I,

Fig. 151, *B*). It is also absent from the very large stem of *C. medullaris*, shown half-size in Fig. 559. Since that stem is more than twice the diameter of the stem of *C. Imrayana* dissected by De Bary, it follows that here complexity is not directly related to size alone.

In the Cyatheaceae the adult leaf-trace is given off directly as a number of disintegrated strands, which spring from the lower margin of the leaf-gap (Fig. 560). This is a highly advanced state as compared with *Gleichenia*; but *Metaxya* and *Lophosoria* suggest intermediate steps in the amplification of the

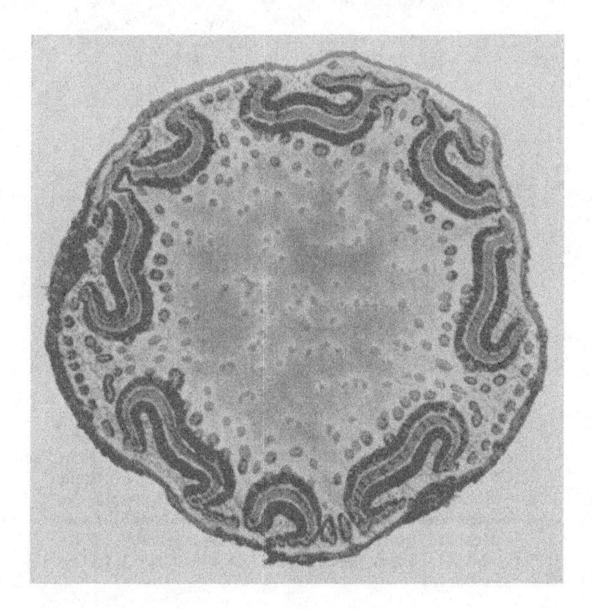

Fig. 559. *Cyathea medullaris*. Transverse section of stem, with adventitious roots and leaf-bases cleared away, showing the vascular system, with numerous medullary strands. ($\frac{1}{2}$ natural size.)

trace, in the formation of the lateral involutions opposite to the pneumatodes, and in its disintegration (compare Vol. I, Fig. 161). There is some variation in the degree of disintegration of the leaf-trace, especially in the upper part of the leaf. It is interesting to note that examples of this are specially evident in *Alsophila*, a genus which on other grounds may be held as the most primitive of the three (Fig. 561). (Compare Bertrand et Cornaille, *La Masse libéro-ligneuse élémentaire des Filicinées*. Lille, 1902.)

The ontogenetic development throws light on the evolutionary history of the complicated vascular system of these stems. Stenzel had already in 1861 shown in *Alsophila aculeata* that the base of the lateral branch may progress through solenostely to the dictyostele of the adult. But the ontogenetic history of the sporeling was first traced in *Alsophila excelsa* by Gwynne-Vaughan (*Ann. of Bot.* xvii, p. 709), and illustrated by a diagrammatic drawing

Fig. 560. *Cyathea Imrayana* Hook. Piece of stem with four leaf-bases, after removal of the outer layers of cortex, seen from without. The margins of the four leaf-gaps, the bundles which spring from them and pass into the leaves, the roots inserted on them (black), and the bundles which run down within the cortex are exposed. The cortical bundles and root-bases are quite free, and the rest are covered by semitransparent parenchyma. (Natural size.) (After De Bary.)

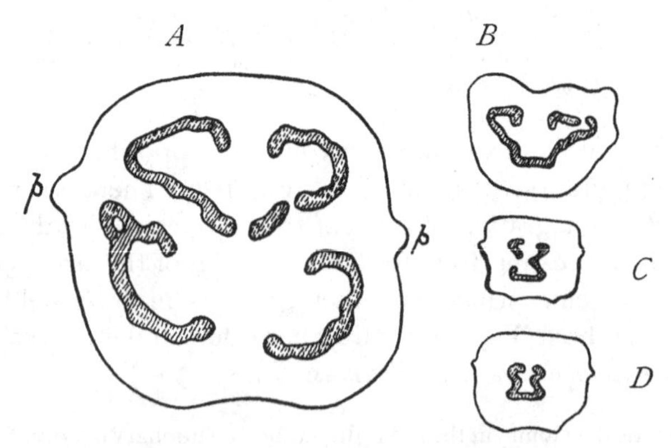

Fig. 561. Sections of the upper part of the rachis of certain related Ferns, showing various degrees of disintegration of the meristele. *A*, *Alsophila atrovirens*; *B*, *Metaxya*; *C*, *D*, *Alsophila Williamsii* Maxon. (All × 10.)

(Fig. 562). At the base of the stem there is a protostele with a solid core of xylem, from which the first leaf-trace departs without disturbance, but the phloem is prolonged downwards in its axil. At the departure of the subsequent leaves this prolongation becomes more pronounced; and subsequent steps follow in a manner similar to what has been seen in the ontogeny of *G. pectinata* (compare Fig. 481, also Vol. I, Fig. 134), with the result that at the level of about the eighth leaf the stem is solenostelic. This stage does not last long in *Alsophila*, for leaf-gaps overlap, and the system gradually becomes dictyostelic. The leaf-trace of the first five or six leaves is undivided, but later two or three strands and at the tenth leaf four strands pass into each leaf-base, two from each side of the leaf-gap. Thus the sporeling of *Alsophila* passes through similar stages to those of *Gleichenia pectinata*, but more abbreviated, and with a divided leaf-trace (compare Vol. I, Fig. 134, p. 144). The first indication of the internal medullary system in *Alsophila* was found by Gwynne-Vaughan at about the tenth leaf. Just below the upper (adaxial) traces of this leaf the xylem projects inwards forming a small ridge: sometimes it separates as a small xylem-strand lying free within the phloem, ending blindly or fusing up again with the main xylem. Later such strands may separate from the main meristele, running as independent vascular strands, and initiating the medullary system, which thus owes its origin to a local thickening at the margins of the leaf-gaps: but the medullary strands do not appear at all till the ordinary cylinder has become dictyostelic.

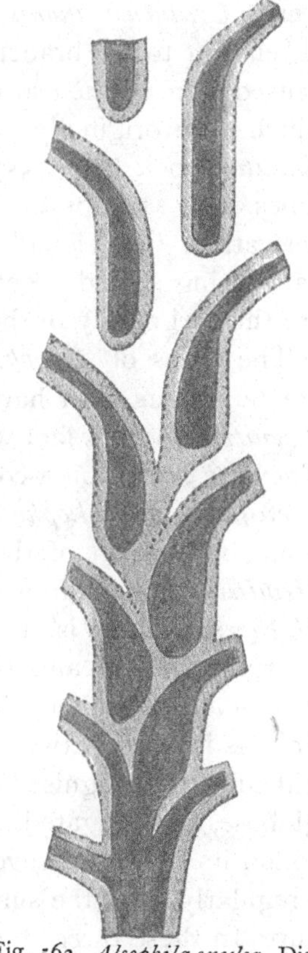

Fig. 562. *Alsophila excelsa.* Diagram of the vascular system of a young plant in median longitudinal section. The xylem is black, the phloem highly shaded, and the endodermis is indicated by a dotted line: the ground-tissue is left white. (After Gwynne-Vaughan.)

The ontogeny thus disclosed for a complex Tree-Fern may be held as a valid suggestion of the way in which the adult condition arrived in descent. It starts from a protostelic state which quickly passes to the solenostelic, and this again to the dictyostelic: lastly by intrusion from the margin of the leaf-gaps the medullary system is produced. All of these are probable steps in the evolutionary story of plants with a massive axis bearing large and closely disposed leaves. The several steps are represented by the protostelic Gleichenias, by the solenostelic *G. pectinata*, and by *Lophosoria* with its transition to dictyostely. The stelar ontogeny thus upholds the sequence suggested.

Sori and Sporangia

The three genera are distinguished from one another by the character of the indusium, the sori being constructed otherwise on the same plan in them all. Their disposition on the leaf or segment is fundamentally the same as in *Gleichenia*, a single series ranging on either side of the midrib. As a rule only one sorus is borne on each vein, but the position relatively to the margin varies, being sometimes near to the midrib, as in most of the highly pinnate types, e.g. *Cyathea serra*: sometimes they lie far out towards the margin, as in most of the condensed types, e.g. *Hemitelia horrida.* Occasionally they may be disposed in two parallel rows on each side of the midrib, a condition seen in *Cyathea brunonis*, or *Alsophila Williamsii* Maxon. Their arrangement in relation to the branching veins is then such as to suggest that the condensed form of the leaf is the result of lateral webbing of pinnae or pinnules which were originally separate: this is seen in the simple-leaved *Cyathea sinuata* Hook. (Fig. 557). Comparison indicates that all such condensed types, with sori approaching the margin, may be held as relatively late and derivative. Occasionally the identity of the sorus is not strictly maintained, pairs being seated together, suggesting either fission or fusion. In no case are the sori of any of the Cyatheaceae actually marginal in position.

The sorus of *Alsophila* is naked, having no indusium. The sporangia are numerous with hairs intermixed, which are specially plentiful in *A. Taenitis* Hook., a fact which is recognised in its old name *Trichopteris* Presl. They are seated on a conically raised receptacle. This distinguishes it from *Gleichenia* and *Lophosoria*, which however *Alsophila* resembles in the usual orientation of the sporangia, and in the presence of the hairs. In *Hemitelia*, which often has the receptacle considerably elongated as it is in *H. capensis*, there is at the base a partial indusium in the form of a scale "varying in size and shape and texture, often indistinct, and often very deciduous" (Hooker). The genus has "the habit of *Cyathea*; a connecting link, as it were, between the latter genus and *Alsophila*, consequently often difficult to recognise" (Hooker). In *Cyathea* the receptacle is elevated, globose, or elongated. The indusium is attached as a cup-like covering round its base, and covers the whole sorus while young, but later it is torn irregularly from the summit, its base remaining as a more or less persistent cup. In these three steps we may see a natural progression from the naked type of sorus found in *Gleichenia* and *Lophosoria* to the fully protected type of *Cyathea*. The characters of the sori of the Cyatheaceae are beautifully presented in Hooker's *Genera Filicum* (for *Cyathea*, Pl. ii, xxiii; *Hemitelia*, Pl. iv, xl, xlii, *A*; *Alsophila*, Pl. ix, xxi). These show in each the elongated receptacle, the crowded sporangia with a high degree of regularity of their orientation so that, as in *Gleichenia*, the distal face is directed basally, and the proximal distally; but this orientation is not always strictly maintained at the apex of the receptacle.

The sporangia of all three genera are of relatively small size, as compared with the larger sporangia of *Lophosoria*, and *Gleichenia* (Fig. 563). The head which is borne upon a four-rowed stalk has an oblique annulus, which is a complete ring encircling a relatively small distal face: in *Hemitelia* it consists of 8 cells, as against 17 in *Metaxya* and over 50 in *Lophosoria*: the ring has 26 cells in *Hemitelia* against 30 in *Metaxya*, and 39 in *Lophosoria*. These are clear indications of steps of simplification: and with it appears an increasing precision of the stomium, which in *Hemitelia*, *Alsophila* and *Metaxya* consists of a compact group of four cells, while in *Lophosoria* it is ill-defined. Thus the sporangium of the Cyatheaceae is more precisely constructed than that of *Lophosoria*, and still more than that of *Gleichenia*, though *G. dichotoma* and *Platyzoma* show features of size and construction that are suggestive for comparison. The spore-output per sporangium is low. In *Alsophila* it is typically 64, as it is also in *Lophosoria* and *Metaxya*. But all of them fall far short of the large numbers in the Gleicheniaceae (excepting the aberrant *Platyzoma*). *Cyathea medullaris* also has 64 spores per sporangium:

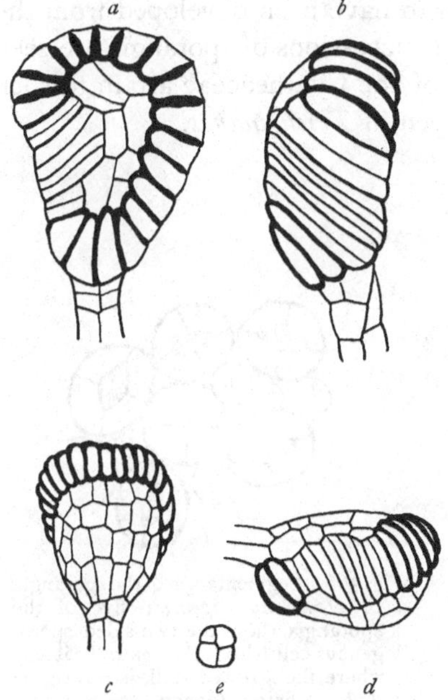

Fig. 563. Sporangia of Cyatheaceae. *a*, *b*, *Hemitelia capensis* Br. *a*, presents the distal or peripheral face. *b*, is seen laterally showing the stomium. (× 100.) *c*, *d*, *e*, *Alsophila excelsa* Br. *c*, presents the basal or proximal face. *d*, shows the stomium with the distal face downwards. *e*, shows a transverse section of the stalk. (× 100.)

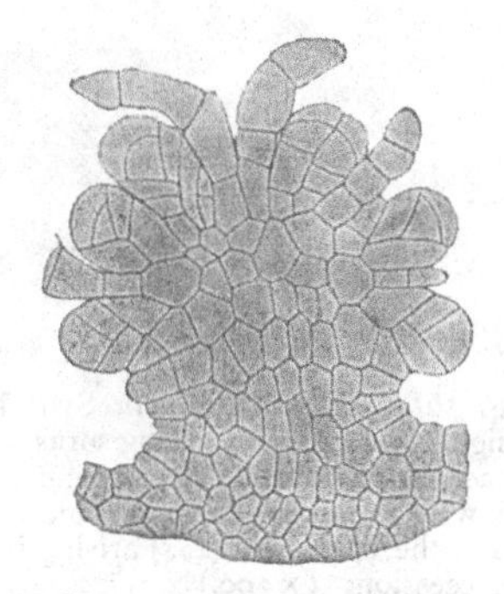

Fig. 564. *Alsophila atrovirens* Presl. A young sorus cut vertically, and showing a slight indication of basipetal succession of the sporangia. (× 200.)

but in *C. dealbata* such low numbers as 16 and even 8 have been recorded from the single sporangium. These facts show that the Cyatheaceae are a family of increasing precision but of reduced complexity of the sporangium, while the extreme is reached in *Cyathea dealbata*. They strengthen the view that this genus is more specialised, while *Alsophila* is less specialised, and that *Hemitelia* takes a middle position.

The development of the sorus has been observed in all three genera. In *Alsophila atrovirens* it first appears as an upgrowth of the leaf-surface (Fig. 564). Upon the receptacle thus formed the sporangia and hairs appear at first distally, while later others are formed lower down. There is in fact a basipetal succession, but it is not long continued, the number of sporangia in the sorus being small in this species. There is no sign of an indusium at the base of the receptacle. All the appearances are as in *Lophosoria*, or in the advanced species of *Gleichenia*, excepting in the number and the basipetal succession of the relatively small sporangia. The origin of the receptacle of *Cyathea* differs in no essential point from that of *Alsophila*, but here a basal indusium appears as a rather massive ring before the first sporangia are formed. These again arise in basipetal succession, and in the instance shown in Fig. 565 the oldest sporangium is seen to have been developed from the extreme tip of the receptacle. The longest successions of sporangia are seen in certain species of *Hemitelia*, but none of the Cyatheaceae attain to such a development of the receptacle as that seen in *Trichomanes*.

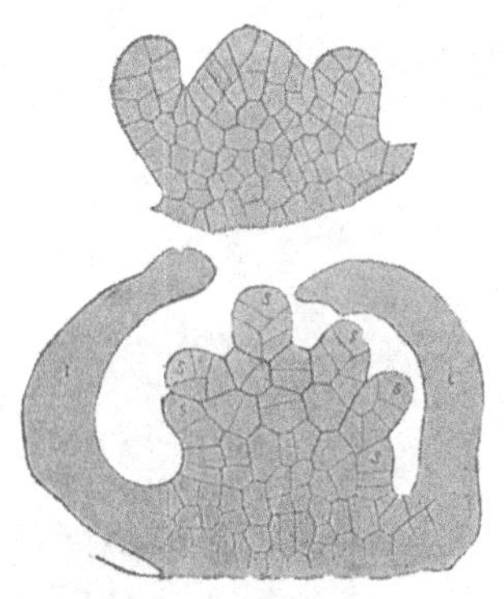

Fig. 565. *Cyathea dealbata* Sw. The upper figure shows a very young sorus, with receptacle and indusium already indicated. The lower shows the indusium (*i*) more advanced, and the sporangia (*s*, *s*) arising in basipetal succession. (× 200.)

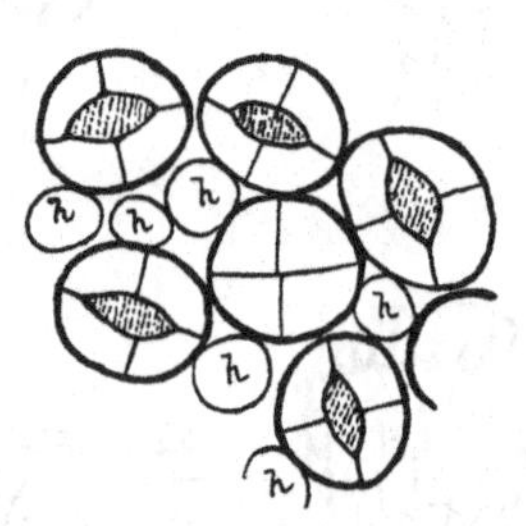

Fig. 566. Segmentation of the sporangia of *Hemitelia capensis*; five of the sporangia show the two-sided sporogenous cell (shaded); centrally is one where the 4-rowed stalk is traversed: *h*, *h* are hairs. (× 200.)

The segmentation of the sporangium presents features of interest. The parent cell frequently has a wedge-shaped base, and the first segment-wall is inserted on one of the oblique lateral walls. This type is thus intermediate between that of the Simplices and that seen in the smaller sporangia of the Leptosporangiates. But the further segmentation is by alternate cleavages in two rows, which are succeeded by the formation of a cap-cell. An examination of the sporangia in Figs. 564, 565, shows this, but the structure is best appreciated in sections tangential to the surface of the receptacle, which cut the individual sporangia transversely (Fig. 566). From these it appears that the internal cell has the shape of a biconvex lens, not of a three-sided pyramid as in most Leptosporangiate Ferns. A natural consequence of this segmentation is that the stalk is composed of four rows of cells, each of the segmental cells having been divided again by a radial wall.

In Fig. 566 the stalk of one of the sporangia has been traversed, showing this structure in the young state, and it persists without further division till the sporangium is ripe. This type of cleavage has been verified for all the three genera of the Cyatheaceae, and is found to be a constant feature for them. It has already been described for *Metaxya*, and it will be seen to exist also in *Dipteris* and *Cheiropleuria*. It thus appears to be widespread among the Superficiales. No instance of it has yet been seen among the Marginales.

If the problem of segmentation in a phyletic sequence with diminishing sporangia be considered, it appears that more than one course is open, just as it is seen to be in the segmentation of a conical apex of stem or leaf or root, where the segmentation may be by four, three, or two rows of segments. A reference to Fig. 243, Vol. I, p. 248 presents the problem in terms of the sporangial stalk, where it is naturally simpler than it would be in the sporangial head itself. A four-sided segmentation has actually been seen in certain sporangia of *Todea*, and the Osmundaceous stalk may perhaps reflect that structure with four internal cells (Vol. I, Fig. 243, *d*). As the sporangia diminish the internal cells may be omitted, so that the cleavages of the initial segmentation become more apparent in the simplified stalk. If the type of the sporangium were one with a relatively short thick stalk, as it is in the gradate Superficiales, it might well be expected that an alternate cleavage would suffice for the formation of the relatively small head, while subdivision of the two rows of segments would readily yield a massive four-rowed stalk. On the other hand, if the sporangial stalk were elongated and thin, and the sporangial head proportionately larger, then a three-rowed cleavage would provide the sporangial head, and would give without sub-division the three-rowed stalk, which is the rule for most Leptosporangiate Ferns, excepting those that are the most advanced. Accordingly it may be held that the short thick stalk and diminished head of the sporangia in the Cyatheoid Ferns, and their kin, may have determined a segmentation that is at least sufficiently unusual to have escaped observation hitherto.

THE GAMETOPHYTE

The gametophyte of the Cyatheaceae presents no special features when grown normally. It is of the usual cordate type. But in old prothalli certain characteristic bristles are produced on both sides of the prothallus, a feature which recurs in other allied Leptosporangiates such as *Diacalpe* and *Woodsia* (compare above, p. 267: also von Goebel, *Organographie*, ii, p. 950). Prof. A. S. Stokey permits me to state that she finds in pure cultures of *Alsophila* and other Cyatheaceae small flattened scales comparable to those of the sporophyte. They appear usually after the archegonia on old prothalli. The chief interest of the gametophyte is, however, centred in the antheridia. Their early segmentations follow those usual in Leptosporangiate Ferns, but with the significant difference that after the distal cap-cell has been formed it may divide further, as has been seen also in *Diacalpe* and *Woodsia*, genera closely associated with the Cyatheaceae (see Heim, 556). The number of spermatocytes traversed in a median vertical section of an antheridium of *Cyathea medullaris* is seen from Bauke's drawings to be 15–18 (*l.c.* Pl. viii, Figs. 1, 8).

This is in near accord with *Woodsia* (Schlumberger, *l.c.* Figs. 2, 3, 4), or with *Dryopteris filix-mas*; such figures compare with the number of spores in the sporangia, according to the table (Vol. I, p. 292). Records are not to hand for *C. dealbata*, though these would be interesting in view of the low spore-output in that species. In these details the Cyatheaceae appear to take an intermediate place between the more primitive Ferns and the Leptosporangiates. The first segmentations of the embryo correspond with those general in Leptosporangiate Ferns (Campbell, *Mosses and Ferns*, p. 391).

COMPARISON

The facts stated in the "Studies in the Phylogeny of the Filicales, II, III," and elsewhere, have been summarised in the preceding chapters. They formed the foundation of a distinction of the Gradatae, and other Ferns holding a middle position between the Coenopterids and ordinary Leptosporangiates, into two large series designated the "Marginales" and the "Superficiales." If such a distinction is to be held valid it must be founded upon consecutive sequences of related forms that are relatively primitive, and the comparisons must refer not to one character or another, but to the sum of such characters as can be observed. The comparison of later types cannot be depended upon to give secure conclusions, owing to the frequency of homoplasy. Since the validity of the two series thus characterised by the position of the sorus has been doubted, the opportunity will now be taken of restating the position, and of marshalling the facts that support it. In particular the reasons for segregating the Cyatheaceae phyletically from the Dicksoniaceae will be reconsidered, for it is round these families that the question chiefly turns (von Goebel, *Organographie*, II Aufl., 2 Teil, p. 1154).

The comparative treatment of the Cyatheaceae naturally starts from the Gleicheniaceae, a Family that dates back certainly to the Mesozoic Period, and possibly it was represented in Palaeozoic time by the Fern *Oligocarpia*. Here the naked sori were clearly superficial, and constructed on a plan closely similar to that of the living species of *Dicranopteris*, such as *G.* (*D.*) *flabellata*. If the ancestry of the Ferns bearing those sori had ever borne their sporangia at the margin (as they probably did in the first instance), the transition from the margin to the surface must have taken place in Palaeozoic times. A comparison of *Oligocarpia*, *G. flabellata*, *G. linearis* and *pectinata*, *Lophosoria*, and *Alsophila* shows a progression from a radiate uniseriate, simple sorus, through an unpractically crowded type, to a gradate state: but always the position was superficial, and the sorus naked. What more probable biological step should then be taken than that, in high-growing tree-ferns with exposed leaves, a basal indusium should be formed as a new structure, to protect the young sporangia at the base of the elongated receptacle? This would then give the type of *Hemitelia*, and finally of *Cyathea*. It is a reasonable view from comparison of the sori alone that the Ferns named form a natural sequence with consistently superficial sori. That sequence should then be tested by comparison in respect of other features. External morphology, anatomy, and dermal appendages should be examined, together with sporangial structure, segmentation, and spore-output: while contributory evidence may also be expected from the gametophyte.

The external morphology seems at first sight unpropitious: for the difference between the creeping *Gleichenia*, with peculiarly constructed leaves, often of unlimited apical growth, seems to differ widely from the Tree-Fern-habit. But it has been shown how *Lophosoria* bridges the difference by its runners, which assume distally the upright pose. *Alsophila aculeata* does the same, though *Metaxya*, which has so often been included in the genus *Alsophila*, retains the creeping habit permanently. Thus the series appears to illustrate the transition from a prone to an upright stem. The branching is characteristic: the bud at the leaf-base being constantly abaxial in *Lophosoria*, *Metaxya*, and *Alsophila*, with its vascular connections median, a fact that strengthens the affinity. It differs from the somewhat similar branching in the Dicksonieae, where the insertion is lateral, and the vascular connection marginal (see Vol. I, p. 77, Fig. 71). The bifurcation of the axis seen so commonly in *Gleichenia* is matched by that occurring occasionally in the Cyatheaceae, while the other branchings may probably be in their origin modifications of dichotomy (Schoute, *Beiträge zur Blattstellungslehre*, ii, Groningen, 1914).

The peculiar features of the leaf in *Gleichenia* depend upon varying localisation of growth combined with interrupted apical activity, carried out in a Pecopterid type of leaf. The leaves both of *Gleichenia* and of *Alsophila* are of the Pecopterid type, and comparison of herbarium-series of *Dicranopteris* and *Alsophila* shows the essential similarity, but disturbed in the former by intermittent intercalary growth of the rachis. In a minor degree this is also habitual in certain Cyatheaceae. The so-called "aphlebiae" of *Hemitelia* are merely basal pinnae left behind by intercalary growth localised above them (Fig. 567). A similar condition is sometimes seen also in *Alsophila* ("Studies, II," Fig. O). Thus the peculiarities which stamp the leaf of *Gleichenia* find their occasional correlative in the Cyatheaceae. The venation in *Gleichenia*, *Lophosoria* and *Metaxya* is always open, and readily referable to a dichotomous source. In the Cyatheaceae it is also open as a rule; but in the more condensed leaf-forms, and especially in *Hemitelia* and some Alsophilas where webbing has produced broad leaf-expanses, occasional vein-fusion is seen. There is never a pronounced reticulum, and the venation is always readily referable to a Pecopterid source. Thus throughout the series there is a general uniformity of the leaf-structure, and even in its modifications. The most advanced states are seen in the Cyatheaceae, as indicated by their occasional vein-fusions.

The anatomy follows suit, the steps of progression being sometimes illustrated in the individual development. The Gleicheniaceae are mostly protostelic, even in the adult state, with highly condensed undivided leaf-trace. But in *G. pectinata*, though the sporeling is protostelic at first, it passes quickly to solenostely: nevertheless the leaf-trace remains undivided. *Metaxya*

Fig. 567. Plant of *Hemitelia setosa* in Edinburgh Botanic Garden. Showing numerous "aphlebioid" pinnae. (From a photograph prepared by direction of Sir Isaac Bayley Balfour.)

is also permanently solenostelic, and its leaf-trace is undivided. But *Lophosoria* shows all stages from solenostely to an imperfect dictyostely in its upward-growing axes, while the leaf-trace is divided in the petiole into three straps. Similarly *Alsophila aculeata* has been shown by Stenzel to pass from protostely, through solenostely, to dictyostely in the same upward-growing shoot: while its leaf-trace becomes here, as in all the larger Cyatheaceae, highly disintegrated. In addition there is a medullary system, and a cortical system is also added in some of the large stems of the Family. These peculiarly consecutive facts lead by gradual steps from the primitive state of the Gleicheniaceae to the highly complex vegetative state of the Cyatheaceae: but always with the same underlying scheme.

A similar progression though not always so consistent characterises the dermal appendages. In the *Dicranopteris* and *Eu-Gleichenia* sections of *Gleichenia* both hairs and scales are present upon the rhizomes and leaves, developed perhaps in relation to a xerophytic habit. But in *G. linearis* and *pectinata*, now separated from the rest of the genus as § *Eu-Dicranopteris*, no scales are found, though in the latter species branched bristles appear at the leaf-bases, borne upon massive emergences. In *Metaxya* and *Lophosoria* hairs only are present; but in the Cyatheaceae broad scales cover the stem and young leaves, while at the leaf-bases they are perched on emergences like those of *G. pectinata*; becoming woody after the scales have fallen away, they form the well-known "armature." The connecting link is *G. pectinata*, whose branched bristles may well be the prototype of the chaffy scales (Fig. 475, p. 196). In this character also the Cyatheaceae appear to be the most advanced.

Turning to the structure of the sorus, the position of which is uniformly superficial, the progression is as in Ferns generally, towards a larger number of sporangia individually smaller, and with a smaller spore-output from each. The Gleicheniaceae, *Lophosoria* and *Metaxya* are all Simplices, with low receptacle and no protection for the sporangia excepting interspersed hairs. It has been shown how an increase in number of the sporangia in *G. pectinata* leads to a mechanical dead-lock, resolved in *Metaxya* by enlarging the area of the sorus. But in the Cyatheaceae it has been resolved by elongation of the receptacle, and the introduction of a gradate sequence of sporangia. In *Alsophila* there is no protection of the young basal sporangia, but in *Hemitelia* an incomplete basal indusium appears on the side of the sorus next the midrib, while in *Cyathea* it is completed as a circular cup. This indusium may be held as a new formation, and the biological value of it is obvious in high-growing Tree-Ferns with exposed leaves. The indusium of the Schizaeaceae was also a new formation: there is no reason to assume that such a development can only happen once. It is this assumption that seems to explain the comparison by von Goebel of the two-lipped indusium of the Dicksonieae with that of the Cyatheaceae (*Organographie*, II. Aufl., p. 1148, Fig. 1142). In view of the sequence here submitted, and supported along so many lines of detailed comparison of form and structure, it would appear more probable that in two distinct types, the one with superficial sori and profuse scales, the other with marginal sori and hairs but no scales, the protection of the sori was secured by independent origin of indusial growths, more or less homoplastic but not homogenetic.

In the series under discussion the sporangium itself shows a progressive diminution in size, together with a difference in dehiscence. This involves also a change from median to obliquely lateral dehiscence: i.e. from von Goebel's "longicide" to his "brevicide" type. Notwithstanding his doubts as to the reality of such a change (*Organographie*, II. Aufl., p. 1180), when a series shows so many points of progressive similarity as the Gleicheniaceae-Cyatheaceae sequence does, the probability appears strong that the phyletic change did actually take place. A similar question arises in regard to the marginal series also. A circumstance that will help towards elasticity of view in such matters is the existence of "looking-glass" images of sporangia from the same plant, such as are seen in *Plagiogyria* (Fig. 546), *Loxsomopsis* (Fig. 524), *Thyrsopteris* (Fig. 529, *E*, *F*), and others. These show that in a "brevicide" type the stomium may lie on either the right or the left side of the ring. If this difference can exist between sporangia from the same individual plant, it shows that the position of the stomium is not a fixed feature. If it can lie either right or left in the closed ring, why not either in a median or in a lateral position? Further, in the organisation of the stomium there is distinct advance within the series from *Gleichenia*, through *Lophosoria*, to the Cyatheaceae, and this goes with the diminution in size and in the spore-output.

As we progress from the Gleicheniaceae to the Cyatheaceae there also follows a change in the segmentation of the young sporangium, which is probably connected with the proportion of the sporangial head to the length and thickness of the stalk. A two-rowed cleavage is constant for *Metaxya*, and for the Cyatheaceae generally. The constancy of this peculiarity here, and its absence (so far as observed) in the Dicksonieae, is a distinctive feature for the Family, which it shares with the Dipterids only, so far as present observation extends. This gives it a special diagnostic value. The spore-numbers have shown the Cyatheaceae to have already fallen to the figure per sporangium usual for Leptosporangiate Ferns, which is so much lower than in the Gleicheniaceae: but a climax is reached in *C. dealbata*, where the numbers may be unusually small. Comparing these results with the numbers of spermatocytes seen in vertical section of the antheridia, their relation comes out similar to that seen in such Ferns as *Dryopteris filix-mas* (see table, Vol. I, p. 292).

The result of such comparison of the series of Ferns with superficial sori, in respect of these various criteria, is to show with impressive uniformity a progression from the ancient type of the Gleicheniaceae to the more modern though still relatively primitive Family of the Cyatheaceae. In all of them the position of the sorus is consistently superficial. As the Gleicheniaceous type is of proved antiquity, and possibly even of Palaeozoic origin, it appears that Ferns with a superficial sorus, linked with the Cyatheaceae by living genera showing intermediate characters, have had a consecutive history as such from the Mesozoic, and possibly from the Palaeozoic Period. On the other hand, the primitive Marginales, as represented by the living Dicksoniaceae, differ not only in the marginal position of their sori, but also in their two-lipped protective indusium, in their dermal appendages being hairs and never scales, and in their sporangial segmentation, so far as yet observed. The dendroid habit, which led earlier systematists to confuse them with the Cyatheaceae, is seen to be a polyphyletic character, as witnessed by the Osmundaceae, and Blechnoideae, etc. It cannot then be held as a valid indication of relationship. The conclusion which follows from this wide comparison is that the primitive Superficiales have been phyletically distinct from the primitive Marginales from very early times, and in fact before the gradate state was adopted: and that their recognition as two parallel series in the phyletic Classification of the Filicales is justified. In addition to habit and soral features the distinction has been supported by palaeontological evidence and by comparative anatomy. These factors seem to have been little regarded by Prof. von Goebel. But here they are held to be essential in leading to a just balance of opinion upon such a question.

A further circumstance that has confused the issue is that in many of the later derivatives of the Marginales the sorus may pass adaptively to a superficial position, while doubtless the converse is possible among the Superficiales. The result of this is that it becomes difficult to pursue the distinction into the more modern Leptosporangiates. This will be found to be a formidable problem in Vol. III. But such difficulties do not justify any negation of the broad conclusion which follows from a comparison of pre-Leptosporangiate types in upward sequence. It is from them that the phyletic lines must be traced, not from their later derivatives.

The Cyatheaceae are so closely related to certain Ferns with mixed sorus that there can be no doubt that the further transition from a gradate to a mixed state has taken place here, as it certainly did in the Dicksoniaceae. The genus *Diacalpe* has a mixed sorus (Davie, *Ann. of Bot.* XXVI, 1912, p. 245). It used to be included in the Cyatheaceae (*Synopsis Filicum*, p. 45), though more recently it has been placed in the Woodsieae. *Peranema*, too, which also has a mixed sorus (Davie, *l.c.*), is included in the Woodsieae, a group which will be held over for detailed treatment later. The genus *Woodsia* itself is clearly related to the gradate Cyatheaceae, but the group bridges over the transition to the mixed state not only by the genera named, but also by the shade-type *Hypoderris*. A ready transition is thus suggested to that large body of Leptosporangiate Ferns that will be styled the Dryopteroideae. On the other hand through *Matteuccia*, which is a gradate genus, a distinct line, characterised by soral fusions as well as by a mixed sorus, leads to the large sequence that may be grouped under the heading of the Blechnoideae. All of these may be held as related by descent to the proto-leptosporangiate Superficiales, constituting lines of further advance which will be seen to run parallel to those derived from the primitive Marginales. They will be held over for detailed examination to Vol. III. They are mentioned here so as to indicate that the Cyatheaceae hold a middle position in the Superficiales comparable to that occupied by the Dicksoniaceae among the Marginales. But the constant feature of soral position stamps the middle terms of the two phyla, represented today by the Cyatheaceae and the Dicksoniaceae, as distinct from the Mesozoic Period onwards.

CYATHEACEAE

(Excl. Thyrsopteridaceae and Dicksoniaceae)

I. ALSOPHILA R. Brown, 1810 ... ... 185 species.
(Excl. *Metaxya* and *Lophosoria*.) Sorus naked.

II. HEMITELIA R. Brown, 1810 ... ... 59 species.
Basal indusium, semicircular.

III. CYATHEA Smith, 1793 ... ... 182 species.
Basal indusium cup-shaped.

BIBLIOGRAPHY FOR CHAPTER XXXIII

559. HOOKER. Genera Filicum. 1842.
560. KARSTEN. Vegetationsorgane d. Palmen. 1847.
561. METTENIUS. Filices Horti Lipsiensis. 1856.
562. STENZEL. Nova Acta. Jena. ii, p. 1. 1861.
563. BAUKE. Pringsh. Jahrb. x, p. 49. 1875.
564. BAUKE. Bot. Zeit. 1880.
565. DE BARY. Comparative Anatomy. Engl. Edn. 1884, p. 291.
566. HEIM. Flora. Bd. 82, p. 355. 1896.
567. BOWER. Studies in spore-producing members. IV. Phil. Trans. 1899, p. 29.
568. ENGLER & PRANTL. Natürl. Pflanzenfam. i, 4, p. 123. 1902. Here the literature is fully quoted.
569. BERTRAND & CORNAILLE. La Masse libéro-ligneuse élémentaire des Filicinées. Lille. 1902.
570. GWYNNE-VAUGHAN. Solenostelic Ferns. II. Ann. of Bot. xvii, p. 709. 1903.
571. SCHLUMBERGER. Flora. Bd. 102, p. 383. 1911.
572. BOWER. Studies in Phylogeny. II, III. Ann. of Bot. 1912, p. 269; 1913, p. 444.
573. SCHOUTE. Ueber verästelte Baumfarne. Groningen. 1914.
574. VON GOEBEL. Organographie. II. Aufl., pp. 950, 1154. 1915–1918.
575. CAMPBELL. Mosses and Ferns. 3rd. Edn. 1918, p. 391.
576. SEWARD. Fossil Plants. ii, p. 366.

CHAPTER XXXIV

DIPTERIDACEAE

THE genus *Dipteris* Reinw., so long merged in the comprehensive genus *Polypodium* on account of its sorus being naked and superficial, has lately been restored to an independent position, and is now held as the sole living representative of the family of the Dipteridaceae (Seward and Dale, *Phil. Trans.* Vol. 194, p. 487, 1901). There is no doubt that this position is justified, though so late as 1902 the genus was still assigned a place among the Aspidieae (Diels, E. and P. i, 4, p. 202): Christensen (*Index*, p. xxvi) accepts the family as distinct, but ranks it under the Aspidieae. It will be shown that it takes its natural place among relatively primitive types in near alliance to the Matoniaceae.

The genus is represented by five living species from the Indo-Malayan Flora. They have creeping rhizomes which show occasional dichotomy. From these arise long-stalked alternate leaves, bearing each a distal lamina repeatedly branched in a dichotomous manner, and with a marked median sinus. The leaves of the different species vary greatly in area, but are alike in plan. The segments may remain narrow, with a marked midrib, and lateral flanges of no great width, as in *D. Lobbiana* Hook., and *D. quinquefurcata* Baker (Fig. 572): or they may be broader, and be more or less webbed into a continuous lamina, which is however still divided by the median sinus into approximately symmetrical halves. This is seen in *D. conjugata* Reinw. (Fig. 568, *A*), and *D. Wallichii* Hook. and Grev. Such leaf-structure is comparable as a whole with that of *Matonia*. But while in both the architecture is based upon helicoid sympodial branching, that in *Matonia* is of the catadromic type while that of *Dipteris* is anadromic (see Vol. I, Chap. V, Fig. 82, *A*, *B*). The same holds also, with peculiar modifications that give a very characteristic appearance, for the leaves of *Dictyophyllum* and *Camptopteris*. These are Mesozoic fossils referred by Nathorst to a relationship with *Dipteris* (see Vol. I, Fig. 82, *C*, *D*. Also Nathorst, *K. Svensk. Vetenskaps-Akad. Hand.* xli, No. 5).

While the primary venation of the leaves of *Dipteris* is dichotomous, the reticulation of the smaller veins is of an advanced type, viz. that described as *venatio anaxeti*. The reticulation is initiated even in the juvenile leaves. In *D. Lobbiana* though the first branchlets may end as "open" veinlets, in the later leaves they are linked together into a reticulum (Fig. 569), while in *D. conjugata* the very first leaves already show the complicated characters of the adult (Fig. 570). Comparing these facts with what is seen in *Matonia* it is evident that in venation *Dipteris* bears characters of advance, and that these are more prominent in *D. conjugata* than in *D. Lobbiana*. This harmonises with the comparison of the lamina as a whole, and particularly with the advanced webbing.

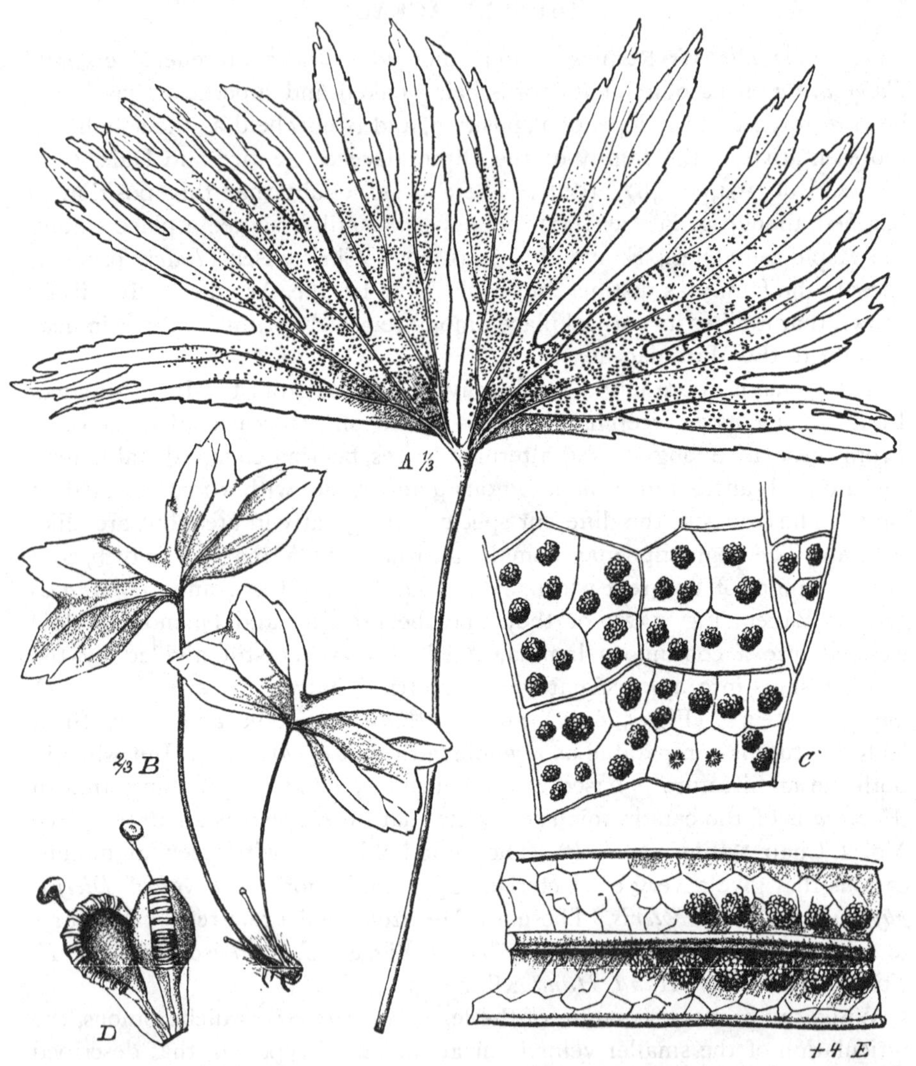

Fig. 568. *Dipteris* Reinw. *A–C*, *D. conjugata* (Kaulf.) Reinw. *A* = leaf of a mature plant. *B* = habit of a young plant. *C* = part of a fertile leaf with venation and sori. *D* = sporangia and paraphyses enlarged. *E* = *D. Lobbiana* (Hook.) Moore. Part of a fertile segment with venation and sori. (*A*, *C*, *D* after Kunze. *B*, *E* after Diels.)

The axis and the bases of the leaves of *Dipteris* are covered with dense brown bristles, which widen conically downwards. They are not actually flattened, though the insertion may be oval as seen in transverse section.

The basal cells are not indurated, and they show signs of intercalary activity; but in the upper parts they are thick-walled, and each bristle runs out into a stiff terminal spine. There is no branching. This state is an advance upon

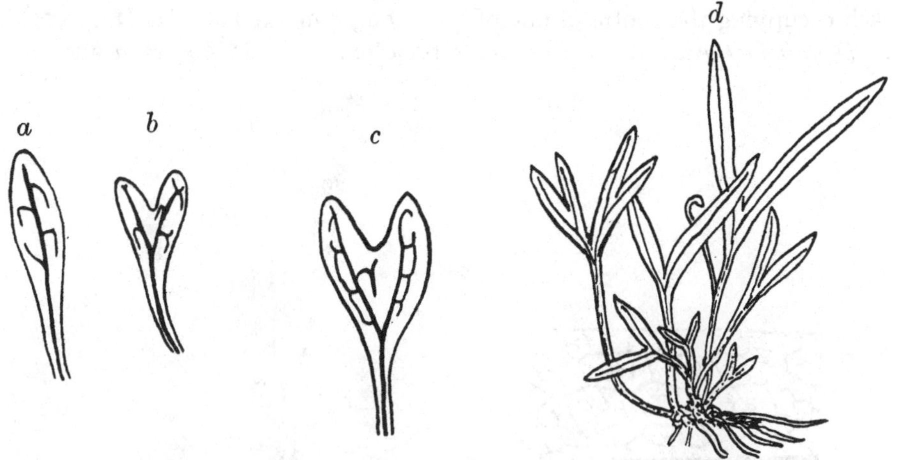

Fig. 569. Juvenile leaves of *Dipteris Lobbiana*. *a*, *b*, from specimens belonging to Dr Lang. *c*, *d*, after drawings by Miss De Bruyn.

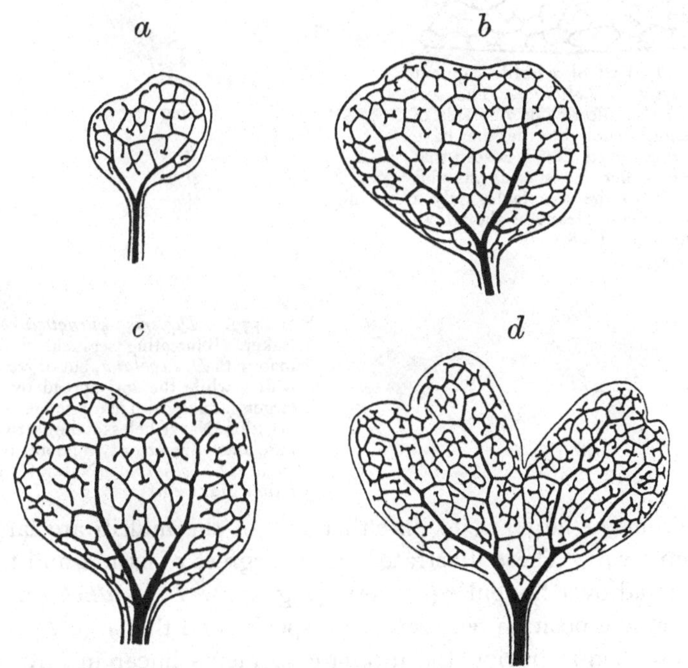

Fig. 570. Juvenile leaves of *Dipteris conjugata*, showing from the first reticulation superposed upon dichotomous branching. (×6.)

the filamentous hairs of *Matonia*: but in both the basal cells are soft and thin-walled, while the distal end is indurated and pointed.

As in *Gleichenia* and *Matonia*, so in *D. Lobbiana* the sori are disposed in regular linear series on either side of the midrib in the narrow leaf-lobes, each occupying the centre of one of the stronger-marked areolae (Fig. 571). In *D. quinquefurcata* the segments are broader than in *D. Lobbiana*, and the

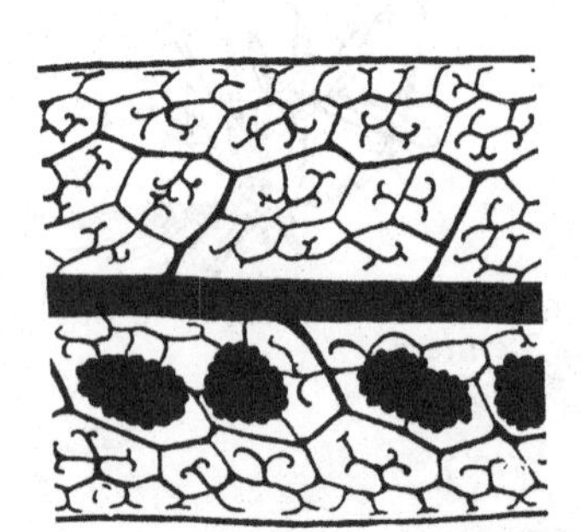

Fig. 571. Part of a fertile segment of *Dipteris Lobbiana*. The sori form a simple row on one side of the midrib: on the other they have been removed, and it is seen that there is here no special vascular supply to the receptacle, only a rather dense plexus of the reticulate veins. (× 8.)

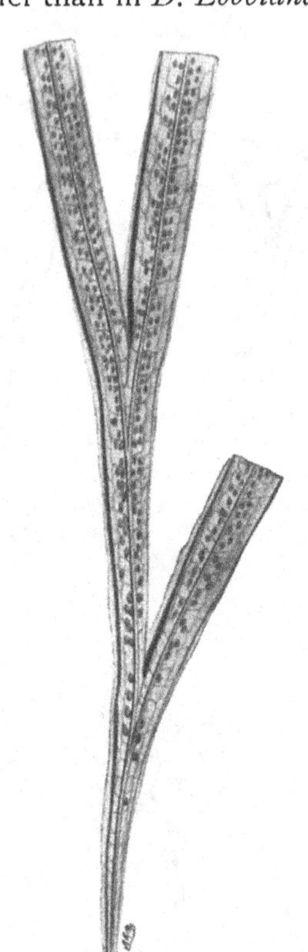

Fig. 572. *Dipteris quinquefurcata* Baker. Bifurcating segment similar in form to *D. Lobbiana*, but of greater width, while the sori spread over a larger surface, with many suggestions of fission. Nevertheless at the narrower base they appear in two lateral rows as in *D. Lobbiana*. (Natural size.) (After Maxwell.)

areolae within the larger veins on either side of the midrib are larger, and contain more sori. These illustrate various degrees of fission, and they thus become spread over the enlarging area (Fig. 572). *D. Wallichii* appears to occupy a middle position between these species and the large *D. conjugata*, for it is described as having the ultimate segments linear in form, and the sori as being similar to those of *D. conjugata* but more numerous than in

D. Lobbiana or *quinquefurcata*. Lastly, in the large-leaved *D. conjugata* the bifurcate lamina is broadly webbed, and the very numerous sori, which are distributed over the wide expanse, may be circular or oval, and not always distinct from one another. They may vary much both in size and shape, and their individuality is often lost, so that nearly the whole of the lower surface of the frond appears as though densely covered with a mass of sporangia (Fig. 573).

There seems to be only one probable way of reading these facts phyletically. Comparison points to *Gleichenia* and *Matonia* as primitive types of leaf to which that of *Dipteris* is related by *D. Lobbiana*. But from this simple narrow-leaved type, with its single row of sori on either side of the midrib, the broader-leaved types of *Dipteris* have broken away as the leaf-area enlarged, and the sori have spread over the extended surface (compare Vol. I,

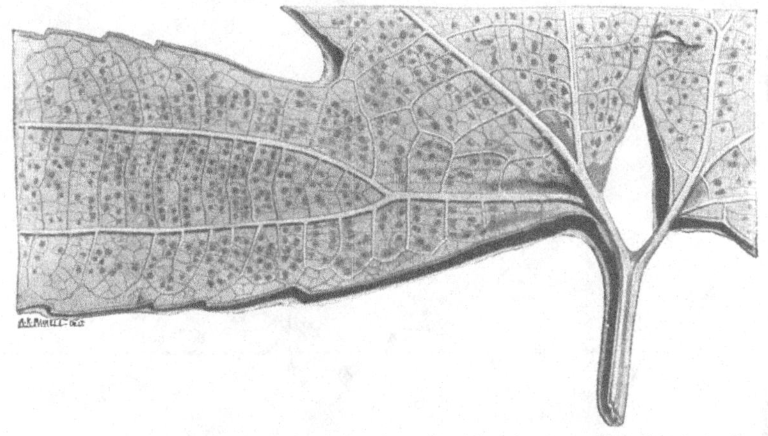

Fig. 573. *Dipteris conjugata* Reinw. Portion of leaf showing its extended surface, the webbing between the pinnae, the venation, and the numerous sori spread over the surface. (Natural size.) (After Maxwell.)

Fig. 222, p. 227). The loss of their individuality suggests one way in which the result may have been brought about, viz. by fission. This process, so clearly seen in the species of this very natural genus, has probably occurred also in other types of Ferns. It is suggested by *Christensenia* among the Marattiaceae, and by *Hypoderris* among the Woodsieae, but much more obviously among certain other advanced Leptosporangiates. It will have to be reckoned with in any general conception of the phylogeny of Ferns.

ANATOMY

The adult rhizome of *Dipteris* is traversed by a simple solenostele the general character of which resembles that of *Metaxya* (Vol. I, Fig. 149, 2). In *D. conjugata* the leaf-traces that arise from it come off each as a single ribbon-like strand, and the leaf-gap soon closes. The margins of this petiolar strand curve inwards to form the usual horse-shoe, which is uninterrupted up to a point close below the lamina (see Vol. I, Fig. 161, 1). These are all relatively primitive characters, and they direct comparison downwards to *Matonia* and *Gleichenia*. But *D. Lobbiana*, notwithstanding its smaller size, shows a somewhat more advanced state, for here the leaf-trace springs from the solenostele of the axis not as one but as two separate straps, given off right and left from the margins of the foliar gap (Fig. 574, *b*). Each of these divides almost at once into two (*d*, *e*), and the four strands arranged in a curve pass outwards into the leaf-base, continuing as separate strands upwards to the further branchings at the base of the lamina. This structure appears as an advance upon that seen in *D. conjugata*, a fact that will be considered later. Notwithstanding this disintegration of the trace the underlying structure is the same as in *D. conjugata*, and in this the genus conforms to the type seen in *Matonia* and *Metaxya*, and ultimately in

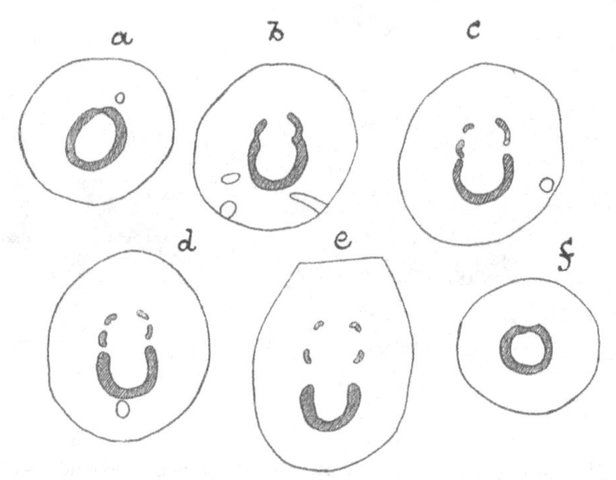

Fig. 574. *a–f*, successive transverse sections from below upwards, showing the separation of the leaf-trace from the solenostele in *Dipteris Lobbiana*. It is disintegrated from the first. (× 3.)

Gleichenia. The latter is the most primitive example of it, being protostelic: but *G. pectinata* and *Platyzoma* show elaborations resulting in solenostely. In *Metaxya* and *Dipteris* a simple solenostele, and in *Matonia* a polycyclic state is seen, but still with undivided leaf-trace. The same appears in *D. Lobbiana*, excepting that here the leaf-trace is disintegrated. Still the anatomical facts for these Ferns, like those from the dermal appendages which are never flattened scales except in *Eu-Gleichenia*, indicate a common and a relatively primitive character for them all.

SORI AND SPORANGIA

The sorus of *Dipteris* is composed of a number of sporangia with which numerous glandular hairs are associated. The sporangia show no regularity of position or of orientation, such as is seen in *Matonia* or *Gleichenia*. There is also an absence of any projecting receptacle. In these features the sorus resembles that of *Metaxya*. The sporangia of the same sorus arise simultaneously in *D. Lobbiana*; but in *D. conjugata* they are formed in succession, those which appear later being distributed without order amongst those first formed. The sorus, in fact, compares with that of the Mixtae, but the succession in time is not long maintained (Fig. 575). It thus appears that within the single genus a direct transition is illustrated between the "simple" and the "mixed" sorus. It is a very natural transition, but its rarity within such near affinity makes it notable.

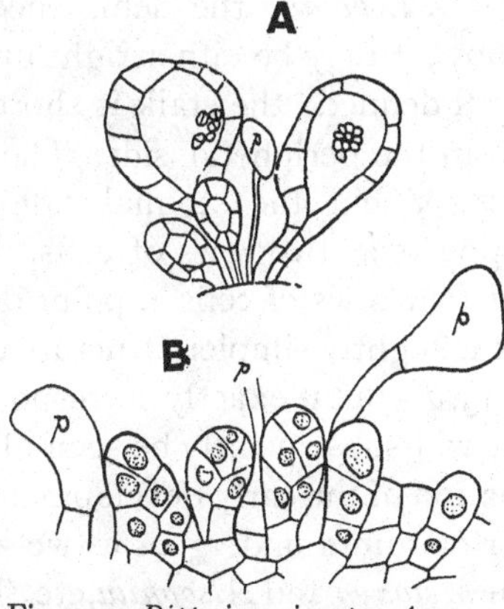

Fig. 575. *Dipteris conjugata*. *A*, young sorus showing sporangia of different ages in juxtaposition. (× 100.) *B*, younger sporangia. *p*, paraphyses. (× 300.) (After Miss Armour.)

The sporangia present details of special interest. Those of *D. Lobbiana* are represented from various aspects in Fig. 576, *a–f*. Seen from the "peripheral" side the annulus appears as a complete ring of cells, though the induration of those that are opposite the stalk is incomplete (*a*, *b*). A comparison may be made with those of *Gleichenia lineata* (= *G. dichotoma*), which show

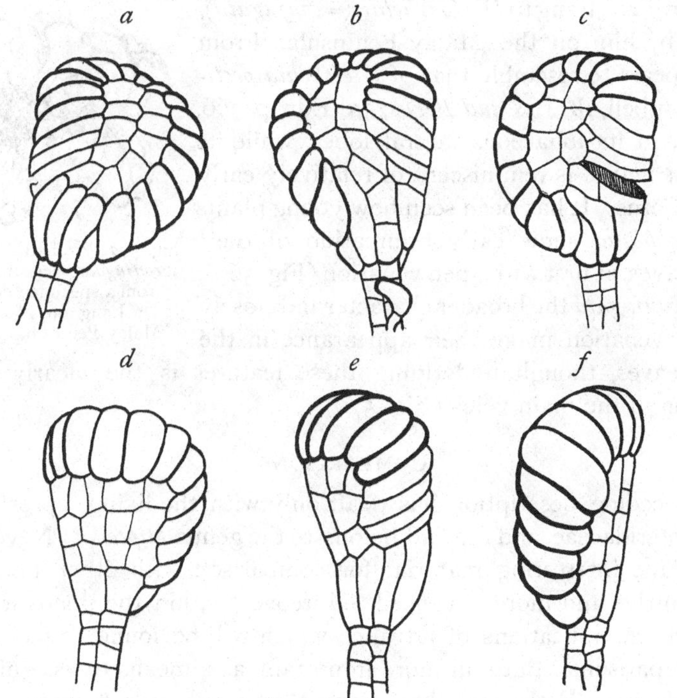

Fig. 576. Sporangia of *Dipteris Lobbiana* seen from various points of view. (× 50.)

similarity of form, and of position of the annulus (Chap. XXIV, Fig. 489, *l*, *m*). But there the induration is complete, and the dehiscence is median and distal. In *D. Lobbiana* the dehiscence is obliquely lateral, and as the figures (*a*, *b*) show, it may be either right or left of the median line. The stomium is not well defined: the stalk is short, and shows two regular rows of cells as seen from the peripheral side. The "central" or basal side is presented in Fig. 576, *d*, but the stomial side of the annulus is hidden: again the stalk appears as two rows of cells. It follows that the stalk is actually composed of four rows of cells, a point definitely proved by transverse sections. This is a slightly simpler structure than that of *Gleichenia lineata* (see Vol. I, Fig. 243, *b*). But it exactly corresponds to that of *Metaxya* and of the Cyatheaceae. It will subsequently be seen also to tally with that of *Cheiropleuria*. Examination of the early development of the sporangium proves that the sporangial primordium undergoes a two-sided cleavage, with two rows of segments, as in *Metaxya* and *Alsophila*, etc. The spore-output as recorded by H. H. Thomas (*Camb. Phil. Soc.* 1922, p. 109) is 64 for *D. conjugata*, while that appears also to be the typical number for *D. Lobbiana*, though the full number is not always developed.

The germination of the spores of *Dipteris* has not been observed, but the structure of the prothallus is represented in Fig. 577, from a specimen ascribed by Prof. Lang to *D. Horsfieldii* (=*conjugata*), collected by him on the Malay Peninsula. From this it appears to resemble that of *Gleichenia pectinata* (Campbell, *Mosses and Ferns*, 3rd Edn. p. 366, Fig. 208), in its foliaceous lateral lobes, while its massive structure is reminiscent of relatively early types of Ferns. It has been seen how young plants of *D. Lobbiana* show early bifurcation of their narrow leaves, at first with open venation (Fig. 569): but in *D. conjugata* the broader character and closely reticulate venation make their appearance in the earliest leaves, though underlying these features is the clearly marked bifurcation of the main veins (Fig. 570).

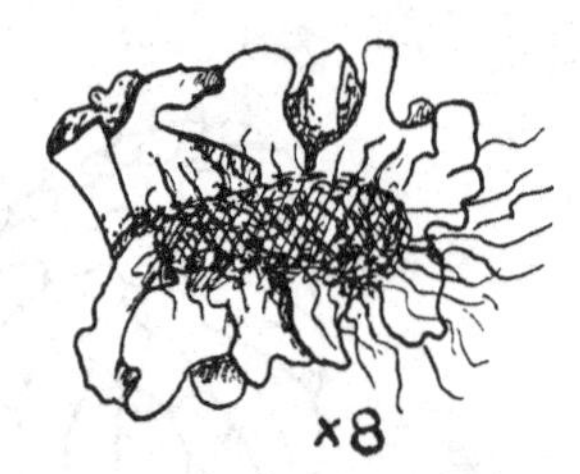

Fig. 577. Prothallus of *Dipteris conjugata* seen from below: from a specimen collected by Dr Lang on Mount Ophir, Malay Peninsula.

COMPARISONS

The preceding description has dealt only with the living representatives of the Dipteridaceae, and they all belong to the genus *Dipteris*. Nevertheless they provide interesting material for comparison with other Ferns, both more primitive and more advanced. Moreover, within the genus itself it is possible to see indications of advance which will be found to facilitate the wider comparisons. But still more important as a means to assigning their phyletic position is the fact that many Mesozoic fossils have with a high degree of certainty been recognised as related to these living Dipteridaceae. The correspondence often appears so close as to justify their being included within the Family. There are hardly any Ferns in respect of which the comparisons with fossils serve so effectively to consolidate the phyletic position of the living types. These are generally regarded as survivals from Mesozoic times, at which period their fossil congeners flourished.

The living species may themselves be seriated according to the characters of leaf-architecture and of sorus, as relatively primitive and relatively advanced. *D. Lobbiana* with its narrow forked lamina, its sori in a simple row on either side of the midrib, and with the sporangia in each of them produced simultaneously, may be held as a relatively primitive type. *D. quinquefurcata* with slightly broader leaf-segments, and with its sori scattered over the enlarged surfaces and showing signs of fission, may take an intermediate place, leading on to the broadly webbed leaves of *D. conjugata*, in which the sori are found to be of the "mixed" type. These features illustrate a progression from a type of sporophyll characteristic of the Gleicheniaceous affinity to one with leaf-surfaces of large area, and a sorus characteristic of many later Leptosporangiate Ferns. They suggest that *Dipteris* is a synthetic type, linking primitive with more advanced Ferns. With this conclusion the vascular construction is in accord: the solenostele and integral leaf-trace of *D. conjugata* readily compare with those of *Gleichenia pectinata* or *Metaxya*, though they fall far short of the special elaboration in the more nearly related *Matonia*. On the other hand *D. Lobbiana* presents in its disintegrated leaf-trace a sign of advance which does not accord with its simpler leaf and sorus. But many instances may be quoted where all the features used in comparison do not progress uniformly. Lastly, the stiff spinous hairs suggest a middle position. They are advanced structurally as compared with the simple hairs of *Matonia*, but they are not widened out into those broad protective scales characteristic of the later Leptosporangiates.

On the other hand the small size of the sporangia, their almost vertical annulus, lateral dehiscence, and the small spore-output appear to indicate an advanced state. But the complete ring of the annulus and the four-rowed stalk point in the direction of the Cyatheaceae and simpler Gleicheniaceae, while the form and structure of the sporangia resemble those of *Gleichenia lineata*. Against this comparison the low spore-output appears as an obstacle. It is here that the recently acquired evidence from fossils comes in with special cogency.

It has long been known that certain Ferns of Mesozoic age resembled the modern Dipteridaceae. Nathorst includes under his separate Family of the Campteridinae the genera *Dictyophyllum*, *Thaumatopteris*, *Camptopteris*, and *Clathropteris*, all of Rhaetic or Jurassic age. They were all large-leaved Ferns, with more or less distinctly ascending, helicoid, dichopodial structure of the lamina, the segments of which were pinnatifid like *Matonia pectinata* (see Vol. I, Fig. 82, *C*, *D*). These he suggested might be placed as a sub-family of the Dipteridaceae. On the other hand, *Hausmannia*, as

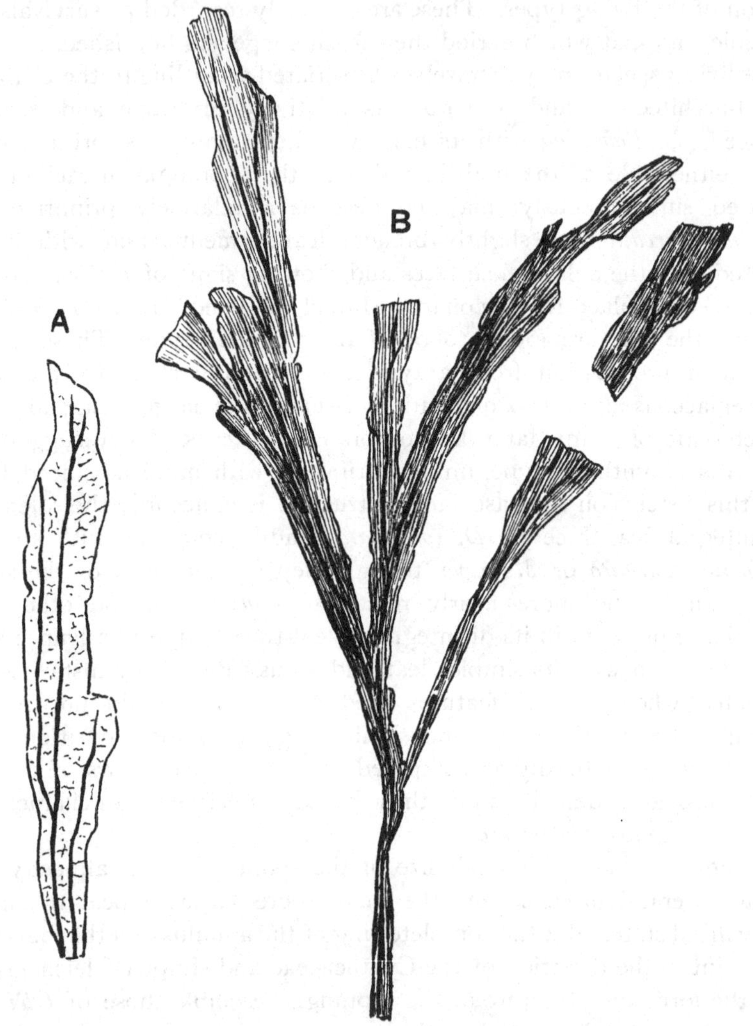

Fig. 578. *Hausmannia dichotoma*, showing the habit-similarity to the narrower-leaved species of *Dipteris*. (Specimens from the late Dr Marcus Gunn's collection of Upper Jurassic plants. Sutherlandshire: very slightly reduced.) (From Seward.)

Halle has recently shown, may safely be more closely related with *Dipteris* itself. In *H. dichotoma* the leaf-structure closely resembles that of *D. Lobbiana* (Fig. 578), while that of an un-named species of *Hausmannia* (Fig. 579), with a more nearly entire lamina, is like that of *D. conjugata*. But it is in the spore-numbers, which go more or less parallel with the size of the sporangia, that the most interesting features of these Mesozoic fossils lie. Halle (*l.c.* p. 23), after remarking on the constancy of the typical number of 64 for *Dipteris*, notes that *Hausmannia Forchammeri* Barth., of Jurassic age, has usually the typical number of 64, but sometimes 128 spores. H. H. Thomas finds in *Dictyophyllum rugosum*, a fossil referred to this affinity, in which the sporangia are smaller than in *D. exile*, the typical number of 128: the fossil comes from the Gristhorpe Beds of the Yorkshire Oolite. In *Thaumatopteris Schenki* Nath., of Rhaetic age, Halle states that the number is probably 128, while in *Dictyophyllum exile* Nath., also of Rhaetic age, the typical number is in most cases 512. We need not, as Halle rightly remarks, assume that these fossils are necessarily all on the same line of descent. But it is a matter for special remark that there is a larger

Fig. 579. *Hausmannia* sp. Upper Jurassic, near Helmsdale, Scotland. From a specimen in the British Museum. (Natural size.) (From Seward.)

spore-output in them than in the living Dipterids. The largest of all is that of *Dictyophyllum exile* which dates from the Rhaetic period, while the smallest among the fossils in question is that of *Hausmannia* from the Jurassic. Moreover the larger number goes along with a greater size of the individual sporangium, and a smaller number of the sporangia in each sorus. Referring to the sori of *H. exile*, Nathorst has been able to state that the number of these in each is probably 4–7 (Halle, *l.c.* p. 16). This number of sporangia in the sorus is not unlike that found in *Matonia*, in *Laccopteris*, *Gleichenia*, and *Oligocarpia*. Such facts when taken with the general morphology and anatomy of all the parts, living and fossil, establish a close nexus between them. They make some degree of actual relationship appear much more probable than has been generally realised: and they suggest that the Dipterids now living form a line of later derivation from a stock essentially Gleicheniaceous in type.

The features of specialised advance, recognised partly on comparative grounds, and not always demonstrated in strict stratigraphical sequence, have been: (i) the development of helicoid dichopodial branching of the lamina; (ii) webbing of the leaf-segments; (iii) fusion of veins to form the *venatio anaxeti*; (iv) advance to solenostely with ultimate disintegration of the leaf-trace; (v) spread of the sori over the enlarged leaf-surface, with a tendency to lose the soral identity in an Acrostichoid state; (vi) increase in number of sporangia in the individual sorus, with indications of transition to the "mixed state"; (vii) diminution in size of the individual sporangium, with reduction of the spore-output of each from such a figure as 512 to 64. All of these are indications of advance as regarded from the point of view

Fig. 580. Photograph by Mr Tansley of *Dipteris conjugata* (=*Horsfieldii*) on the edge of Padang Batu, Mt Ophir, showing its native habit (much reduced).

detailed in Vol. I, and the palaeontological evidence tends, so far as it goes, to support the general progression in time from the Rhaetic period onwards to the present. It is specially significant that, while certain archaic features still survive in the living species of *Dipteris*, none of them show either a radiate uniseriate sorus with few sporangia, as *Matonia* does, nor a spore-output higher than 64 from each. But it remains to be seen how far this result is borne out by the living Ferns that comparison indicates as further derivatives from the Dipterid-stock; and in particular *Cheiropleuria*.

The occurrence of *Matonia* and *Dipteris* as living Ferns in the Malayan region, and particularly their survival together on the heights of Mount Ophir, has long been a fact of commanding interest (Fig. 580). But palaeogeography shows that in Mesozoic times the spread of the *Matonia-Dipteris* alliance was very wide. In his Hooker Lecture, p. 238, Seward remarks: "There can be no doubt that the genera *Matonia* and *Dipteris* belong to a section of the Filicales which in former days rivalled in its geographical range the cosmopolitan Bracken-Fern of to-day: their present restricted range is not an indication of relatively recent origin (p. 233)....The problem of the original home of the *Dipteris-Matonia* stock is not easy of solution.... By the Rhaetic period they were thoroughly established in the Tonkin region, also in Germany and Scania....There is no good reason to suppose that this alliance was more widely represented in Tertiary floras than it is at the present day."

A further question may be the relation of *Matonia* to *Dipteris*. The similarities are so strong that a general relationship must be admitted. But *Matonia* stands apart both in stelar structure and in the sorus. The former is in advance of that of *Dipteris* in complexity, and size alone will not explain the difference, as is proved by comparison of Vol. I, Fig. 149, 2, 3. On the other hand, the sorus of *Matonia* appears to be conservative in respect of the uniseriate arrangement of the few large sporangia, though specialised in respect of the indusium, and the low spore-output. In view of the low spore-counts of the modern Dipterids, and the high counts of the Rhaetic *Dictyophyllum exile*, it would appear more desirable than ever to know the spore-output for *Laccopteris* and *Matonidium*. Pending such data *Matonia* may be held as a very ancient type, related on the one hand to the Gleicheniaceae, on the other to the Dipteridaceae. It is difficult to place the two genera as terms of a sequence. The probability seems to be that they represent independent lines of specialisation from a stock related to the Gleicheniaceae.

DIPTERIDACEAE

(Aspidieae-Dipteridinae of Diels)

Genus I. DIPTERIS Reinwards, 1824 ... ... 6 species

Mesozoic Fossils.
- *Hausmannia* Dunker, 1846.
- *Dictyophyllum* Lindley and Hutton, 1834.
- *Thaumatopteris* Goeppert, 1841.
- *Clathropteris* Brongniart, 1825.

BIBLIOGRAPHY FOR CHAPTER XXXIV

577. SEWARD & DALE. *Dipteris.* Phil. Trans. Vol. 194, p. 487, 1901, where references to fossils are fully given.

578. DIELS. Natürl. Pflanzenfam. i, 4, p. 202. 1902.

579. NATHORST. Ueber *Dictyophyllum* und *Camptopteris.* K. Svensk. Vetenkaps-Akad. Hand. Bd. xli, No. 5. 1906

580. BOWER. Origin of a Land Flora. p. 618. 1908.
581. SEWARD. Fossil Plants. ii, p. 380. 1910.
582. BOWER. Studies in Phylogeny. V. Ann. of Bot. xxix, p. 109. 1915.
583. BOWER. Leaf Architecture. Trans. Roy. Soc. Edin. li, p. 687. 1916.
584. HALLE. Sporangia of some Mesozoic Ferns. Arch. Bot. K. Svensk. Vetenkaps-Akad. Hand. Bd. 17, No. 1. 1921, where there are full references to the fossil literature.
585. HAMSHAW THOMAS. *Dictyophyllum rugosum.* Proc. Camb. Phil. Soc. 1922, p. 109.
586. SEWARD. Hooker Lecture. Linn. Soc. Journ. xlvi, p. 219. 1922.

CHAPTER XXXV

GENERAL REVIEW OF THE PRIMITIVE FERNS

IN Chapters XVIII—XXV the Ferns styled collectively the Simplices have been examined and compared according to the criteria laid down in Volume I of this work. In Chapter XXVI a general survey of the Simplices was given, and their relation to the palaeontological record discussed, which demonstrates the correctness of the view that in those having a simple sorus we see the most primitive types of Ferns. They are strongly represented in the Palaeozoic Age, though representatives of some of them persist to the present day. Their comparison showed further that according to the position of the sorus two types might be distinguished as represented among them: First, those where the sporangia are borne in close relation to the tip or margin of the leaf, or segment. This feature is held on comparative grounds to be primitive, and the Ferns that show it have been styled the Marginales. Secondly, there have been recognised those in which the sporangia are borne upon the surface of the expanded blade, a feature that is held to be relatively derivative: and these were styled the Superficiales. In the Schizaeaceae a pronounced marginal type is seen: in the Gleicheniaceae a pronounced type of the Superficial Ferns.

In Chapters XXVII—XXXIV certain Families have been examined which comparison indicates as more advanced than the Simplices, in respect of the sum of their characters, as judged along the lines adopted in Volume I. But since they possess features reminiscent of those seen in the Simplices, though still falling short of the full development seen in the Leptosporangiate Ferns, they have been accorded an intermediate place in the system. Many of them are characterised by having a Gradate sorus, which may be either marginal or superficial. Their sporangia tend towards a smaller size, and the spore-output from each is as a rule smaller than in the more primitive Ferns. They belong for the most part to types which were represented in the Mesozoic Period, while some of the living species correspond very closely with the fossils of Jurassic or even Triassic age. Phyletically they may be regarded as the probable predecessors of the bulk of the modern Leptosporangiate Ferns, a view which detailed comparative study amply confirms.

These Ferns of the Mesozoic types fall into several distinct Families, and these again segregate themselves naturally into two groups, the one suggesting relationship to the Schizaeaceae, especially on the ground of their marginal sori, the others suggesting relationship to the Gleicheniaceae, and having superficial sori. It has already been recognised that as the evolution of the Ferns proceeded there has probably been a general tendency to a phyletic slide of the sorus from the margin to the surface of the broadening leaf. But this did not happen once for all. In certain Ferns the transition was effected early, in others late. Hence the distinction is not an absolute one. But as the sori of the Schizaeaceae themselves are always marginal in primary origin, and those of the Gleicheniaceae are as constantly superficial, the distinction in respect of this character dates from Palaeozoic time. The families closely related to these two ancient types show for the most part a similar constancy in position of their sori. Thus the distinction between the ancient Marginales and Superficiales must be held as a real one and phyletically early.

The families that group themselves naturally with the Schizaeaceae on the sum of their characters, including that of soral position, are the Hymenophyllaceae, Loxsomaceae and Dicksoniaceae (including the Thyrsopterideae, Dicksonieae, and Dennstaedtiinae), and they may be held as relatively primitive Marginales. Those that group themselves naturally with the Gleicheniaceae on the sum of their characters, including that of soral position, are the Protocyatheaceae (*Lophosoria* and *Metaxya*), the Cyatheaceae (excl. Dicksonieae), and the Dipteridaceae. The curiously isolated family of the Plagiogyriaceae, represented only by the living genus *Plagiogyria*, is difficult to place. It certainly is a relatively primitive type, and may very possibly represent a direct offshoot from the Osmundaceae, of the type of *Todea*, with which it shares the upright habit, open venation, certain points of anatomical structure, and the superficial sorus. All these Ferns may be held as relatively primitive Superficiales.

This grouping separates the Dicksoniaceae from the Cyatheaceae, a separation which is long overdue. Their grouping collectively under the title of the Cyatheaceae was in the first instance based upon their sharing the dendroid habit. Strong indications of their real distinctness lie not only in the constant difference of their soral position, but also in the two-sided segmentation, with a four-rowed stalk, seen in the sporangia of the Cyatheaceae, a point which they share with the Dipterids, but which is absent so far as yet observed in the Dicksoniaceae. Further, there is the difference in the indusium, which is absent in *Alsophila* as it is also in *Lophosoria* and *Gleichenia*; but constantly present in two-lipped form in the Dicksoniaceae. Again the consistent presence of dermal hairs in the Dicksoniaceae, as against the broad scales of almost all the Cyatheaceae, is a strong indication of distinctness. Finally, the near relation of *Lophosoria* on the one hand to *Gleichenia* and on the other to *Alsophila*, all of them having superficial sori, is strong evidence of a transition from the simple superficial type to the gradate without any relation to the marginal series at all. If it had not been for the common feature of a dendroid habit it is improbable that Ferns differing in so many material features would have been classed together. The differences leave little doubt of the propriety of separating the Dicksoniaceae from the Cyatheaceae. Accordingly the latter name is here used in a restricted sense, as excluding the marginal Dicksoniaceae.

Some of these ancient Families of Ferns are represented to-day only by a single genus (*Matonia*, *Dipteris*), or even by a solitary species (*Loxsoma*, *Thyrsopteris*, *Lophosoria*, *Metaxya*, *Stromatopteris*, *Platyzoma*). This immediately suggests antiquity. It therefore becomes a question of interest to see what is the distribution of such Ferns to-day, and how it compares with that of the past. The facts for several of the Families here treated have been supplied in succinct map-form by Prof. Seward, in his Hooker Lecture of 1922 (*Journ. Linn. Soc.* xlvi, p. 219). The Map *A* shows approximately the present distribution of the Gleicheniaceae, as a shaded area. It extends throughout the tropics as a broad belt. But Seward remarks that the absence of *Gleichenia* from Northern Africa, the whole of Europe, Western Asia, and practically the whole of North America is a surprising fact, to which the geographical distribution of the fossils affords a striking contrast. The recorded Mesozoic distribution of the family is indicated on the same map by enclosed unshaded areas, which show that it then extended widely in N. America and Europe, and even into the Arctic Circle, while the specimens themselves appear closely similar to the living forms. There has then been a notable skrinkage of the area occupied, from the Mesozoic period to the present time.

The Map *B* embodies similar results for the *Matonia-Dipteris* alliance, the living representatives of which are now narrowly restricted to the Malayan region, as shown by the shaded areas. But in Mesozoic times their range was world-wide, as indicated by the enclosed but non-shaded areas on the map. These naturally record only the areas in which the fossils happen to have been collected. The imperfection of the record probably accounts for their isolation.

The distribution of the Schizaeaceae at the present time is indicated on Map *C*, by the shaded area, showing that the family now extends throughout the tropics. But again the records of the Mesozoic period indicate that the family then extended northwards into Europe, and into other northern areas where they are not now found. Again the present day area is restricted as compared with that of Mesozoic time.

A similarly constructed Map, *D*, relates to the Marattiaceae, of which "the oldest examples of fronds with fertile pinnae agreeing generally in habit with *Angiopteris*, *Macroglossum*, and *Archangiopteris* are from the Upper Triassic beds of Switzerland, Austria, and France" (Seward, *l.c.* p. 235). "It is in the older Mesozoic rocks that we first encounter Ferns which agree closely in habit as also in character of the sori with recent representatives of the family" (Seward, *l.c.* p. 236). Again this Map shows

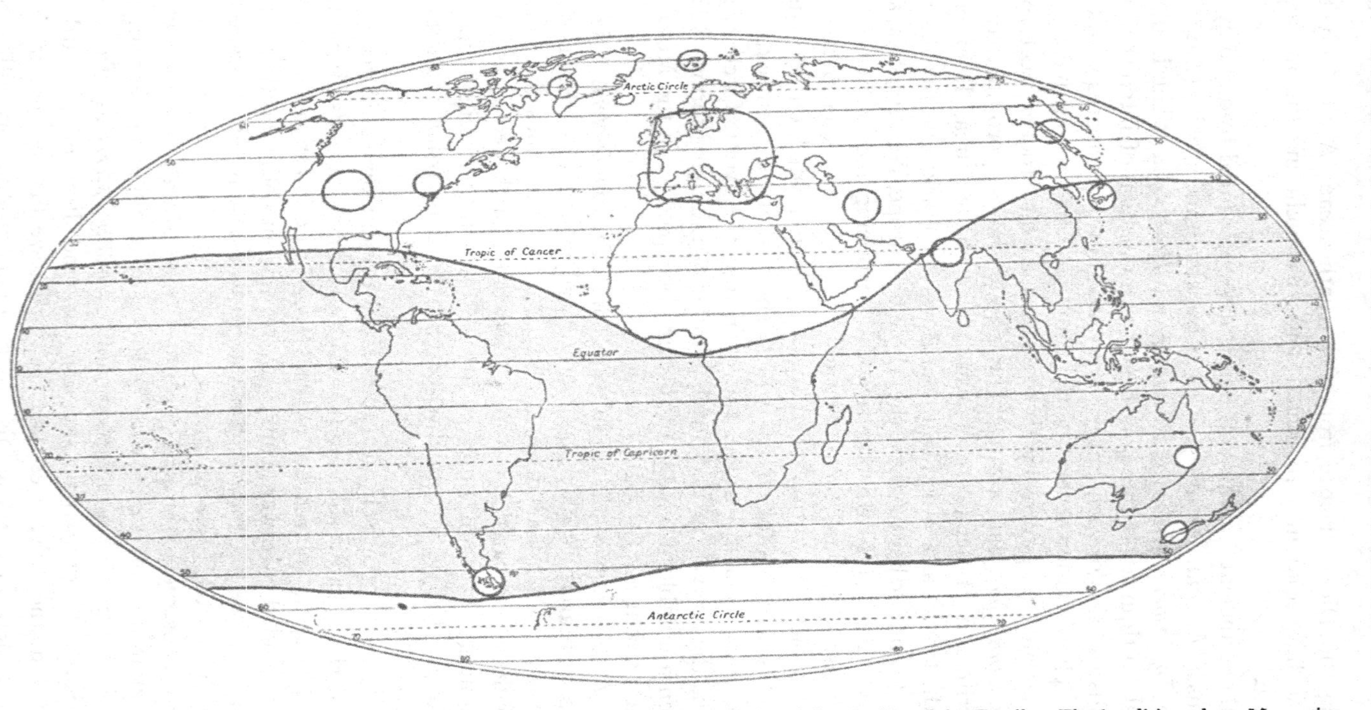

Map *A*. GLEICHENIACEAE. The shaded area shows approximately the present limits of distribution of the Family. The localities where Mesozoic representatives have been found are within the enclosed unshaded areas. From Prof. Seward's "Hooker Lecture," *Linn. Journ. Bot.* xlvi, Pl. 16, 1922.

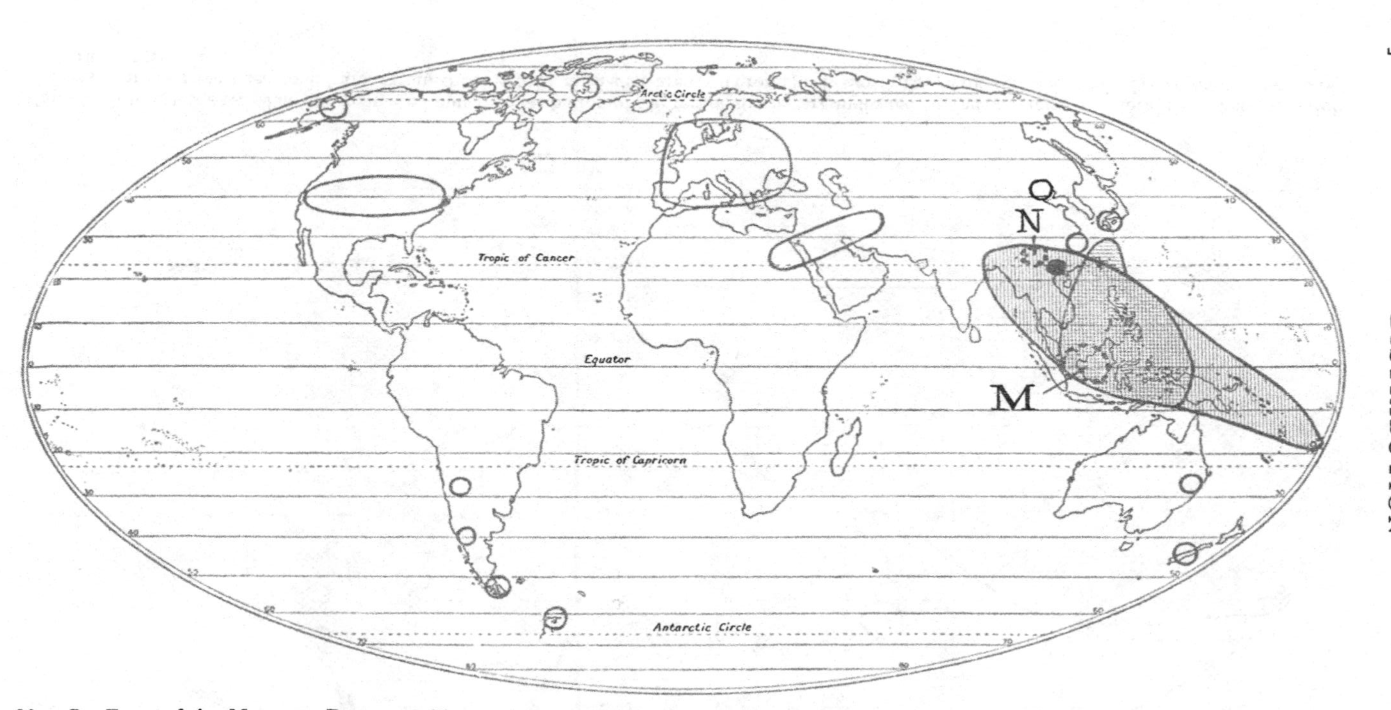

Map *B*. Ferns of the Matonia-Dipteris alliance. The area indicated by vertical lines is approximately that now occupied by *Dipteris*: the horizontal lines show the distribution of *Cheiropleuria*: the area *M*, within the dotted boundary, is that occupied by *Matonia*: the area *N* shows the home of *Neocheiropteris*. The other areas, including the Tonkin region (black dot), enclose localities from which Mesozoic Ferns mentioned in Seward's lecture have been obtained. (From Prof. Seward's "Hooker Lecture," *Linn. Journ. Bot.* xlvi, Pl. 17, 1922.)

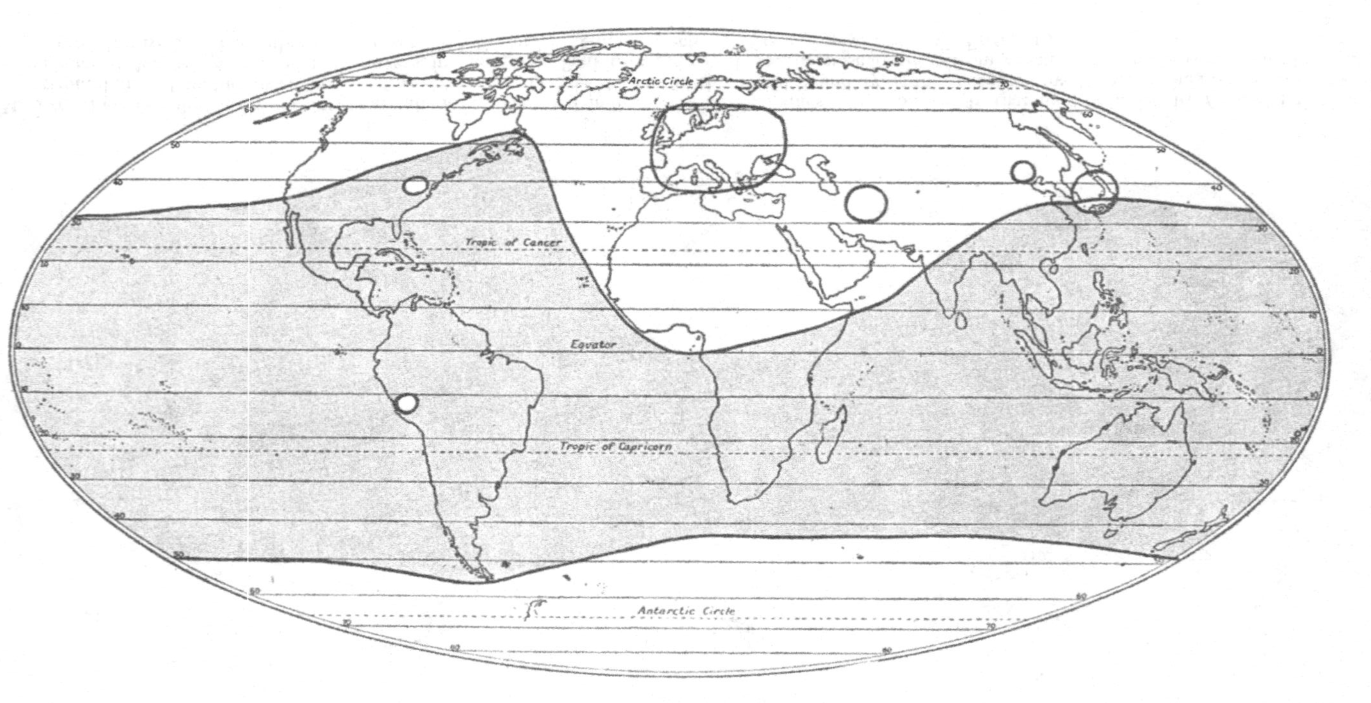

Map *C*. SCHIZAEACEAE. The approximate limits of distribution of the recent species are indicated by the shading. The localities where Mesozoic representatives have been found are within the enclosed unshaded areas. (From Prof. Seward's "Hooker Lecture," *Linn. Journ. Bot.* xlvi, Pl. 18, 1922.)

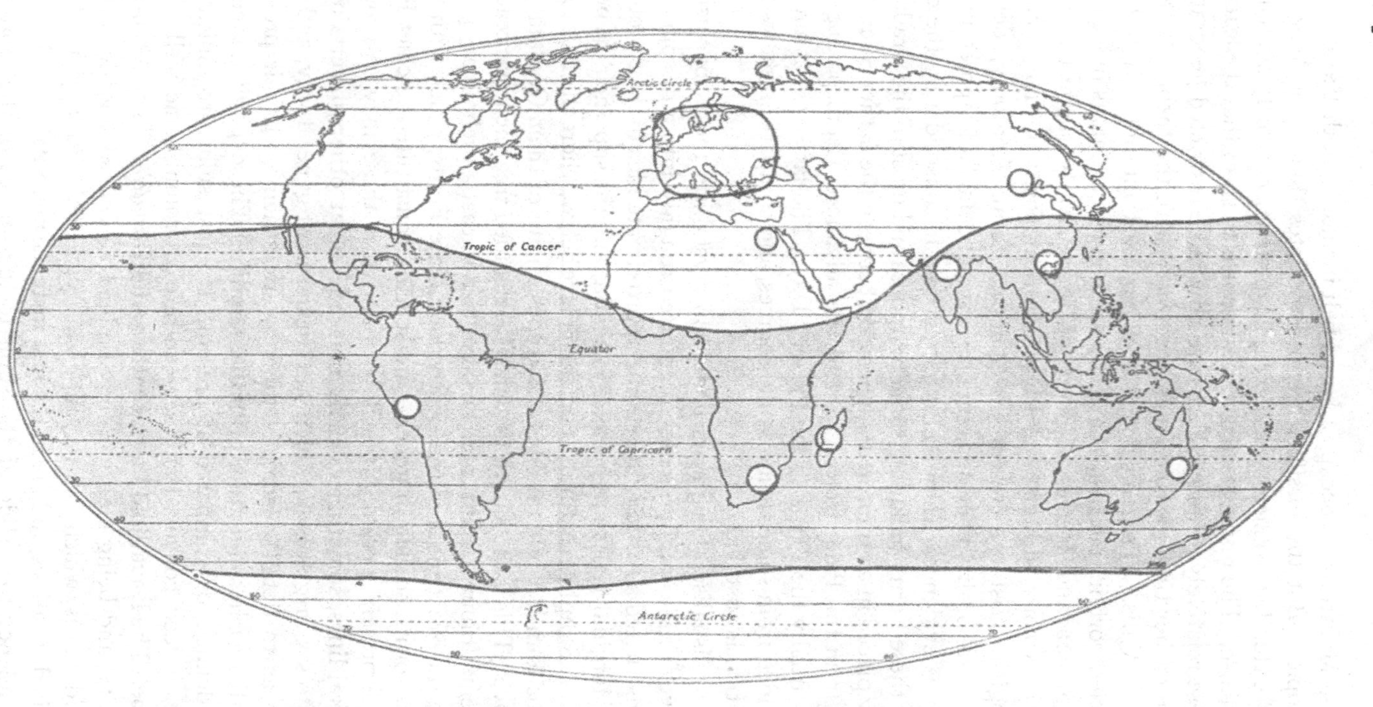

Map *D*. MARATTIACEAE. The shaded area shows approximately the present distribution of the family. The localities which have yielded Mesozoic species are within the enclosed unshaded areas. (From Prof. Seward's "Hooker Lecture," *Linn. Journ. Bot.* xlvi, Pl. 19, 1922.)

how the living Marattiaceae occupy a broad belt including both Eastern and Western Tropics. But the fossil record demonstrates that in the Mesozoic Period they extended farther northward, in a manner curiously similar to the Schizaeaceae. The result of such comparisons is to show that the leading types of these relatively primitive Ferns are now more restricted geographically than in earlier times, a fact which accords ill with the theory of "Age and Area" of Willis. The geological record proves them not only to have been of ancient origin, but also that they are now decadent, as indeed the paucity of genera and species of many of them clearly suggests. Such results are in accord with our general theoretical position. The facts of Palaeontological History, present distribution, Form, Anatomy, dermal appendages, soral position and structure, sporangial detail, and spore-output, all converge as evidence. They all take their part in consolidating the conclusion that these Ferns, including some Simplices and most of the Gradatae, are properly placed in an intermediate position between the earlier Simplices, which date from the Palaeozoic Age, and the advanced Leptosporangiates that are essentially the Ferns of the Present Day. The latter we assume to have descended from some such original sources, and comparison shows that the assumption is justified.

In order to visualise, though not unduly to crystallise, these conclusions, the primitive families of Ferns may be plotted roughly into a disconnected scheme, constructed so as to convey some approximate idea of their probable relations. Marginal types are placed to the left, and the superficial types to the right. The Simplices naturally take their place below, where the Palaeozoic types will be found: while the Mesozoic types, many of them Gradatae, are ranged above. This is the length to which the detailed treatment has reached in the present Volume. But it may be permitted to cast a preliminary glance onwards to the great mass of the Leptosporangiate Ferns, which with their multifarious forms, so rich in genera and species, constitute the chief Fern-Flora of the present period. It will be found that these may be grouped along natural lines, sometimes clearly marked but at others uncertain and obscure, into six large phyla, which are (in the most general way, and with reservation as to detail) placed in the phyletic scheme, in their probable relation to the earlier and more primitive families. These phyla cannot be discussed, defined, or compared at present. It must suffice to indicate, as their names indeed convey, that they centre each round some well-known generic type. The Davallioid Ferns centre round *Davallia*. They retain their soral identity, and being marginal may be held as Dicksonioid derivatives. The Pteroid Ferns, which centre round the type of *Pteris*, appear also as Dicksonioid derivatives, and are in the first instance marginal: but they show a strong tendency to a slide of the sorus to the surface of the sporophyll. Moreover in them the sori are apt to be fused into linear sequence, which becomes a leading feature. The Gymnogrammoid Ferns centre round *Gymnogramme*, and though this phylum is less definite than any of the others, it may perhaps be traced as derivative, in part or in whole, from forms such as the living *Plagiogyria*, and ultimately from an Osmundaceous source. The Blechnoid Ferns centre round *Blechnum*, and have in common the fusion of superficial sori of a primitively Cyatheoid type into linear sequence, but along evolutionary lines quite distinct from the fusion seen in *Pteris*. The Dryopteroid Ferns centre round *Dryopteris*, and differ from the Blechnoids in

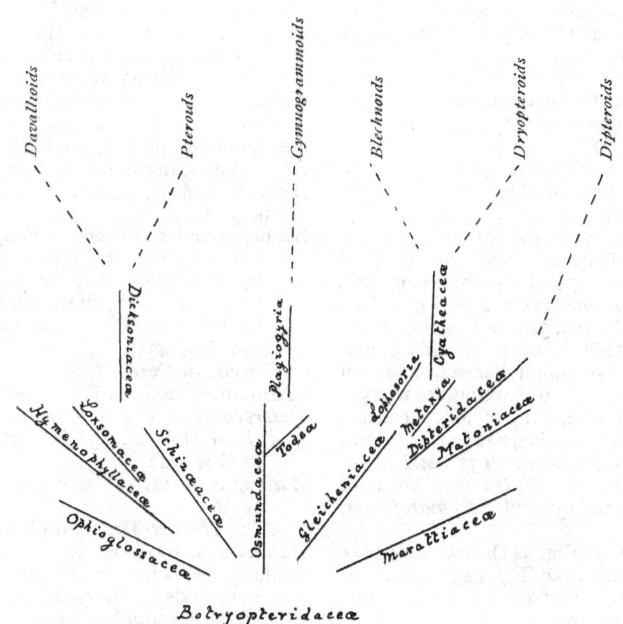

Phyletic Scheme for the more Primitive Filicales.

maintaining the individuality of their superficial sori, which are linked with those of the Cyatheoids by unmistakeable intermediate types. The Dipteroid Ferns may be held as derivatives from the type of *Dipteris*, a genus which already shows in *D. conjugata* the advance to a broad leaf-surface, and a mixed type of sorus, advancing towards the acrostichoid state. The comparative working out of the natural affinities of the Leptosporangiate Ferns along such lines, and the discussion of their grouping in relation to these suggested central types, must be held over as the subject for Volume III.

INDEX

CAMBRIDGE: PRINTED BY W. LEWIS, M.A., AT THE UNIVERSITY PRESS

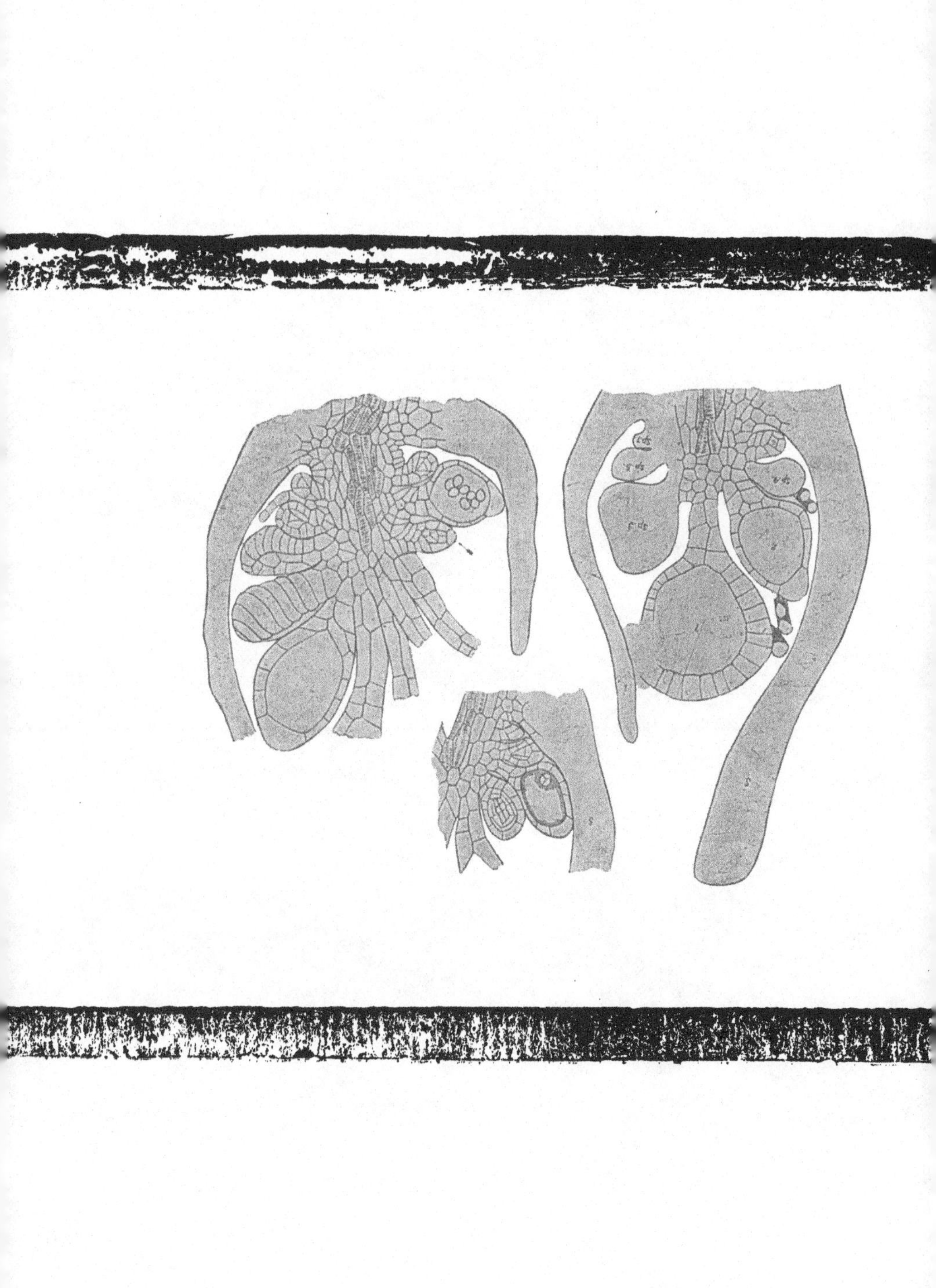

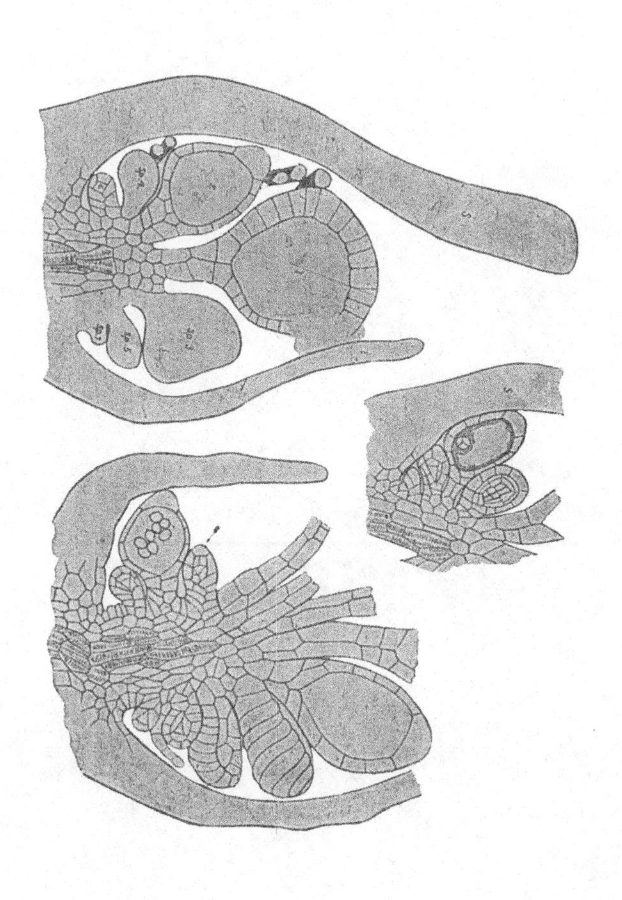

Printed in the United States
By Bookmasters